Lead-free Soldering Process Development and Reliability

Wiley Series in Quality & Reliability Engineering

Dr. Andre Kleyner
Series Editor
The Wiley Series in Quality & Reliability Engineering aims to provide a solid educational foundation for both
practitioners and researchers in the Q&R field and to expand the reader's knowledge base to include the latest
developments in this field. The series will provide a lasting and positive contribution to the teaching and practice of
engineering.
The series coverage will contain, but is not exclusive to,

- Statistical methods
- Physics of failure
- Reliability modeling
- Functional safety
- Six-sigma methods
- Lead-free electronics
- Warranty analysis/management
- Risk and safety analysis

Wiley Series in Quality & Reliability Engineering

Lead-free Soldering Process Development and Reliability
by Jasbir Bath (Editor)
2020

Thermodynamic Degradation Science: Physics of Failure, Accelerated Testing, Fatigue and Reliability
by Alec Feinberg
October 2016

Design for Safety
by Louis J. Gullo, Jack Dixon
February 2018

Next Generation HALT and HASS: Robust Design of Electronics and Systems
by Kirk A. Gray, John J. Paschkewitz
May 2016

Reliability and Risk Models: Setting Reliability Requirements, 2nd Edition
by Michael Todinov
September 2015

Applied Reliability Engineering and Risk Analysis: Probabilistic Models and Statistical Inference
by Ilia B. Frenkel, Alex Karagrigoriou, Anatoly Lisnianski, Andre V. Kleyner
September 2013

Design for Reliability
by Dev G. Raheja (Editor), Louis J. Gullo (Editor)
July 2012

*Effective FMEAs: Achieving Safe, Reliable, and Economical Products and Processes Using Failure Modes
and Effects Analysis*
by Carl Carlson
April 2012

Failure Analysis: A Practical Guide for Manufacturers of Electronic Components and Systems
by Marius Bazu, Titu Bajenescu
April 2011

Reliability Technology: Principles and Practice of Failure Prevention in Electronic Systems
by Norman Pascoe
April 2011

Improving Product Reliability: Strategies and Implementation
by Mark A. Levin, Ted T. Kalal
March 2003

Test Engineering: A Concise Guide to Cost-Effective Design, Development and Manufacture
by Patrick O'Connor
April 2001

Integrated Circuit Failure Analysis: A Guide to Preparation Techniques
by Friedrich Beck
January 1998

Measurement and Calibration Requirements for Quality Assurance to ISO 9000
by Alan S. Morris
October 1997

Electronic Component Reliability: Fundamentals, Modelling, Evaluation, and Assurance
by Finn Jensen
1995

Lead-free Soldering Process Development and Reliability

Edited by

Mr. Jasbir Bath
Bath Consultancy LLC
11040 Bollinger Canyon Road, #E-122
San Ramon
CA 94582, USA
Email: Jasbir_Bath@yahoo.com

Registered Office
John Wiley & Sons, Inc., 111 River Street, Hoboken, NJ 07030, USA

Editorial Office
111 River Street, Hoboken, NJ 07030, USA

For details of our global editorial offices, customer services, and more information about Wiley products visit us at www.wiley.com.

Wiley also publishes its books in a variety of electronic formats and by print-on-demand. Some content that appears in standard print versions of this book may not be available in other formats.

Library of Congress Cataloging-in-Publication Data

Names: Bath, Jasbir, editor.
Title: Lead-free soldering process development and reliability / edited by Jasbir Bath, Bath Consultancy LLC.
Description: First edition. | Hoboken, NJ : John Wiley & Sons, Inc., 2020. | Series: Wiley series in quality & reliability engineering | Includes bibliographical references and index.
Identifiers: LCCN 2020004969 (print) | LCCN 2020004970 (ebook) | ISBN 9781119482031 (hardback) | ISBN 9781119482048 (adobe pdf) | ISBN 9781119481935 (epub)
Subjects: LCSH: Electronic packaging. | Solder and soldering.
Classification: LCC TK7870.15 .L434 2020 (print) | LCC TK7870.15 (ebook) | DDC 621.381/046–dc23
LC record available at https://lccn.loc.gov/2020004969
LC ebook record available at https://lccn.loc.gov/2020004970

Cover Design: Wiley
Cover Image: © Paul Krugman/Shutterstock

Set in 9.5/12.5pt STIXTwoText by SPi Global, Chennai, India

Printed in the United States of America

V10019067_061120

Contents

List of Contributors

Raiyo Aspandiar
Intel Corporation
Hillsboro, OR
USA

Nilesh Badwe
Intel Corporation
Hillsboro, OR
USA

Jasbir Bath
Bath Consultancy LLC
San Ramon, CA
USA

Silvio Bertling
Mesa
Arizona, USA

Peter Borgesen
Integrated Electronics Engineering
Center
Binghamton University, State
University of New York, NY
USA

Kevin Byrd
Intel Corporation
Hillsboro, OR
USA

Richard J. Coyle
Nokia Bell Laboratories
Murray Hill, NJ
USA

Travis Dale
School of Mechanical Engineering
Purdue University
West Lafayette, IN
USA

Gerjan Diepstraten
Vitronics Soltec
Oosterhout, The Netherlands

Carol Handwerker
School of Materials Engineering
Purdue University
West Lafayette, IN
USA

Shantanu Joshi
Koki Solder America
Cincinnati, OH
USA

Jason Keeping
Celestica Inc.
Toronto
Canada

Ning-Cheng Lee
Indium Corporation
Clinton, NY
USA

Elizabeth McClamrock
School of Materials Engineering
Purdue University
West Lafayette, IN
USA

Jennifer Nguyen
Flex
Milpitas, California
USA

Rick Nichols
Atotech
Berlin
Germany

Ganesh Subbarayan
School of Mechanical Engineering
Purdue University
West Lafayette, IN
USA

Karl Sauter
Oracle Corporation
Santa Clara, California
USA

Maxim Serebreni
Department of Mechanical
Engineering
University of Maryland
College Park, MD
USA

Brian J. Toleno
Microsoft
Mountain View, California
USA

Alyssa Yaeger
School of Materials Engineering
Purdue University
West Lafayette, IN
USA

Hongwen Zhang
Indium Corporation
Clinton, NY
USA

Introduction

With the movement to lead-free soldering in electronics manufacturing production, there is a need for an updated review of various topics in this area for prac-
ticing process, quality and reliability engineers and managers to be able to use to
address issues in production.

The book gives updates in areas for which research is ongoing, and addresses
new topics which are relevant to lead-free soldering. It covers a list of key topics
including developments in process engineering, alloys, printed circuit board (PCB)
surface finishes, PCB laminates, and reliability assessments.

Chapter 1 discusses lead-free surface mount technology (SMT) with review of
the surface mount process for lead-free soldering, including printing, component
placement, reflow, inspection, and test.

Chapter 2 covers lead-free selective and wave soldering in terms of flux and preheat processes as well as the similarities and differences of these processes.

Chapter 3 discusses the issues during lead-free rework for the assembled components based on the higher lead-free soldering temperatures and the range of small
and large components to rework as well as temperature and moisture sensitivity with components and boards. The chapter reviews updates in lead-free rework
technology including hand soldering, ball grid array/chip scale package BGA/CSP
rework, and PTH (Pin Through Hole) rework.

Chapter 4 discusses lead-free solder paste and flux technology and the characteristics needed for these materials to ensure good solder paste performance. It also
reviews the defects which can occur during electronics manufacturing, including micro solder balls, voiding, tombstoning, bridging, opens, head-on-pillow, and
non-wet opens.

Chapter 5 covers low temperature lead-free alloys and pastes, with an emphasis
on the Bi-Sn system and the development work ongoing in this area.

Chapter 6 discusses solder materials, silver-sintering materials, and TLPB (transient liquid phase bonding) materials as the three potential candidate types for
lead-free, high-temperature, die-attachment materials.

Chapter 7 covers the drivers, benefits, and concerns associated with the development and implementation of third generation, high reliability lead-free solders.

Chapter 8 reviews lead-free surface finish alternates to the electrolytic nickel immersion gold surface finish in relation to performance characteristics and cost, with selection of a final board surface finish being of significance to the assembly and reliability of the product.

Chapter 9 discusses several critical factors relating to PCB laminate materials and describes why they are important to ensuring that the finished board performance requirements are met.

Chapter 10 reviews the use of adhesives in the manufacturing of high-density lead-free surface mount assemblies, with a discussion of the two adhesive applications used widely for increased reliability: underfills and encapsulants.

Chapter 11 overviews thermal cycling reliability in relation to lead-free solder joints, with the reliability of solder interconnects influenced by all aspects of the electronic assembly, ranging from the component package style and circuit board construction to the solder composition.

Chapter 12 discusses intermetallic compounds (IMCs) formed in the lead-free solder joints with a discussion of the roles that IMCs play in determining solder joint reliability, and how those roles change as a result of aging or damage induced by thermal cycling. It covers the performance of common lead-free solder alloys in combination with metallizations and surface finishes to understand what to expect in these specific systems and the problems that may arise when combining new solder alloys and surface finishes/metallizations and the methodologies that can be used to separate out the different possible root causes.

The final chapter (Chapter 13) covers industry updates in the use of conformal coatings and their use in electronics manufacturing and their effect on reliability. Various aspects of conformal coatings are discussed, including Environmental Health and Safety (EHS) requirements, the five basic conformal coating types and new emerging materials, preparation, application, cure, and inspection of conformal coatings, repair and rework, and design guidance on when and where coatings are required, and which physical characteristics and properties are important to consider.

1

Lead-Free Surface Mount Technology

Jennifer Nguyen[1] and Jasbir Bath[2]

[1] Flex, Milpitas, California, USA
[2] Bath Consultancy LLC, San Ramon, CA, USA

1.1 Introduction

Surface mount technology (SMT) involves the assembly or attachment of surface mount devices (SMDs) onto the printed circuit board (PCB). Today, the majority of the products are built using surface mount technology and lead-free process. This chapter will review the surface mount process for lead-free soldering, including printing, component placement, reflow, inspection, and test. The chapter also discusses some advanced miniaturization technologies used in the SMT process.

1.2 Lead-Free Solder Paste Alloys

Today, there are a variety of lead-free solder paste alloys available in the market. SnAgCu (SAC) materials with 3.0–4.0% Ag and 0.5–0.9% Cu and remainder Sn are widely accepted within the industry. Among them, Sn3.0Ag0.5Cu (SAC305) is still the most common alloy used in the SMT process. These SnAgCu alloys have the liquidus temperature of around 217 °C. As the cost of Ag has increased over the past years, the use of low Ag alloy materials such as Sn0.3-1.0AgCu or SnCu/SnCuNi has increased. These alloys have approximately 10 °C higher melting temperature than SAC305 and may need to be processed at slightly higher temperature during the reflow process.

Low temperature lead-free alloys which contain SnBi/SnBiAg are also used. These alloys have melting temperature around 140 °C and can be processed at 170–190 °C. These low temperature alloys usually have high bismuth content and they create some reliability concerns, especially on mechanical reliability. These

Lead-free Soldering Process Development and Reliability, First Edition. Edited by Jasbir Bath.
© 2020 John Wiley & Sons, Inc. Published 2020 by John Wiley & Sons, Inc.

low temperature alloys are used on certain applications such as light-emitting diode (LED)/TV products. In recent years, there is a desire for low temperature lead-free alloy alternatives with better reliability. The drivers for these low temperature alloys include component warpage, low energy consumption, and component or board sensitivity to the higher temperature lead-free process. These alloys typically have higher liquidus temperature than traditional SnBi/SnBiAg alloys, but they still have lower liquidus temperature than SAC305. These alloys have gained a lot of interest in the industry in the recent years, and some are available in the market and used in production.

1.3 Solder Paste Printing

1.3.1 Introduction

One of the most important processes of the surface mount assembly is the application of solder paste to the PCB. This process must accurately deposit the correct amount of solder paste onto each of the pads to be soldered. Screen-printing the solder paste through a foil or stencil is the most commonly used technique, although other technique such as jet printing is also used.

There is no major change to solder paste printing for lead-free processes. The same printer can be used for tin-lead and lead-free printing. In general, the same stencil design guidelines can be used for lead-free process.

1.3.2 Key Paste Printing Elements

Solder paste printing process is one of the most important processes in surface mount technology. This process can account for the majority of the assembly defects if it is not controlled properly. For effective solder paste printing, the following key factors need to be optimized and controlled:

- PCB support
- Squeegee (type, speed, pressure, angle)
- Stencil (thickness, aperture, cleanliness, snap off, separation speed)
- Solder paste (including type, viscosity)

PCB support is important to the printing process. Good PCB support holds the PCB flat against the stencil during the screen-printing process. PCB support is generally provided with the screen-printing machines. If the board is not properly supported, solder defects such as bridging, insufficient solder, and solder smearing can be seen. For fine pitch printing such 0.3/0.4 mm pitch chip scale package (CSP), 0201/01005 (Imperial) chip component, a dedicated custom-made fixture for printing or vacuum support should be used.

Squeegees, squeegee pressure, and speed are other critical parameters in the screen-printing process. Metal squeegees are commonly used for printing solder paste, and rubber or polyurethane squeegees are used for epoxy printing. A squeegee angle of 60 °C to the stencil is typically used [1]. Squeegee speed and squeegee pressure are critical for good printing. The speed of the squeegee determines how much time the solder paste can roll and settle into the apertures of the stencil and onto the pads of the PCB. In the beginning of lead-free conversion, a slower printing speed was used because the lead-free solder paste was stickier than tin-lead solder paste. Today, many lead-free solder pastes can print well at high speed.

The speed setting is widely varied from a typical range of 20–100 mm/s^{-1} depending on the size of the aperture, the size of PCB, and the quantity of boards being assembled, etc. Printing speed used depends on the solder paste supplier or is optimized by a Design of Experiment (DOE). It is typically between 40 and 80 mm s^{-1}. During the solder paste printing, it is important to apply sufficient squeegee pressure and this pressure should be evenly distributed across the entire squeegees. Too little pressure can cause incomplete solder paste transfer to the PCB or paste smearing. Too much pressure can cause the paste to squeeze between the stencil and the pad.

Stencil is another key factor in the solder paste printing. Metal stencils are used in solder paste printing. Stainless steel material is commonly used; however, metal stencils can be made of copper, bronze, or nickel [2]. There are several types of screen-printing stencil, including chemical etch, laser cut, and electroformed [2]. The thickness of the stencil is typically 125 μm (5 mil) or 150 μm (6 mil). Stencils with the thickness of 100 μm (4 mil) or thinner have become more popular with the high density and fine pitch components such as 0201/01005 (Imperial) chip components or 0.4/0.3 mm pitch CSP or quad flat no-leads/bottom termination component (QFN/BTC) components. Thicker stencils than 150 μm are typically used when more paste is needed. Stencil thickness and aperture size determine the amount of paste deposited on the pad. In general, stencil aperture must be three times and preferably five times the diameter of the solder particles. To ensure the proper paste release and efficient printing, the aspect ratio should be greater than 1.5, and the area ratio should be greater 0.66.

The aspect ratio is defined by Eq. (1.1), and the area ratio is shown in Eq. (1.2).

$$\text{Aspect ratio} = \text{Aperture Width/Stencil Thickness} \tag{1.1}$$

Area Ratio

$$= \text{Area of the Aperture } (L \times W)/\text{Area of Aperture Walls } (2 \times (L + W) \times T) \tag{1.2}$$

Snap off and stencil separation speed are also important for good printing quality. Snap off is the distance between the stencil and the PCB. For metal stencil

Figure 1.1 Example of tailing at the edge of the paste due to high separation speed.

printing, the snap off should be zero. This is also called contact printing. A high snap off will result in a thicker layer of solder paste. Stencil separation speed is the speed of separation between the stencil and PCB after printing. Traditionally, high separation speed will result in clogging of the stencil apertures or tailing at edges around the solder paste deposited (Figure 1.1). However, lead-free pastes tend to have a higher adherence than tin-lead pastes and may prefer high separation speed than tin-lead solder paste. Separation speed varies depending on the solder pastes and its supplier, and the supplier's recommendation should generally be followed.

Last but not least, the correct solder paste type and material should be used. The correct type of solder paste should be selected based upon the size of the apertures within the stencil. Type 3 was commonly used in the tin-lead process; however, Type 4 has become a more common lead-free solder paste type in the recent years due to the increase in miniaturized components on the printed circuit board. The

Table 1.1 General solder paste type and particle sizes.

Paste type	Particle size (μm)
3	45–25
4	38–20
5	25–15
6	15–5

release from the apertures of the stencil is affected by the particle size within the selected solder paste. Table 1.1 lists the particle size of different solder paste type.

Both tin-lead and lead-free solder paste should be refrigerated while being stored to maintain its shelf life but must be brought to room temperature before use to maintain quality. Some new lead-free solder pastes require no refrigeration and can be stored at room temperature. The solder paste should be mixed properly before use to ensure even distribution of any separated material throughout the paste. It is recommended to follow the solder paste manufacturer's recommendations for storage and handling conditions.

1.4 Component Placement

1.4.1 Introduction

After the correct amount of solder paste is applied, components are placed on the PCB at the specific locations. The component placement process includes board loading and registration, fiducial vision alignment, component pick-up, component inspection, and alignment and placement. The component placement must be precise and in accordance with the schematics. Pick and place machines are used in this process. There are different types of pick and placement machine available in the market. Some machines are designed specifically for speed whereas others are more focused on flexibility. The machines designed for speed are generally referred to as "chip shooters" and can achieve component placement rates of up to 100 000 cph (components per hour). The flexible pick and place machine can handle components ranging from 01005 (Imperial) chips to large components such as ball grid arrays (BGAs), connectors, etc. Flexible machines typically have slower pick and place speed than the chip shooter. The machines are selected depending on the types, sizes, and volumes of the surface mount components. The same pick and place equipment can be used for tin-lead and lead-free components.

1.4.2 Key Placement Parameters

Component placement is an important factor in surface mount assembly. It affects not only the assembly time but also the reliability of the solder joint. Placement accuracy and placement speed are critical in this process. To achieve accurate placement and high output, the following factors need to be considered:

- Nozzle
- Vision system
- PCB support
- Component size, packaging
- Feeder capacity

1.4.2.1 Nozzle

It is very important that the correct nozzle be selected for each different part to be placed to ensure accurate and consistent placement. There are many different types of nozzle for pick and place components. Most nozzles use a vacuum to hold the components. For handling small components, positive pressure is often supplied in addition to vacuum at the moment of placement so that the component would be completely released from the nozzle. Component flatness at the top surface is important for the pick and place process. Certain components such as connectors that do not have a flat top surface can have a pick-up pad inserted or pre-attached by the supplier for pick and place purposes. Some alternative nozzles have a gripper, which grips the component sides instead. The gripper is typically for placing some odd-shaped components. However, the placement speed is typically slower as compared to the nozzles that hold the component by vacuum. In addition, extra space is required between the components to accommodate the grippers.

1.4.2.2 Vision System

The vision system inspects every component before placement. It checks the part dimensions and any component damage before placement. It is important to program each component with the correct tolerance parameters to allow the machine to determine if an incorrect part has been loaded and also not to reject acceptable components.

1.4.2.3 PCB Support

The PCB needs to have adequate support during component placement. Improper PCB support can cause component misalignment or missing components.

1.4.2.4 Component Size, Packaging, and Feeder Capacity

The surface mount components on the PCB will differ in size. It is common to have small components positioned close to large components in high density design. All small components need to be placed before larger components so that the larger components do not get disturbed and misaligned during placement.

The surface mount components are supplied in different ways. The most common component packages are tape and reel, tubes, and trays.

1.4.2.5 Feeder Capacity

Feeders are used to feed components to a fixed location for the pick-up mechanism. Feeder types include tape and reel feeder, matrix tray feeder, bulk feeder, and tube feeder. The tape and reel feeders come in different sizes and are the most common feeder for placing large quantities of small components. The number of tape feeders that can be loaded into the machine at a time will play an important role in determining the speed of component placement. The matrix tray feeders are typically used for large and/or expensive components such as BGA or QFN/BTC components. The tray holds the components securely without damaging the body or leads. However, the pick and place process for the tray feeder is often slower than the tape feeder.

1.5 Reflow Process

1.5.1 Introduction

In reflow soldering, the solder paste and solder balls for the case of a BGA component must be heated sufficiently above its melting point and become completely molten, in order to form reliable joints. In the case of components with leads, the solder paste must wet the plating on component leads to form the desired heel and toe fillets.

There is no one best reflow profile for all board assemblies. Ideally, a reflow profile must be characterized for each board assembly using thermocouples at multiple locations on and around the component devices and board. The solder paste type, component, and board thermal sensitivity must be considered in reflow profile development.

Lead-free solders typically process at higher temperature than tin-lead solder due to the high melting temperature of typical lead-free solders. Lead-free solder such as SAC305 (Sn3Ag0.5Cu) have an initial melting point of 217 °C and a final melting point of 220 °C. Lead-free reflow typically has a narrower process window than tin-lead reflow due to the component or board maximum temperature limitations.

1.5.2 Key Parameters

Solder joint formation depends on temperature and time during reflow. There are four phases of a reflow process, including preheat, soak, reflow, and cooling. In addition, reflow atmosphere plays an important role in the reflow process. The key parameters for reflow will be discussed in the following sections.

1.5.2.1 Preheat

The preheat phase prepares the PCB and components for actual reflow. It helps to reduce the thermal shock and temperature difference between the PCB and components and reflow temperature. A quick ramp rate during the preheat can damage the component. In general, a ramp rate between 1.0 and 3.0 °C/s^{-1} is recommended, and the temperature change should be evenly distributed throughout the PCB. Preheat also removes some flux volatiles and prepares the solder paste material for reflow.

1.5.2.2 Soak

Soak is also known as the pre-reflow phase. In this phase, the flux in the solder paste gets activated, and this helps to remove oxidation on the component leads, PCB pads or on the solder particles' surface. Also, soaking phase allows the thermal gradient across the PCB to equilibrate prior to reflow. In this way, the entire assembly sees nearly the same reflow conditions to form consistent solder bonds. For large boards or boards with a large range of component sizes, a longer soak time is usually helpful to achieve successful assembly to help ensure the delta Temperature across the board is reduced. Soak profiles are also used to minimize voiding when assembling such components as BGA, land grid array (LGA), and QFN/BTC.

1.5.2.3 Reflow

As the solder reaches the solder melting temperature, the board enters the reflow phase. Peak temperature and time above liquidus temperature are important factors in this phase. The peak temperature is generally 20–30 °C above the liquidus temperature of the alloy, and reflow time is typically 30–90 seconds in order to form a good solder joint and proper intermetallic formation at the interfaces.

A typical reflow profile chart is shown in Figure 1.2, and typical profile parameters are listed in Table 1.2.

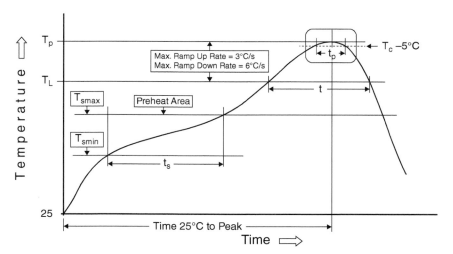

Figure 1.2 Reflow profile chart [3]. Copyright 2019 by IPC International, Inc. and is used with IPC's permission. This image may not be altered or further reproduced without the prior written consent of IPC.

Table 1.2 Typical lead-free profile parameters for lead-free.

Parameters	Typical lead-free profile
Preheat ramp rate	1–3 °C s^{-1}
Preheat and soak temperature range	110–210 °C
Preheat and soak time	60–180 s
Reflow time	30–90 s
Peak temperature	235–255 °C (for alloy liquidus temperature of ~217 °C)
Cooling rate	~3–4 °C s^{-1}

1.5.2.4 Cooling

Cooling affects the grain structure of the solder joint. Fast cooling rate results in fine grain structure which is assumed to have a more reliable solder joint and bond. However, too fast a cooling rate can exert thermal stress on the solder joint, which can result in fractures or tears on the solder joint. In general, cooling rate should not exceed 3–4 °C s^{-1}.

1.5.2.5 Reflow Atmosphere

Lead-free reflow can be done in both air and nitrogen environment. Reflow soldering in an inert atmosphere such as nitrogen reduces the solder oxidation during reflow and results in better wetting and appearance of the solder joint. However, nitrogen adds additional cost to the reflow process. Today, most lead-free reflow in manufacturing can be done in an air environment. For fine pitch components and some advanced assembly such as flip chip assembly and package on package (PoP) components, a nitrogen atmosphere is recommended

1.6 Vacuum Soldering

Lead-free soldering results in more voiding in the component solder joint as compared to tin-lead soldering. In addition, the assembly of BTCs generates more voids at the thermal pad solder joints and causes some quality and reliability concerns. Reflow soldering using vacuum or vacuum soldering becomes an attractive solution. Vacuum soldering uses gases and pressure to create a vacuum environment for reflow. Vacuum soldering helps to eliminate or minimize voiding

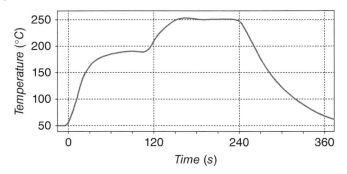

Figure 1.3 Example of lead-free profile using a vacuum reflow oven.

to a few percent. Hydrogen atmosphere is also used. Vacuum soldering is not a new technique and it has been used for a long time for various applications such as brazing (hard soldering). The use of vacuum in soldering electronics has increased in recent years.

Although vacuum soldering can significantly reduce voids for most electronics components, vacuum soldering has its own concern. Vacuum soldering is typically a batch process which can limit the throughput of the assembly. However, some in-line vacuum soldering equipment is also available. In addition, vacuum soldering requires a higher temperature rising rate and a longer reflow time which generates some concerns regarding the components and solder joint reliability. An example of a lead-free profile generated using vacuum soldering is shown in Figure 1.3.

1.7 Paste in Hole

Paste-in-hole or pin-in-paste technology is also known as through-hole reflow. Paste in hole is a method for inserting and soldering plated-through-hole (PTH) components during the surface mount process. The process involves screen-printing in and around PTH barrels, inserting PTH components after SMT placement, and reflowing the solder paste in the reflow oven to form the SMT and PTH solder joints simultaneously. Paste-in-hole reflow eliminates the wave solder process on selected PTH components and may reduce manufacturing costs and improve cycle times.

However, there are some restrictions with the paste-in-hole process. First, the plated-through hole component must be able to survive the reflow process temperature. Second, all PTH components for pin-in-paste process must be on one side of the PCB. In addition, the pin and paste process typically works well with only thin boards. The board thickness for pin-in-paste assembly should be less than 2.4 mm (0.093″), and the ideal thickness for this process should be ≤1.6 mm (0.062″).

A large amount of solder paste is required to fill the hole for a larger board thickness. When the pin-in-paste technology is used for thicker printed circuit boards (≥2.4 mm), an overprinted stencil aperture and/or solder preform can be used to be able to achieve good hole fill for PTH components.

1.8 Robotic Soldering

Robotic soldering is a soldering method which uses robots to solder surface mount or pin through hole components onto the printed circuit board. A soldering robot is an automated system that performs soldering tasks with precision and repeatability. Robotic soldering can help to eliminate human error from manual soldering and help to improve throughput and yield. Robotic soldering is typically used for point to point soldering applications, but it is also available for line soldering applications. Robotic soldering is used in many industries, such as Automotive Electronics and Solar Modules, Aerospace, Military and Medical. An example of a soldering robot is shown in Figure 1.4. Both benchtop and in-line robotic soldering systems are available in the market.

Figure 1.4 Example of a soldering robot. Source: Courtesy of Japan Unix. https://www.japanunix.com/en/ products/automation.

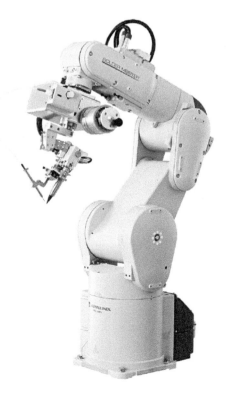

1.9 Advanced Technologies

Advanced surface mount technologies relate to advanced packaging and product miniaturization. Miniaturization is the trend in surface mount technology today. Both PCB and component sizes are getting smaller and smaller while the number of signal inputs and outputs is increasing along with better performance expected.

The following sections will discuss some advanced technologies that are getting popular in surface mount technology such as Flip Chip and PoP.

1.9.1 Flip Chip

Flip chip assembly is the direct connection of active die onto printed circuit boards, regardless of interconnect material and method. In typical component packaging, the interconnection between the die and the carrier is made using wire. The die is attached to the carrier face up. A wire is bonded first to the die, then looped and bonded to the carrier. In contrast, the interconnection between the die and carrier in flip chip packaging is made through a conductive bump that is placed directly on the die surface. The bumped die is then "flipped over" and placed face down, with the bumps connecting to the carrier directly. A bump is typically 60–100 μm high, and 80–125 μm in diameter. The flip chip connection is formed by using solder or conductive adhesive. By far the most common flip chip interconnect method is solder, and will be discussed in the next section.

The flip chip assembly to the board typically requires no paste. Instead the bumped flip chip is dipped in flux before placing and reflow. Flux is used to hold the die in place during the assembly and to help remove oxide from the pads and component bumps during the reflow process. Nitrogen atmosphere is required for flip chip assembly. The nitrogen prevents oxidation during reflow and promotes good wetting of the pads. The bare die is typically underfilled for protection and enhancement of thermal and mechanical reliability.

1.9.2 Package on Package

PoP is a method to solder two or more BGA packages vertically. This allows higher component density in devices. This component packaging is commonly used in consumer products such as mobile phones, digital cameras, etc. An example of package on package is shown in Figure 1.5.

Package on package technology provides many benefits. First, PoP uses less PCB real estate of the motherboard. Second, PoP results in better electrical performance of devices because PoP shortens the length between different interoperating parts,

Figure 1.5 Package on package (PoP) component.

such as the controller and the memory. Shorter routing between the circuits yields faster signals and reduces noise and cross talk. In addition, PoP allows the memory package to be separate from the logic packaging. The user can test these components separately, change the components based on the product requirements or select different suppliers for the package.

Package on package can be processed using standard surface mount machines equipped with a dip-fluxing unit. The bottom package, along with the other SMT components on the board, is placed first. The top part is then dipped in flux (or solder paste) and placed. The flux (or paste) material must be carefully selected, and the flux amount and thickness must be controlled to get a high yield. Nitrogen is highly recommended for better yield.

1.10 Inspection

1.10.1 Solder Paste Inspection (SPI)

Since most PCB solder joint defects relate to the solder paste printing process, it is critical to have a good solder paste printing inspection method. In the beginning, solder paste height was manually inspected and tracked for solder paste printing control. However, solder paste volume was then shown to link to the solder joint quality and reliability, and thus it needs to be monitored closely. Automatic solder paste inspection systems were then developed to address the need for manufacturers to be able to monitor solder paste alignment and volume during the printing process. Solder paste inspection can be 2D or 3D inspection. 2D solder paste inspection checks the area of the paste deposit only while 3D inspection captures the height of the solder paste and enables the equipment to accurately measure the total volume of the paste deposited. The 3D Solder Paste Inspection (SPI) is commonly used to monitor and control the paste printing process and helps to correct any defects that are caused by solder printing early in the assembly process before the solder joint is formed. An example of 3D solder paste inspection image is shown in Figure 1.6.

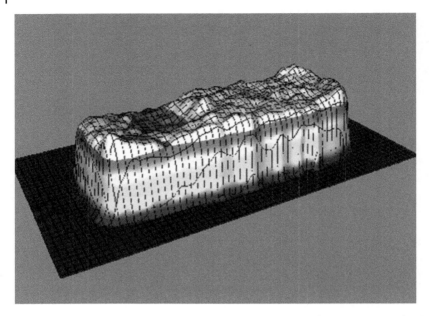

Figure 1.6 Solder paste inspection image. Source: Courtesy of Koh Young America, Inc.

1.10.2 Solder Joint Inspection

Solder joint inspection is necessary to ensure the quality and reliability of the solder joint. There are different ways to inspect the solder joint such as visual inspection (by the human eye or by using optical devices (magnifying glass, microscope, etc.)). Some solder joints are hidden under the components and cannot be visually inspected and need to be X-rayed for inspection of assembly defects. For example, the solder joints of BGA or BTC components are inspected using this technique. Automated optical inspection (AOI) and automated X-ray inspection (AXI) are common systems used in manufacturing to ensure the quality and reliability of the solder joint.

1.10.2.1 Automated Optical Inspection (AOI)

AOI is an automated visual inspection system used in the SMT process to identify failure and quality defects such as wrong component placement, missing component, component misalignment, bridging, missing solder, etc. Automated optical inspection can be done in line or off line, and it can be used pre or post reflow process. AOI is much more reliable and repeatable than manual visual inspection. AOI uses camera and machine vision systems to provide images for defect analysis. The surface of the board is visually scanned using several light sources and high definition cameras to create the picture of the board. The captured image is then

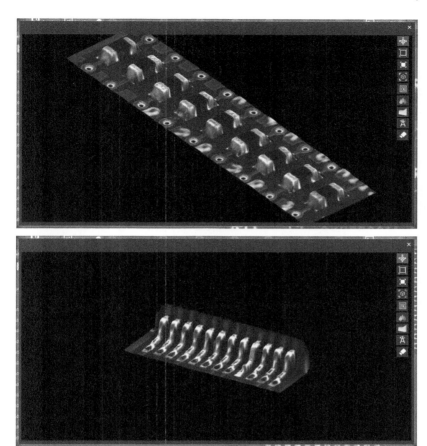

Figure 1.7 Example AOI inspection images of lead-free soldered chip (top) and lead-frame components (bottom). Source: Courtesy of Cyberoptics Corporation.

processed and compared with the knowledge the machine has of what it should look like. Using this comparison, the AOI system can detect defects and generate a report. A golden good board or known status board and design information is needed for generating the database of what the solder joint should be. Examples of AOI images for lead-free soldered chip and lead-frame components are shown in Figure 1.7.

1.10.2.2 X-ray Inspection

Automatic optical inspection works well for printed circuit boards where the solder joints are visible. When the solder joint is under the package and hidden such as in the case of BGA, CSP, and BTC components, X-ray inspection is

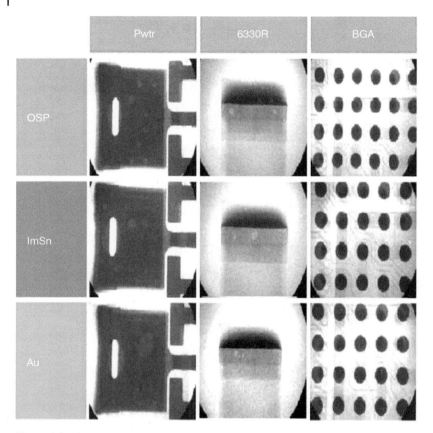

Figure 1.8 Example of power transistor/BTC, chip, and lead-free Sn3Ag0.5Cu BGA component solder joints soldered with lead-free Sn0.7Cu1.1Ag0.06Ni1.8Bi solder for three different board surface finishes (organic solderability preservative, immersion tin, NiAu) under X-ray inspection [4].

needed. X-ray inspection can also reveal other solder joint defects such as voiding or head-on-pillow (HoP). Manual or automated x-ray inspection is used. AXI is typically used in the production environment. Example images of X-ray inspection for lead-free soldered Power Transistor/BTC, 6330 chip and BGA components are shown in Figure 1.8 for three different board surface finishes.

1.11 Conclusions

In general, surface mount technology for lead-free soldering is similar to tin-lead soldering. Lead-free soldering requires higher process temperatures and has a narrower process window than tin-lead. Therefore, the process parameters should be closely controlled and optimized for high yield.

Besides traditional pick and place and reflow methods, the surface mount components can also be assembled using some alternative methods such robotic soldering or vacuum soldering. As the products get smaller while the performance is increased, the use of advanced miniaturized technologies such as fine pitch components (0.3/0.4 mm pitch component), smaller chip components (0201/01005 Imperial), PoP components and flip chip component has increased with surface mount technology becoming the major and dominant technology to assemble most electronics components today.

References

1 Mohanty, R., Claiborne, B. and Andres, F. (2010) Effect of squeegee blade on solder paste print quality. SMTnet, 17 June. https://smtnet.com/library/index .cfm?fuseaction=view_article&article_id=1630\ignorespaces2010 (accessed 3 February 2020).

2 Gopal, S., Rohani, J., Yusof, S., and Bakar, Z. (2006). Optimization of solder paste printing parameters using design of experiments. *Journal Teknologi* 43 (A): 11–20.

3 IPC/JEDEC J-STD-020E (2015) *Moisture/reflow sensitivity classification for non-hermetic surface mount devices.* January 2015. IPC International.

4 Nakatsuma, M., Wada, T., Mori, K. et al. (Koki Company Limited) (2013) A study of lead-free low silver solder alloys with nickel additions. SMTAI conference.

2

Wave/Selective Soldering

Gerjan Diepstraten

Vitronics Soltec, Oosterhout, The Netherlands

2.1 Introduction

This chapter covers selective and wave soldering. Both processes have a lot in common, but there are significant differences that should be understood to run them successfully. Flux and preheat processes are described in general, but where necessary the differences are pointed out in the following sections.

2.2 Flux

2.2.1 The Function of a Flux

To understand what a flux does, one should wave solder a board without flux on it. The result would be solder dross on the solder mask, bridges, solder balls and no hole filling and solder webbing all over. The following makes clear what a flux does for the soldering:

- It is a film between solder mask and solder (avoiding adherence of the solder to the board)
- Cleans metal surfaces by removing oxides and makes soldering possible
- Promotes the wetting of the solder (for better hole fill)
- Eliminates solder bridges and solder balls

The functions of the flux are identical for wave and selective soldering. However, there is a significant difference in the flux spreading. In wave soldering the flux should cover the whole bottom side of the assembly that runs over the liquidous solder. In selective soldering the flux should only be applied on the areas that

Lead-free Soldering Process Development and Reliability, First Edition. Edited by Jasbir Bath.
© 2020 John Wiley & Sons, Inc. Published 2020 by John Wiley & Sons, Inc.

contact the small solder fountains in the nozzles. There should be no flux in the surrounding or SMD areas of the assembly. For this reason one should not use the same flux for wave and selective soldering. The selective flux should spread less, have a higher solids content, and be inert after soldering.

2.2.2 Flux Contents

More detailed information about flux can be found in the IPC standard J-STD-004 [1]. Flux is a liquid that consist of >90% of a solvent. The solvent is the carrier for applying the "solids" (chemistry dissolved in the solvent) to the bottom side of the board. Popular solvents are alcohol and demineralized water. During the preheating of the assembly the solvents will evaporate. This is required to avoid spattering when the board enters the liquidous solder. There are criteria to classify a flux. "No-clean" fluxes are less aggressive and do not need to be cleaned after soldering. The rosins or resins will cover the remaining activators after soldering and the residue is safe enough to leave on the board. Water-soluble fluxes are stronger/more active and make soldering easier. However, their residues need a cleaning process. There is a trend to clean the "no-clean" flux also for products that require a conformal coating or for high reliability products. Keep in mind that these fluxes are not designed to be cleaned, so the cleaning is harder to do.

The other 5–10% of chemistry in the flux is activators that clean the metal surfaces and improve the wetting speed of the solder, and surfactants. The surfactants are a type of soap that changes the surface tension and lets the flux flow more easily into the small barrels of the through-hole components. Special water-based fluxes need surfactants to make them flow. The surfactants for wave soldering fluxes are different from selective wave fluxes.

The selective soldering fluxes are typical "no-clean" rosin-based fluxes that have a higher solid contents (10–20 wt%). For this process water-based VOC-free fluxes are not preferred since these fluxes are hygroscopic – attracting water molecules from the environment. This makes them sensitive for electromigration reliability issues.

2.3 Amount of Flux Application on a Board

If you are fortunate, the flux supplier has a number on the datasheet of the flux in terms of the amount of flux to apply. However, the majority of the datasheets does not say anything about the amount of flux required to make a good solder joint. Those datasheets that contain flux concentrations indicate about 500–2000 $\mu g/in^2$. This is a wide tolerance which also is good since the amount itself is not so critical. Twenty percent more or less from the target will not show insufficient soldering or excessive flux. The main reason for there not being an exact number for all applications is because the amount is very much dependent on the solderability of the materials and the thermal mass of the assembly. There should be care taken to avoid applying too much flux since this may generate more solder balls. In selective soldering the application location of the flux is critical as there needs to be enough flux in the solder area and in the barrels.

There are some process indicators that will tell if more flux or less flux is required. In case of open joints, spikes, solder webbing or bridging, there might be more flux required since all these issues indicate a lack of flux activity. Too much flux may result in solder splashes, solder balling, or electromigration reliability issues after soldering. The end user may not like flux residues for cosmetic reasons or pin testability. Most flux processes apply just enough flux to have a robust process without excessive flux residues. It should be kept in mind that more flux will not always solve a process issue.

2.4 Flux Handling

There are some major differences between an alcohol-based and a VOC-free water-based flux. The alcohol will not freeze where the VOC-free water-based flux should be kept above the freezing point. The alcohol flux is a hazardous chemistry that requires special shipping documents and safety precautions. It should be ensured that the flux is not expired before using it. In case of doubt one can check acid content and density to verify its properties. If a water-based flux was frozen, the supplier should be contacted for handling guidelines.

All solid particles should be dissolved in the solvent liquid. However, shaking the jar before using is always recommended. The user needs to be aware that there can be contamination at the bottom of the jar. Therefore it is preferred to have a filter in the supply tubes. For proper storage conditions the datasheet of the flux should be reviewed.

2.5 Flux Application

2.5.1 Methods to Apply Flux (Wave Soldering)

There are different methods to apply a flux to the printed circuit board (PCB), but spray systems are commonly used for wave soldering. The foam and drum fluxer used in the earlier days of wave soldering have been reduced, with less interest because of two reasons:

1. Difficult to change the flux amount. Process window was tight.
2. Open storage system; the alcohol could evaporate and thus a density control procedure or tooling was needed.

Some high-volume production lines in Asia still have some foam flux units running. It applies an amount of flux which is hard to control, but the benefit is that the flux penetrates very well into the copper barrels, which results in good hole filling. If the flux density is not properly controlled, the flux will have a higher solid content when a part of the alcohol is evaporated. This will improve the solderability but leave more flux residues behind.

For the majority of the wave soldering applications spray fluxers are used. This can be an ultrasonic or an atomized air spray application. Ultrasonic generates a fine mist of flux. The disadvantage of this is it has no force to penetrate the flux into the barrels. The advantage of the foam fluxer – the capillary force of the flux – is missing in the ultrasonic fluxer. All other properties of the ultrasonic fluxer are excellent. The flux amount control, the low deviation of the flux, no clogging, and a closed system are all benefits to have in a robust and consistent fluxing process. The more advanced systems have a cleaning cycle after production to remove the flux residues in the ultrasonic head.

An air atomized spray system mixes the air and flux just before applying it. The spray nozzle has a body and an air cap. The air cap provides the atomized air at a pressure of approximately 0.3–0.7 bar. The higher the air pressure, the more bouncing of the flux on the board. Most nozzles have different settings for water-based fluxes due to the different density of water compared with alcohol-based fluxes. Due to the air pressure there is a better capillary power for flux penetration. Some applications are optimized to have a better flux deposition; with more equal deposition over the PCB.

Spraying can be done when the nozzle moves from front to rear side and backwards which generates more overlap or only in one direction (front to back) to get a more equal deposition of the flux. Flux amounts can be varied by pump speed or air/nitrogen pressure depending on the kind of supply system which is installed. The system is more sensitive to clogging of flux. There are systems in the market that have a flux flow control to identify any clogging of the nozzles. Advanced systems purge solvent through the air lines after the production is stopped to blow out all remaining flux.

Servo-jetted spray is an alternative for spraying. Multiple small jets are installed next to one another and the flux is jetted through these nozzles to the bottom side of the board. When aggressive OA (organic acid) type fluxes are used, compatible jets should be selected. Air knifes help to support the barrel fill (Figure 2.1).

The multiple jets on a unit apply a wide band of flux. The transfer speed of the fluxer can be reduced because of the wide flux width. Due to the lower speed the flux will penetrate much better into the PCB copper barrels which will benefit the hole filling during soldering. Since the jets have a very small orifice, proper cleaning after fluxing is required to prevent clogging. The start and stop of the fluxing can be adjusted to selectively spray the board or pallet.

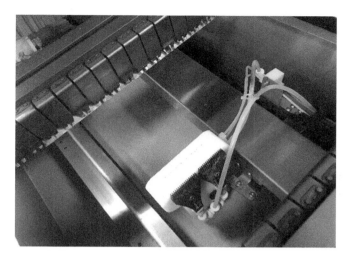

Figure 2.1 A head servo jet fluxer with air knife.

2.5.2 Methods to Apply Flux (Selective Soldering)

Selective soldering machines use a different technique to apply the flux on the PCB. The main reason for this is the small amount of flux required to be applied very precisely on the area where the solder is touching the bottom side of the board. The systems use dropjet nozzles that are capable of applying very fine amounts of flux. Typical for a dropjet is an orifice opening of 100–300 μm, which defines the flow range.

The system pressure, orifice diameter, valve on-time, and frequency are the parameters that define the drop size and flux volume applied. Most of the applications have limited the parameters by fixing the pressure and/or frequency. Dropjets are calibrated and require a software module that is capable of making corrections to ensure they all spray the same amount under the same conditions (Figure 2.2).

The amounts of flux applied for selective soldering are very small, sometimes up to 20 times less than in wave soldering. The flux concentration (percentage of solids) may be higher to get better soldering results. Solder temperatures are higher, which require more active/stronger flux and more concentration of flux. The flux deposition position and amounts are more critical compared to wave soldering. In order to have a robust fluxing process, flowmeters and laser sensors are used to control flux volume and deposition. Clogging and improper spray direction can be detected by these sensors.

Programming of the fluxing recipe is different from wave soldering. In wave soldering the fluxer settings may be general for multiple boards, but in selective

Figure 2.2 Drop jet with robot unit and flux supply hoses.

soldering each recipe has its own fluxing program. Solder areas can have a flux line dragged or have only a dot sprayed to the solder spot (without movement of the robot). The droplet size is bigger when the on-time of the valve is longer. The flux supply is a closed system and therefore there is no evaporation of solvents and there is no necessity for monitoring of the flux density.

2.6   Preheat

2.6.1   Preheat Process-Heating Methods

For both wave and selective soldering processes preheating is an important part to activate fluxes and evaporate the solvents in the flux chemistry. In addition, it brings heat into the assembly necessary for good hole filling and reduces the thermal shock when the board contacts the solder. There are three methods that are commonly used to heat up a PCB. The method used is the result of the thermal mass of the PCB and the required throughput. Available methods are:

- IR lamps
- Electric tubular heaters
- Forced convection

An infrared lamp transfers energy to the lower temperature PCB through electromagnetic radiation. Depending on the temperature of the emitted body, the wavelength of the peak of IR radiation is between 750 and 3000 nm. A part of the transferred energy will be reflected by the assembly and another part will be absorbed. How much is dependent on the wavelength and the heated material.

There are two ways to control the IR lamps: by changing the power to create another wavelength or by switching the lamp continuously on and off with a certain frequency. The last method requires more power and reduces the lifetime of the lamps. The IR lamps are often used in the full tunnel wave machines since they do not interfere with the nitrogen system. The lamps are typically mounted on the bottom side of the conveyor but can also be applied as a support heating system on the topside. In selective soldering the lamps can be mounted above the soldering station to keep the assembly at a consistent temperature during the whole soldering cycle (Figure 2.3).

IR lamps respond very fast to temperature setting changes. The reaction time of an IR lamp is between 1 and 3 seconds. This makes it very suitable for mixed production lines. In a barcode-driven line it is a fast way to have different temperature settings for different products. A disadvantage for the IR heaters is the temperature difference across the board (dT). The PCB can have areas that absorb more heat than other parts and additionally the IR lamps are not consistent in temperature over the length of the tube.

Electric tubular heaters are used widely for the purpose of producing radiant heat in wave soldering machines. These heaters are typically mounted underneath

Figure 2.3 IR lamps on bottom and top side.

the conveyor only. The heaters can come in different shapes, with heating easily and evenly distributed. The heating up time of these heaters is approximately 90 seconds. The heating elements are very robust and durable. The heating is consistent over the width of the conveyor (resulting in a small dT across the board). The power is only one parameter that defines the temperature of the heaters.

Forced convection heaters blow hot air from the top or bottom side of the board. Advanced systems also have the ability to change fan speed for more efficient heat transfer. Hot air distributes the heat more equally across the PCB. The result is a smaller dT over the assembly. The topside heating also brings significantly more energy into the assembly, which improves the hole filling. The risk of blowing hot gas to the bottom side of the board is that the high temperatures may degrade the flux if the flux is not robust enough. Depending on the activation system and rosins that are used, some fluxes cannot withstand high hot air gas temperatures (Figure 2.4).

The total length of the preheating is important when products are soldered in high volume or if the product has a high thermal mass. Some of the wave solder machines are offered with multiple preheat modules which are easily exchangeable. This process flexibility might be needed for factories that deal with a large variety of products (high and low thermal mass). For a high mix production, using IR lamps in the last part of the preheat gives the flexibility for running in high volume without large changeover times.

Preheating in selective soldering is different than for wave soldering. Some low-cost small selective soldering machines do not have a preheat option. On this

Figure 2.4 Forced convection preheater wave solder machine.

type of soldering machine only low thermal mass products can be soldered with low throughput. The board is fluxed and the solvents of the flux will evaporate when the board is transferred above the solder pot. The heat of the pot increases the flux temperatures just high enough to have the alcohol evaporated before the joint is soldered.

Preheating the assembly before and after fluxing is another possible process configuration. The background of this is to minimize the flux spreading. A pre-heated board will reduce the spread of flux when it is an alcohol-based flux, but the spreading will increase for water-based fluxes since water becomes more mobile when it is heated. The drawback of the minor spreading is that the flux may not penetrate enough into the barrels of the through holes, resulting in hole filling issues.

The other difference between preheating in selective soldering versus wave is that the conveyor is not moving in selective soldering. In a wave application the conveyor does not stop, but in selective soldering the board hits a blocking device so the position of the soldered part toward the heating elements is always the same. This affects the delta T (dT).

The majority of the selective soldering lines have IR lamps for preheating. The lamps are mounted in parallel to the conveyor underneath the conveyor. For smaller boards not all lamps have to be heated, but only those located underneath the board. The heat distribution over the lamps is not consistent. The middle part of the lamps heats more efficiently. There is a potential risk that part of the assembly is overheated and part of the assembly has not enough heat to get good soldering performance.

The higher-end selective soldering machines may have one or two forced convection heating stations. The principle of these units is identical to the wave soldering units, with top and bottom side heating as well as adjustable fan speed.

2.6.2 Preheat Temperatures

The preheat temperature depends on the flux used in combination with the thermal mass of the assembly. Typically the flux supplier defines the required temperature and documents this in the flux datasheet. It is recommended that the topside board temperature be listed for reference. What setting and time is required to achieve this temperature depends on the thermal mass of the assembly. A multilayer 0.135″ (3.2 mm) thick board requires higher settings versus a thinner 0.062″ (1.6 mm) thick board. For good hole fill, higher preheat temperature is better. The solder will solidify during soldering when the topside is too cold; there is not enough energy in the board. Also, the thermal shock will be less when the board enters the liquidous solder when the temperature is higher. On the other hand, too high a gas temperature may degrade the flux or some components on the topside of the board which are sensitive to high temperatures. Typically a flux can withstand gas temperatures not higher than 160 °C. The hot gas will damage the flux activation system and poor soldering (bridging, poor hole filling) will be the result.

2.6.3 Preheat Time

In general, the preheat profile should have a linear slope to the required topside board temperature. The slope should be not higher than 3 °C s^{-1}. A faster heat may damage components. For a typical topside temperature of 120 °C and ambient temperature of 20 °C, the shortest preheat time is 33 seconds. In a high-volume selective soldering application where the solder time is approximately 30 seconds, there should be at least two preheat stations of a maximum 30 seconds preheat cycle – including transport – to avoid the preheating becoming the bottleneck. If the assembly has a lot of copper or thermal mass, there should be an additional topside heater to keep the topside temperature consistent during soldering. The longer the board is heated to this temperature, the less active the flux will be. The activators in the flux may evaporate slowly, even though their activation temperature is around 180 °C.

In a wave solder line the throughput is defined by the preheat length and the contact time during soldering. For a typical preheat slope of 1.5 °Cs^{-1} and 120 °C topside temperature, this requires a preheat length of 1000 mm when the conveyor speed is 90 cm min^{-1}. A typical larger wave soldering machine will have three preheat modules of 600 mm, so in total a 1800 mm heated length. Smaller machines will have two preheat modules, which affects the throughput.

2.6.4 Controlling Preheat Temperatures

Preheat modules may drift in temperature over time. During start-up the machine will be cold. After many production hours the temperature inside the machine may drift slightly when the exhaust is not sufficient enough. In order to have a consistent topside temperature a pyrometer can be installed that reads the topside board temperature on a predefined spot. Depending on the measured value, the power of the heaters can be adjusted to maintain the same temperature during the day. Modern solder machines record this data in the database and temperatures are stored for each board for quality traceability.

In selective soldering machines with a robot that picks up the board to bring it to the solder position, it is possible that the board has to wait because of unexpected events. In that case, the pyrometer is used to control the power of the IR lamps in such a way that the temperature of the topside of the board is maintained constant and the board is ready to be picked up at anytime.

2.6.5   Board Warpage Compensation (Selective Soldering)

The PCBs that need to be soldered have already one or two heating cycles (during reflow) and may warp after preheating. This warpage is very critical for point to point soldering. If the board bends too much during dragging of the board, the nozzle may touch the pins and lift the component. A board warpage compensation is available in most selective machines. This unit (typically a laser sensor) measures the bending over the length and width of the board and the software will compensate the nozzle position during soldering if required.

2.7   Selective Soldering

2.7.1   Different Selective Soldering Point to Point Nozzles (Selective Soldering)

In all liquid solder applications, the solder is pumped through a wave former or nozzle. A point to point solder application is similar to hand soldering but instead of manual soldering the robot determines the position and contact time of the soldering. The wave formers – nozzles – in point to point soldering can have many different shapes depending on the application and assemblies to be soldered. The nozzles can be divided into two groups:

1. Wettable nozzles
2. Non-wettable nozzles

A wettable nozzle is made out of cold rolled steel. The characteristic of this material is that it can be tinned. The advantage of tinned nozzles is that they overflow in all directions, which allows soldering front to back or sideways without turning the PCB or solder pot. They need to be cleaned properly to avoid oxidation. The base material will wear out – steel walls will get thinner – which is why they need to be replaced periodically. Wetted nozzles can come closer to surface mount devices during soldering without washing them off. The wave height of wettable nozzles is limited. Special designs like a "bullet" nozzle have increased wave heights. The critical aspect is the nozzle material. Wettable nozzles are made from a ferrous alloy that has a limited amount of wear and still has good wetting properties. To keep them wettable some machines offer automatic tinning systems. Some use a liquid or adipic acid to clean while others use mechanical methods to remove oxides (Figure 2.5).

Non-wettable nozzles are made from stainless steel and have a special treatment to be compatible with lead-free alloys. Their lifetime can be more than a decade without cleaning since the solder does not interact with the metal. Non-wettable nozzles make the solder flow in one direction, just like in wave soldering

Figure 2.5 Wettable nozzle with de-bridging tool.

applications. Therefore the selective soldering equipment with a non-wettable nozzle needs a rotating device for soldering.

2.7.2 Solder Temperatures (Selective Soldering)

The solder temperatures for selective soldering are slightly higher than for wave soldering (260 °C). However, they are much lower than lead-free hand soldering temperatures for rework wire (400 °C). Typically the solder temperature is in between 280 and 300 °C. The smaller nozzles need more heat to get enough energy into the assembly and have complete hole fill. Some engineers do not want to have temperatures above 300 °C because of fillet/pad lifting or flux activation loss. In addition, the oxidation becomes more critical at higher temperatures, resulting in more dross and an increased risk for solder joint bridging.

What is different from wave soldering is that there is no significant difference in solder temperature for a typical Sn3Ag0.5Cu alloy versus a low silver Sn1Ag0.5Cu lead-free alloy with a higher melting point. For both alloys the superheat (difference between operating and solidification temperature) is high enough to get good solder joints for both types of lead-free alloys. A more critical issue would be for lead-free six-part alloys (such as Sn3.7Ag0.65Cu3.0Bi1.43Sb0.14Ni) for high reliability applications which will lose the specific as-received alloy composition during production soldering because of the high temperature and oxidation level and small amounts of solder used in the solder pot. The composition of the alloy will drift and leave mixtures of high melting areas. Elevated soldering temperatures are needed, which makes these alloys not suitable for a robust selective soldering process.

2.7.3 Dip/Contact Times (Selective Soldering)

The contact time in selective soldering depends on the wetting speed of the solder joint, component lead, and board pad. The thermal mass of the lead, board thickness, and copper layers, board surface finish, and solder alloy are material properties that have an impact on the dip/contact time. Machine parameters like nozzle diameter, wave height, solder temperature, and dip time affect the wetting. A low thermal mass solder joint will have good hole fill after 2 seconds. In general a dip time of 8 seconds is the maximum to avoid overheating and board pad or solder fillet lifting. For the pad the Cu thickness is typically 0.018 mm and Cu dissolution rate is 0.001 mm s^{-1} for SAC305 [2]. Longer lead lengths also influence the solder hole filling since solder contact is the best way to transfer heat into the solder area. Typically longer leads will have a better hole fill.

2.7.4 Drag Conditions (Selective Soldering)

Many times, a connector with multiple leads needs to be soldered. This can be a double or even triple row connector up to 32 leads per row. For cycle time and heat transfer the nozzle dimension is a critical parameter. If there is enough space a wider nozzle is preferred. A wider nozzle comes with more energy (faster wetting) and multiple rows are soldered in one soldering drag. It should be noted that a connector may lift or the board may warp when soldering a very long connector. For those long pin connectors it is better to fix the connector first by dipping the front and rear leads in the solder pot before dragging the complete connector. The drag speed for a non-wettable nozzle can be slightly higher because of the flow energy to one side of the nozzle. There are two output characteristics that define the drag speed: hole filling and solder bridging. For wettable nozzles a lower drag speed is better, to avoid solder bridging. Using non-wettable nozzles in combination with a de-bridging device is faster. Typical drag speeds for this type of process are 2–10 mm s^{-1}. The drag speed can be the robot moving the PCB over a fix positioned solder pot or a moving solder pot under a fixed PCB.

2.7.5 Nitrogen Environment (Selective Soldering)

Whether a wettable or non-wettable nozzle is used for point to point soldering, the environment around the nozzle should be inert. The solder temperatures at the nozzle are so high that oxidation will occur in air and solder bridging cannot be avoided. An air environment will also generate too much dross in the small solder pot. It is important that the nitrogen flow over the pot is smooth with a low laminar flow. If the nitrogen is blowing too fast, the fine tin oxide film that is generated will be blown onto the bottom side of the PCB and adhere to the solder mask. A well trained engineer is able to see if the oxygen level at the solder area is low enough. Typically the solder joint will be shinier when the oxygen level is low. Some machines have a separate nitrogen supply to the nozzle and solder area. Preheating nitrogen before injecting it over the solder pot is also an alternative. It is hard to measure oxygen levels during soldering, but close to the nozzles the oxygen level should be in the 50–100 ppm O_2 range for a robust soldering process.

2.7.6 Wave Height Controls (Selective Soldering)

The contact time depends on the wave height of the solder. It is important that the wave height is stable over time to avoid soldering defects. There are different methods to verify the wave height. This can be done with a needle checking at which Z-height there is contact with the solder. Another method is using a camera that checks the wave shape when there is no board in the wave. Most common is to move the solder pot and wave former to a photoelectric fork-style sensor that checks the height with a beam.

When a needle makes contact the software is able to define the wave height. Critical with this type of measurement is the oxide film on the solder, which may act as an insulator. It is also possible that tin oxides adhere to the needle (although it is made of titanium), resulting in an incorrect reading. A camera can define the highest point of the solder in between two production boards. There is a minimum loss in cycle time when a camera is used. It is different from a beam photoelectric fork sensor that combines sender and receiver in a single housing. For this sensor the soldering point has to move to the measuring position, which affects the cycle time.

When a magnetic solder pump is used, the wave height is influenced by the temperature of the coil. A lower coil temperature results in a higher solder wave. Temperature compensation or closed loop wave height measurement are methods to maintain the wave height consistently for a more robust soldering process. The wettable nozzles have a lower wave height compared to non-wettable nozzles. Longer leads may require dip soldering instead of dragging the connector over the wave.

2.7.7 De-Bridging Tools (Selective Soldering)

Selective soldering is sensitive to solder bridging. Solder bridges occur when the solder does not separate from two or more leads before it solidifies. Solder temperatures are high and lead-free solder tends to bridge when flux activation is lost. Solder bridging can be avoided by applying the correct design rules. Short leads of components, wider pitch and small pads should be considered. In order to overcome solder bridging the wettable nozzle should have good peeling properties. The attraction of the steel to the solder makes the solder drain from the leads.

For finer pitch components with longer leads the wettability of the nozzles may not be enough to peel off the solder without bridging. For those applications a small tube blowing hot nitrogen over the leads can remove the bridge and make a more stable process. The gas temperature, gas flow, and the angle the gas is blown toward the leads with the solder in between determine the yield. The gas temperature must be well above the melting point of the solder otherwise the gas will solidify the solder and generate a bridge instead of removing the solder. The flow should be higher than the capillary force between the pins that hold the solder. Typical values are 2–10 lpm nitrogen. Too high a nitrogen flow may disturb the flow of the solder and cause other process issues, like solder depression.

2.7.8 Solder Pot (Selective Soldering)

In selective soldering the position of the overflowing nozzle (solder pot) toward the PCB must be accurate and repeatable. The distance from the pin pad to be soldered toward adjacent components is often not more than 1.00–2.00 mm. For this reason, the solder pot is mounted on an accurate guiding system that guarantees a consistent position. Not only the X and Y axis should be accurate but also in the Z-direction the solder pot position must be accurate. In the Z-direction the distance from the solder nozzle toward the bottom side of the PCB is defined. This distance is, however, often influenced by the warpage of the PCBs. In most applications selective soldering is the second or third heating process after reflow soldering. Due to these reflow cycles the assemblies are warped when positioned in the soldering area. A board warpage compensation can be established in the software to maintain a consistent distance in the Z-direction and keep the contact time identical for all soldered assemblies. When non-wettable nozzles are used or de-bridging with hot gas the solder direction (which is the move direction of the robot) is dependent on the orientation of the component on the board. In these applications there is an additional rotation required of the solder pot or PCB. For wettable nozzle applications this rotation is not required since the solder is circumferentially overflowing. Some selective solder machines have the solder pot fixed and move the PCB in a gripper system. Most of today's machines leave the PCB in the conveyor and move the solder pot to the solder positions to reduce cycle time.

Although soldering under an angle like in the traditional wave solder machines has a preference mostly the in-line selective soldering machines cannot tilt the board. There are two different methods to pump the solder. This can be done traditionally with a pump and impeller like in wave soldering. Electromagnetic pumps are the alternative. Some machines have more than one solder pot. This can be for several reasons: to reduce cycle time, to use different alloys for different products, or to have different nozzle diameters for different components.

2.7.9 Topside Heating during Soldering (Selective Soldering)

Selective soldering applications can be very complex and product boards can have a high thermal mass. Some assemblies are mounted in housings that absorb a lot of heat. Other assemblies have multiple copper layers connected to the board barrels that make the heat flow away rapidly from the solder area. To have enough energy in the board a topside heater above the solder nozzle is a common option. For most of the applications this is an IR lamp heater that keeps the topside board temperature equal during the soldering. A pyrometer is used to control it. Also, for applications with many connectors where the cycle time can be over 10 minutes it is useful to have an additional heater. So, one should apply a topside heater during soldering if the process cycle is long or when the assembly has a high thermal mass and it is hard to get enough solder hole fill. A disadvantage of the topside heater is that it takes away the view on the soldering processing. A process camera is a good feature to have to watch the soldering on the screen or on a remote computer.

2.7.10 Selective Soldering Dip Process with Nozzle Plates (Selective Soldering)

The soldering dip process with nozzle plates solders all joints in one dip. The solder time is independent of the number of solder joints. The application is designed for high volumes (short cycle times) of the same product. To make this process possible each product has its own dedicated plate with special nozzles on the solder areas. To maintain a low oxygen level in this area, each product has its own cover. The PCB design should have some keep out areas where the through-hole components are located. Typically, there must be 1.00 mm more space around the rim outside of the nozzle. The nozzles should be made from a material that is compatible with the high lead-free solder temperatures. It should have a coating or diffusion layer that prevents the material from dissolving into the solder. Otherwise the lifetime of the nozzles is limited, and the solder becomes impure, which leads to soldering issues.

The PCB is positioned above the nozzles and the solder is pumped to fill the board barrels with solder. Several parameters are important to achieve a proper hole fill and avoid solder bridging. The contact time with the solder defines how well solder wetting will be. The program can be setup such that the solder speed is reduced so that the solder drains from the soldered pins while the board is still in place. The deceleration of the soldering speed is critical. The other method is that the solder pump speed remains constant, and the board is lifted out of the solder with a defined speed. Here the lifting speed of the robot for the board is critical. To keep the board flat some of the nozzle plates have position pins on which the board sits, and that defines the distance between topside of the nozzles to the bottom side of the board (Figure 2.6).

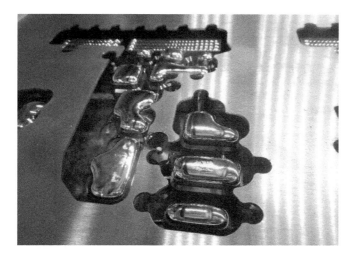

Figure 2.6 Dip solder process with nozzles covered by shield to generate nitrogen solder environment.

2.7.11 Solder Temperatures for Multi-Wave Dip Soldering (Selective Soldering)

For good wetting the solder temperature should be high. In this dipping process it depends on the design of the nozzle. If the nozzle has overflowing solder, the temperature can be slightly lower. If not, the temperatures for lead-free solders are 280–320 °C. Although it is preferred to stay below 300 °C, some high thermal mass board can only be soldered properly at 320 °C.

Nevertheless, there are applications where soldering can be achieved with lead-free solder at 260 °C, with similar temperatures to wave soldering processes. This can only be done in combination with a very well-performing flux. Manufacturers that run these types of application typically solder with rosin-based high solid content (10–20% solids) no-clean fluxes.

2.7.12 Nitrogen Environment (Selective Soldering)

As with all selective soldering applications, the solder area should be inert to avoid oxidation. An inert environment also reduces the dross formation of the solder. A minor issue of the low oxygen content is that overflowing solder will splash and generate small solder balls. Special shields can be installed to avoid solder balls bouncing and adhering to the solder mask of the board. Preferably the nitrogen is supplied through a porous tube, which gives a very smooth flow. The amount of flow depends on the dimensions of the solder pot. Typically, around 75 lpm is supplied for the solder pot that handles up to 300×250 mm dimension boards. Soldering engineers can evaluate by visual inspection of the soldered fillets whether the nitrogen environment is low or not. In the case of low oxygen levels (below 100 ppm O_2) the solder joints are shinier.

2.7.13 Wave Height Control (Selective Soldering)

The wave height is a function of several parameters and conditions. The amount of solder in the pot, the pump speed, the amount of dross, soldering temperature, and inert environment all have influence on the actual solder height. There are automatic solder feeder systems that calculate how much solder is consumed and automatically add more wire to the pot if necessary. Keeping the solder level in the solder pot consistent does not guarantee a consistent wave height. Due to the fact that every multi-wave plate is different and customized for the assembly, the easiest method to verify wave height is using a needle that is mounted on the robot. The needle goes down in the robot and contacts the solder. The needle determines when there is contact with the solder. When there is contact with solder there is electrical contact. The Z-position of the robot is known and can be verified. In the software the frequency of wave height check can be set. Some assemblies require a wave height check after each board, but in most cases wave height checking after 10 boards is accurate enough. Wave height verification by a needle requires time and will affect cycle time. Other applications have a dummy nozzle with a sensor for continuous wave height control.

2.7.14 Dip Time – Contact Time with Solder (Selective Soldering)

When soldering areas that are free of oxides and the solder temperature is high enough, a contact of 2–3 seconds should be long enough to get proper solder hole filling. Unfortunately, selective soldering is the second or third heating cycle after reflow soldering. The reflow conditions can be very critical and influence the soldering results on selective soldering afterwards. Studies in the industry show that reflow under nitrogen atmosphere and peak temperatures lower than 245 °C are recommended to keep critical board finishes like copper organic solderability preservative (OSP) and immersion Sn still solderable. Also, the time between the last reflow cycle and selective soldering should be limited. Other studies in the industry have showed the importance of the flux. When applying enough flux, the thermal mass of the board is less critical. However, when the flux activation is not enough to remove the surface oxides, board hole filling will be insufficient.

In general, a contact of more than 8 seconds will not be accepted as components will be damaged, or the soldered board material may show board pad or solder fillet lifting. Eight seconds is acceptable for 0.018 mm thick Cu pads.

2.7.15   Solder Flow Acceleration and Deceleration (Selective Soldering)

Multi-wave soldering can be done in different ways. The solder can be overflowing when the board is dipped into the nozzle. Most of the processes have the solder level low and increase the pump speed when the board is positioned above the nozzles. The acceleration of the solder is critical to get enough heat to the solder area. The vertical force of the solder can be used for fast wetting and good hole filling. The acceleration of the solder pump is defined in rpm/s. The target is to purge solder into the copper board barrels without overflowing the nozzles otherwise solder oxides will adhere to the solder mask of the board.

Deceleration of the wave is even more important. The target of an optimized deceleration is to eliminate solder bridging. The preferred deceleration depends on the connector to be soldered. Component lead length and pitch are critical parameters. Most of the assemblies require a low deceleration rate. Faster deceleration of the solder generates more solder bridges.

2.7.16   De-Bridging Tools (Selective Soldering)

Most applications have no overflowing solder. This makes it very sensitive for solder bridging. The solder must be free of oxides before dipping otherwise the pins of the connector will bridge. To make sure that all oxides are removed from the molten solder, applications have a flush cycle prior to the soldering. For some seconds the pump speed is increased to make the oxides drain off from the solder.

When a flush cycle is not enough to eliminate bridging, there are dedicated tools to reduce the defects. A good PCB design with good solderable connectors will not require additional tooling. However, also in selective soldering the board layouts are not always properly designed, and the materials used are lower cost, which reduces the process window. Lead protrusion length of the connectors is another factor that influences bridging but it is very hard to change without adding additional costs. Bridge-free soldering requires tools like screens or wettable strips in the nozzles to drain the solder from the leads. Screens are made from non-wettable material. The metal sits in between the pins when the board is positioned for soldering. The screen is a barrier between solder and the solder mask of the PCB. When the board is soldered and transported away from the nozzle, the screen will drain the solder and eliminate bridging. Screens can only be used in accurate solder machines that use robots that are able to position the board in a consistent position. Screens are very fragile due to the small pitch between the leads. Sticky flux and solder oxides need to be removed to maintain a robust process.

A wettable strip mounted in the nozzle lower than the leads of the connector will drain the solder from the component pins when the solder level in the nozzle decreases. The physics behind these strips are identical to that of the wettable nozzles as previously discussed. The wettable material attracts the solder. These wettable strips have a tendency to oxidize when they are exposed to air. Since they are wettable, there is an intermetallic layer that grows over time and results in wear of the strip. Wettable nozzles need maintenance and replacement after time.

2.7.17   Pallets (Selective Soldering)

Selective soldering can be done with or without pallets. If the board dimensions allow it, it is more convenient to run without pallets. Pallets require more handling, need to be cooled down after soldering and need to be cleaned.

Pallets have benefits too, in that a pallet is easier to keep the board flat or it is possible to add tooling to keep it flat during preheat and soldering. In a pallet it is also possible to make shielding to avoid overheating of surface mount device areas on the bottom side of the PCB. In a multi-wave dip soldering process, the pallet can also be used for positioning during soldering. A guiding pin will fit into the hole located in the pallet which guarantees a very consistent position. Tools to keep components in place can also be part of the pallet.

During preheating the pallet will absorb a lot of heat. The preheat configuration must allow the use of pallets. The pallet material is exposed to excessive heating cycles. For a consistent process it is important that the pallet is cold when entering the machine to have similar conditions over time. The material used should be able to remain flat and be compatible with the multiple heating cycles. Most pallets are made from fiber-reinforced plastic materials.

2.7.18   Conveyor (Selective Soldering)

The conveyor in a selective soldering machine must transfer the board from one sub-process to the other. This should be done quickly since it influences the cycle time. If the acceleration or speed of the conveyor is too fast, components may drop, so there is a speed limitation.

There are different concepts used. For transportation of the PCB/pallet the pin chain is commonly used. For heavy pallets roller chains are used to reduce the wear of the chain. In the solder area a pin conveyor requires a large amount of space. A nozzle cannot solder close to the pin because of the conveyor construction. Disk conveyors require less keep out area and are therefore used in solder stations and robots.

2.8   Wave Soldering

2.8.1   Wave Formers (Wave Soldering)

A wave solder machine will have a single or dual wave. The dual wave is introduced to overcome open solder joints of surface mount device components on the solder side. The shadowing effect of surface mount device components in molten solder may result in solder opens and by introducing a turbulent or chip wave before the main wave this problem is eliminated. The ideal main wave starts with a turbulent wave that pushes solder into the board barrels and ends with a laminar wave that eliminates solder bridging.

After the introduction of lead-free soldering the wave formers and solder pots were made from titanium, cast iron, or specially treated stainless steel to avoid erosion of the material. Several different designs are available in the market with their specific characteristics.

The first wave can either be a turbulent wave former or a chip wave. The chip wave is a laminar wave that flows in the transport direction of the product to be soldered. This chip wave makes the first contact so brings the heat into the assembly and solders the components including the shadow side of the surface mount devices. In case a turbulent wave is used this will have a small amount of contact with the board. The turbulent wave will also cover the shadow side. For both wave formers the contact length will be short, in the range of 1–2 cm. The contact time should be limited, otherwise all flux activity is gone before the board enters the main wave (Figure 2.7).

The main wave or second wave former should be installed as close as possible to the first wave. If the distance between both waves is too large, the solder will solidify in between, and it will be hard to achieve an acceptable board hole fill. The main wave has a longer contact time necessary for good hole fill. Several solutions are available for generating a turbulent wave. There is a "smart"-wave that uses

Figure 2.7   Double wave former with turbulent first and laminar main wave.

a hexagon shaft in front of the wave to make the solder turbulent. Another wave former that generates a turbulent wave is the "Worthmann" wave. A thick massive plate with a defined hole pattern generates small solder fountains. The latest designs of wave formers have an extended solder contact area. Since the wetting of lead-free soldered products is more difficult more thermal energy and a longer dwell time capability is required.

Some engineers prefer only one wave former for lead-free applications. The reason is twofold: the flux activation might be lost after the first wave, and in between the waves the solder will cool down. For easy-to-solder applications this can be a laminar wave that flows to front and backside. For more thermally challenging boards the wave former should have a turbulent part. Some of the machines can increase the pump speed when the board is running over the wave so that more thermal energy and solder penetrates the small diameter copper board barrels.

2.8.2 Pallets (Wave Soldering)

Using pallets in a wave soldering process can have advantages when design rules are adhered to. The pallet material can cover surface mount device reflowed components on the solder side, which creates a selective soldering process on a wave solder machine. Pallets do absorb a lot of heat, which requires more energy during preheating. There should be no flux underneath the edges of the PCB where it sits in the pallet as the presence of flux may wick the solder in between PCB and pallet. The advantage of a pallet on the wave is that the solder can be pumped with more power without the risk of overflowing (solder on top of the board) which benefits hole filling. There should be channels on the bottom side of the pallet that allow the solder to drain out and enough space surrounding a copper board barrel so that the solder can contact the component pin and board pad.

There should be enough pallets for each product that has to be soldered. The pallets need to be cooled down before reuse. Pallets should be cleaned in a wash machine after a period of time.

2.8.3 Nitrogen Environment (Wave Soldering)

Wave soldering in an inert environment benefits solder wetting which results in less solder hole fill issues. In a nitrogen atmosphere the solder may have more solder balls because of the different surface tension of the solder in nitrogen. It is preferred to have hot gas injected into the solder area as cold nitrogen may freeze the solder and may generate more solder bridging. Apart from the inert environment nitrogen also reduces the solder dross formation of lead-free solder dramatically. The advantage of less dross is not only to reduce the alloy consumption cost, but also result in less maintenance and soldering defects.

There are many different alternatives for wave soldering in nitrogen. Some machines are complete inert tunnel systems for the preheat and solder area. Other options have only a nitrogen shroud over the solder pot. Nitrogen consumption therefore is different from 10 to 30 m^3 h^{-1} for a tunnel concept. There is no need to go below 100 ppm O$_2$ with oxygen levels in the wave solder environment. With 200 ppm O$_2$ a robust soldering process is feasible. Wave soldered joints assembled in a nitrogen environment are shinier.

2.8.4   Process Control (Wave Soldering)

It is recommended to have a wave height measurement in the solder pot. When the wave height is controlled the contact time will be consistent. The level in the solder pot should be checked and solder must be added when required. Not only does the solder level in the pot define the wave height, but also the dross has an influence. Too much solder dross would reduce the solder flow and the pump speed would need to be increased.

The contact time depends on the assembly and solder temperature. A higher temperature requires less contact time to achieve the same soldering quality. Common contact times for low mass products are 2–3 seconds. For multilayer boards and high thermal mass assemblies in lead-free processes, 10 seconds of contact time is about the maximum. Solder temperatures are limited during wave soldering. Where Sn37Pb tin- lead soldering requires about 250 °C, for lead-free 260–265 °C is a common solder temperature for wave soldering. Temperatures above 270 °C may damage components or result in solder fillet lifting or board pad lifting and copper board dissolution. The solder temperature for SnCu-based alloys will be slightly higher (5–10 °C) due to the higher melting point. SnAgCu are preferred for thicker boards with high thermal mass. SnCu alloys are selected for less challenging assemblies. The lack of silver (Ag) in the lead-free SnCu solder results in less potential for metal erosion (damage of the stainless steel parts in the solder pot) compared with SnAgCu lead-free solder and lower alloy cost.

2.8.5   Conveyor (Wave Soldering)

Wave solder machines have finger conveyors with the exception of some that run with pallets and uses a chain. The fingers can have different shapes and depending on the load they must carry they need to be more robust and strong. Some conveyors run small low mass PCBs whereas others transport heavy pallets. There are V-shape fingers for PCBs up to 5 kg. For heavy pallets up to 13 kg typical L-shape type fingers are used. Some fingers are made of stainless steel, but others are titanium which has a better compatibility with the lead-free alloys.

Special tools and options are in the market to keep the fingers clean and free of solder. Some are brushes, but also alcohol or other liquids are used to have fingers free of solder and flux residues. Critical with L-shaped fingers is that it is possible to get solder in between the L finger and the pallet. This solder may lift the pallet and affect soldering performance. It may cause open solder joints. Another available option for a conveyor is crash detection. If the outfeed conveyor is blocked and a board may crash for some unexpected reason these units will switch off the conveyor and solder wave to avoid more damage. The conveyor must run smoothly, and especially when the board is soldered there should be no external stress applied on the molten solder.

2.9 Conclusions

In liquidous soldering the flux is the important chemistry to clean the surfaces and support the wetting of the metal surfaces to be soldered. Preheating is the process prior to soldering that evaporates the solvents of the flux, activates it and applies more thermal heat into the assembly. During soldering the wetting is defined by solder materials, temperature, and contact time. After the contact with the liquidous solder, bridging can be avoided by blowing hot gas over these locations. Pallets are used to cover surface mount devices and keep the boards flat during soldering.

Selective soldering is different from traditional wave soldering. Because it is selective, the flux should be supplied in specific solder areas and not spread further. The small solder fountains have higher solder temperatures to wet the surfaces. Nitrogen should support the soldering by eliminating oxidation and improving the wetting at these higher process temperatures.

References

1 IPC J-STD-004. *Requirements for soldering fluxes*. IPC International.

2 Kennedy, J., Hillmann, D., and Wilcoxon, R. (2013) NASA DOD Phase 2: SAC305 and SN100C copper dissolution testing. International Conference of Soldering and Reliability, Toronto (16 May).

3

Lead-Free Rework

Jasbir Bath

Bath Consultancy LLC, San Ramon, CA, USA

3.1   Introduction

In addition to challenges in lead-free soldering assembly, there are issues during rework for the assembled components based on the higher soldering temperatures and the range of small and large components to rework as well as temperature and moisture sensitivity with components and boards.

The chapter will review updates in lead-free rework technology including hand soldering, (ball grid array) BGA/CSP (chip scale package) rework, and PTH (Pin Through Hole) rework, which will include lead-free alloys used, rework temperatures encountered, and equipment used.

3.2   Hand Soldering Rework for SMT and PTH Components

3.2.1   Alloy and Flux Choices

3.2.1.1   Alloys

There are many alloy choices for lead-free hand soldering rework based on the alloys used for the assembly operations. Typically the lead-free rework alloys used are Sn3-4Ag0.5Cu (mp: 217 °C), Sn3.5Ag (mp: 221 °C), and Sn0.7Cu (mp: 227 °C) based alloys. Based on assembly operations for surface mount technology (SMT) being Sn3AgCu, the hand soldering rework wire would typically be Sn3Ag0.5Cu.

For assembly operations for wave, Sn3AgCu or Sn0.7Cu based alloys could be used, with the hand soldering rework wire typically being Sn3Ag0.5Cu or Sn3.5Ag. Sn0.7Cu based hand soldering rework alloys could also be used, with

Lead-free Soldering Process Development and Reliability, First Edition. Edited by Jasbir Bath.

the advantage of lower cost with the reduced silver content. This could potentially require the need for higher soldering iron tip temperatures, especially for thicker, more thermally demanding boards.

3.2.1.2 Flux
In terms of hand soldering rework flux trends, there are choices in terms of the use of tacky flux versus liquid flux versus cored rework wire with flux contained in the rework wire. The general trend for hand soldering rework is to use cored rework wire or tacky flux. There are reliability concerns on the use of liquid rework flux due to hand soldering heat activation questions with the use of liquid flux, which requires liquid fluxes to be validated to be reliable if not exposed to sufficient heat during rework soldering. The same concerns in terms of flux reliability would also need to be addressed for tacky fluxes.

3.2.2 Soldering Iron Tip Life

One area which should be considered for hand soldering applications is the soldering iron tip life. This is more of a cost consideration if there is increased of use of soldering irons in production, with the higher erosion rates of the lead-free high tin solder alloys versus tin-lead solder alloys during the hand soldering operation. The soldering iron tip, which typically has a layer of iron over the copper, is subject to erosion during high temperature hand soldering rework operations. An example of a soldering iron tip cross-section is shown in Figure 3.1.

Studies have shown different erosion rates of copper and iron in high tin containing lead-free solders compared to tin-lead solders as shown in Figures 3.2 and 3.3. The copper plate erosion test results are shown in Figure 3.2. Based on the results, the anti-erosion alloy (Sn0.3Ag0.7Cu0.04Co) has less erosion than Sn0.7Cu, which has less erosion than Sn0.7Cu0.3Ag, which has less than Sn3Ag0.5Cu. This shows the benefit of the cobalt addition to the lead-free solder to reduce copper erosion.

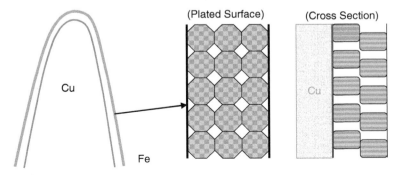

Figure 3.1 Soldering tip cross-section [1].

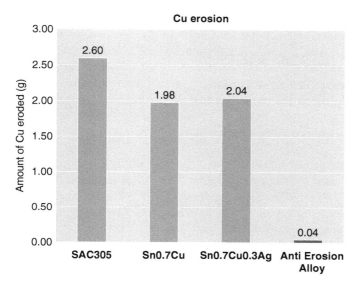

Figure 3.2 Erosion results of copper plate on different alloys (anti-erosion alloy is Sn0.3Ag0.7Cu0.04Co) [1].

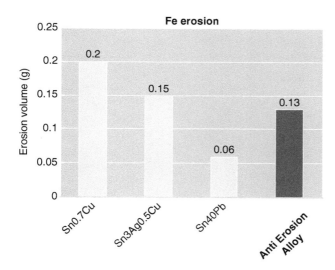

Figure 3.3 Erosion results of iron plate on different alloys (anti-erosion alloy is Sn0.3Ag0.7Cu0.04Co) [1].

The iron plate erosion test results are shown in Figure 3.3. Based on the results, the 60Sn40Pb tin-lead solder has less erosion than the anti-erosion lead-free alloy (Sn0.3Ag0.7Cu0.04Co), which has less erosion than Sn3Ag0.5Cu which has less than 99.3Sn0.7Cu. This shows the benefit of the cobalt addition to the lead-free solder to reduce iron erosion and indicates that the iron erosion is more dependent on the tin content of the solder alloy, with higher tin content giving higher iron erosion.

These studies have been confirmed during hand soldering operations with soldering irons as shown in Figures 3.4 and 3.5.

The results of the soldering iron tip erosion test in terms of iron plating erosion thicknesses are shown in Figure 3.4, which indicates that the Sn0.3Ag0.7Cu0.04Co solder alloy has a much lower solder iron tip erosion rate than lead-free Sn3Ag0.5Cu alloy and is similar to tin-lead solder.

After 10 000 shots (uses), the soldering iron tips used with the three solder alloys were cross-sectioned as shown in Figure 3.5. Based on the results, solder iron tip erosion is large with Sn3Ag0.5Cu alloy but is reduced with the Sn0.3Ag0.7Cu0.04Co alloy which is similar or better than the solder iron tip erosion rate with tin-lead solder.

Based on analysis of the cross-sections, it was found that for the Sn0.3Ag0.7Cu0. 04Co lead-free alloy there was a barrier layer being formed as a result of the reaction between the solder iron tip and the melted solder. This barrier layer was

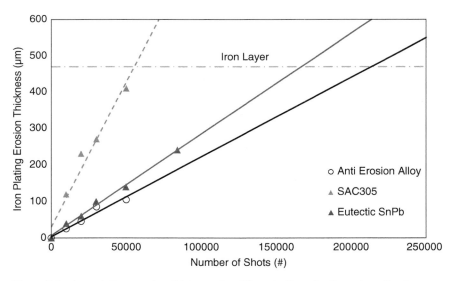

Figure 3.4 Iron plating erosion thickness for different alloys. (anti-erosion alloy is Sn0.3Ag0.7Cu0.04Co) [1].

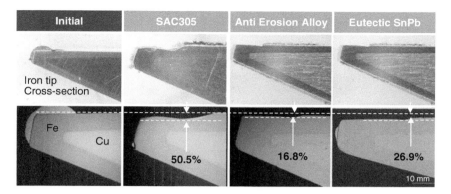

Figure 3.5 Cross-section of solder iron tips used with different alloys (Sn3Ag0.5Cu, Sn0.3Ag0.7Cu0.04Co, Sn40Pb) after 10 000 shots (uses). [1].

Figure 3.6 Schematic view of reaction between solder iron tip and melted solder (Sn3Ag0.5Cu compared with Sn0.3Ag0.7Cu0.04Co alloy) [1].

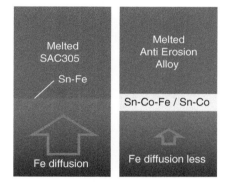

found to be Sn-Co-Fe/Sn-Co, which helped to reduce the solder iron erosion rate compared with Sn3Ag0.5Cu solder, as shown schematically in Figure 3.6 [1].

3.2.3 Hand Soldering Temperatures and Times

Soldering temperatures and times for lead-free hand soldering rework depend on the type of component to rework and thermal mass of the board being soldered to. The maximum soldering iron tip temperatures and times are indicated for tin-lead and lead-free soldering in Tables 3.1 and 3.2.

Tables 3.1 and 3.2 show the maximum peak soldering tip temperatures and times should be similar for both tin-lead (mp: 183 °C) and lead-free Sn3Ag0.5Cu (mp: 217 °C) or Sn0.7CuNi (mp: 227 °C) hand soldering, even though the melting temperature of the lead-free solder used is higher. This helps to ensure that there is minimal thermal damage to the components or board during hand soldering rework.

Table 3.1 Tin-lead hand soldering technology rework trend [2].

Soldering process	Parameter	Units	2015	2017	2019	2021	2027
SnPb	Soldering iron peak temperature used	°C	375	375	375	375	400
	Total contact time	s	6	6	6	**6**	**8**
	Smallest pitch to be reworked by hand	mm	0.35	0.3	0.275	0.25	0.25
	Smallest type of discrete being reworked imperial/[metric]	—	0201/[0603]	01005/[0402] and [03015]	01005/[0402] and [03015]	008004/[0201]	008004/[0201]
	Type of wire alloy used	—	Sn37Pb	Sn37Pb	Sn37Pb	Sn37Pb	Sn37Pb

Table 3.2 Lead-free hand soldering technology rework trend [2, 3].

Soldering process	Parameter	Units	2015	2017	2019	2025	2027
	Soldering iron peak temperature used	°C	375	375	375	375	400
	Total contact time	s	6	6	6	6	8
	Smallest pitch to be reworked by hand	mm	0.35	0.3	0.275	0.25	0.25
Pb-free	Smallest type of discrete being reworked imperial/[metric]	—	0201/[0603]	01005/[0402] and [03015]	01005/[0402] and [03015]	008004/[0201]	008004/[0201]
	Type of wire alloy used	—	SAC305/SnCuNi (low tip dissolution alloys)	SAC305/SnCuNi (low tip dissolution alloys)	SAC 305/SnCuNi (low tip dissolution alloys)	SAC305/SnCuNi (low tip dissolution alloys)	SAC305/SnCuNi (low tip dissolution alloys)

Table 3.2 above shows the trend to use Sn3Ag0.5Cu and Sn0.7CuNi lead-free cored wire for lead-free hand soldering operations. The nickel addition to SnCu lead-free solder helping to reduce soldering iron tip dissolution similar to cobalt additions to lead-free tin-based solder.

3.3 BGA/CSP Rework

3.3.1 Alloy and Flux Choices

3.3.1.1 Alloys
Alloy choices for lead-free BGA/CSP rework are based on the alloys used for the assembly operations. The lead-free rework alloys used would be chosen from Sn3-4Ag0.5Cu (mp: 217 °C), Sn3.5Ag (mp: 221 °C), and Sn0.7Cu (mp: 227 °C) based alloys. Lead-free Sn0.7Cu based alloys are lower in cost but are higher in melting temperature than Sn3Ag0.5Cu. Due to the higher board and component temperatures encountered during lead-free BGA/CSP rework, Sn3Ag0.5Cu solder is generally used.

3.3.1.2 Flux
In terms of BGA rework flux trends, there are choices in terms of the use of solder paste versus tacky flux. The general trend for BGA rework is to use Sn3Ag0.5Cu solder paste for larger-sized BGA component rework where solder joint volume and reliability requirements would be more of a concern with tacky flux used for smaller CSP components based on ease of use and speed in rework operations. The discussion to use solder paste or tacky flux depends on the type of BGA component being reworked and the end-use reliability requirement of the product.

3.3.2 BGA/CSP Rework Soldering Temperatures and Times

Soldering temperatures and times for lead-free BGA rework depend on the size of the component to rework and the thermal mass of the board being soldered to. The maximum BGA component package and solder joint temperatures and times are indicated for tin-lead and lead-free soldering in Tables 3.3 and 3.4.

Tables 3.3 and 3.4 show the maximum peak component package and solder joint temperatures and times for tin-lead (mp: 183 °C) and lead-free Sn3Ag0.5Cu (mp: 217 °C) BGA rework. The maximum component package temperature and time closely follow the maximum component package temperature and time ratings used by component suppliers who qualify their moisture-sensitive components according to IPC/JEDEC J-STD-020 standard [4]. This helps to ensure that there is minimal thermal damage to the components or board during BGA/CSP rework.

Table 3.3 Tin-lead BGA rework [2].

Soldering process	Parameter	Units	2015	2017	2019	2025	2027
	Maximum package temperature	°C	220–240	220–240	220–240	220–240	220–240
	Minimum reworkable pitch	mm	0.35	0.3	0.25	0.25	0.2
	Target solder joint temperature	°C	205	205	205	205	205
	Target delta T across solder joints	°C	<10	<10	<10	<10	<10
	Typical rework profile length (time)	min	6–8	6–8	6–8	6–8	6–8
	Typical time above liquidus at solder joint (TAL)	s	45–90	45–90	45–90	45–90	45–90
	Number of allowable area array reworks at a specific location	#	3	3	3	3	3
	Type of rework (Conv./IR/other) (other is laser and vapor phase rework)	%	80/15/5	80/15/5	70/15/15	70/10/20	60/10/30

Table 3.4 Lead-free BGA rework [2, 3].

Soldering process	Parameter	Units	2015	2017	2019	2025	2027
	Maximum package temperature	°C	245–260	245–260	245–265	245–270	245–270
	Minimum reworkable pitch	mm	0.35	0.3	0.25	0.25	0.2
	Minimum solder joint temperature	°C	235	235	235	235	235
	Target delta T across solder joints	°C	<10	<10	<10	<10	<10
	Typical rework profile length (time)	min	8	8	8	8	8
	Time above liquidus (TAL)	s	60–90	60–90	60–90	60–90	60–90
	Number of allowable area array reworks at a specific location	#	3	3	3	3	**3**
	Type of rework (Conv./IR/other) (other is laser and vapor phase rework)	%	80/15/5	80/15/5	70/15/15	70/10/20	60/10/30

It should be noted in Table 3.4 that there is a trend going forward for some components to exceed the current maximum component temperature requirements during lead-free BGA rework which are indicated in IPC/JEDEC J-STD-020 standard, with temperatures over 260 °C being forecasted.

3.3.3 Component Temperatures in Relation to IPC/JEDEC J-STD-020 and Component/Board Warpage Standards

3.3.3.1 IPC/JEDEC J-STD-020 Standard

During lead-free BGA/CSP rework, the biggest challenge is to maintain a low delta temperature between the solder joint and component package during the rework operation. As already indicated, the standard that the industry uses to help define the component temperature and time limits during lead-free BGA rework is IPC/JEDEC J-STD-020 [4]. This standard indicates that for lead-free SMT assembly processing the moisture-sensitive components are rated to 3x 245 °C or 3x 260 °C peak reflow temperature, which is related to package thickness/volume. In addition, a 1x 260 °C rating is to be conducted for BGA/CSP parts not rated to 260 °C based on package thickness/volume to simulate area array/BGA rework. The time above 217 °C for this 260 °C rating profile should be 60–150 seconds to simulate the time that the component could reach during rework.

3.3.3.2 Component Warpage Standards

In addition to concerns during lead-free BGA/CSP rework of component temperature ratings being exceeded, which could cause damage to the component and boards, there are concerns with warpage of components during rework. Warpage of components can cause solder joint assembly and reliability issues.

JEDEC JEP 95 publication [5] previously only referred to measurements of component coplanarity/flatness at room temperature. The maximum package warpage at room temperature was 3 mils (75 μm) to 8 mils (200 μm), dependent on component ball pitch from 0.4 to 1.27 mm. Component coplanarity and flatness at room temperature can be different than at SMT reflow and during BGA/CSP rework, which can lead to SMT yield loss due to component part or board warpage. As a result of this, work done has been done by JEITA and JEDEC to update component coplanarity specifications to include coplanarity requirements during SMT reflow and BGA/CSP rework. Based on the JEITA standard [6] and JEDEC publication [7], the maximum component package flatness during reflow would be 3.6 mils (90 μm) to 10 mils (250 μm), dependent on component size, ball diameter, and ball pitch (from 0.4 to 1.5 mm).

3.3.3.3 Board Warpage Standards

In addition to concerns of component warpage during lead-free BGA/CSP rework, there are concerns with warpage of boards during rework leading to solder joint and reliability issues [8]. In terms of the level of warpage allowed on the board, IPC standard IPC-6012 [9] (for rigid boards) indicates that board surface mounting up to 0.75% board bow or twist is an acceptable condition (7.5 mils in.$^{-1}$) at room temperature. From IPC-6012 standard [9], for space and military avionics (Class 3/A), up to 0.5% board bow or twist is an acceptable condition. IPC-6016 (high density interconnect [HDI] boards) [10] indicates that up to 0.6% board bow and twist for harsh environments (automotive and space) is an acceptable condition. These tolerances could lead to significant board warpage which would lead to solder joint and reliability issues. Many companies are requesting tighter tolerances to 0.5% of the board length. There is a need for development of standards to assess and set limits for board warpage during reflow.

3.3.4 Equipment Updates for Lead-Free BGA/CSP Rework

Rework equipment will need to be capable of processing large, thermally massive printed circuit boards (PCBs) with lead-free solder. The trend has been to increase the amount of bottom side preheat in BGA rework equipment to reduce the amount of top nozzle preheat needed, which would reduce the delta temperature between the top of the component and the solder joint. There are needs during lead-free area array rework for increased board temperatures with up to 175 °C preheat with increased board temperature testing requirements. For tin-lead soldering, board preheat temperatures of 125 °C are being encountered during area array rework. Temperature repeatability studies on different rework equipment have been done in the past to understand the temperature variation between the top of the component and the solder joint during lead-free rework and how this can be minimized [11].

Since the process window for rework has narrowed, there may be a need to develop special tooling (e.g. baffles in nozzle designs) and software control on rework temperature profiles. To further increase the productivity of the rework process, methodologies need to be developed which provide higher and more efficient heat transfer rates.

3.3.5 Adjacent Component Temperatures

Keep out spaces are minimal in the higher density boards such as cell phones and laptops. For larger-sized boards keep out spaces are typically larger. The tighter the spacing, the more challenging for rework. Challenges encountered during development work have some adjacent components being partially reflowed during rework which has led to component reliability issues. Work on investigation of this issue has shown that the use of increased bottom side preheat and reduced topside nozzle heat have helped to reduce the issue [11].

Thermal management during rework is critical with a need for design standardization on recommended board spaces or "keep out areas" around these area array components without damaging adjacent components. Some cell phone and laptop boards have already moved to zero adjacent spacing, with no keep out areas for consumer products.

3.4 Non-standard Component Rework (Including BTC/QFN)

3.4.1 Alloy and Flux Choices

3.4.1.1 Alloys

In terms of alloy choices for lead-free non-standard component rework which would include BTC/QFN components, they would be similar to those alloys used for BGA/CSP rework and based on the alloys used for the assembly operations. The lead-free rework alloys used would be chosen from Sn3-4Ag0.5Cu (mp: 217 °C), Sn3.5Ag (mp: 221 °C), and Sn0.7Cu (mp: 227 °C) based alloys. Sn0.7Cu based alloys are lower in cost but are higher in melting temperature than Sn3Ag0.5Cu. Due to the higher board and component temperatures encountered during lead-free non-standard component rework similar to lead-free BGA/CSP rework, Sn3Ag0.5Cu is generally used.

3.4.1.2 Flux

In terms of lead-free non-standard component rework flux trends, there are choices in terms of the use of solder paste versus tacky flux. Sn3Ag0.5Cu solder paste would generally be used for rework where solder joint volume and reliability requirements would be more of a concern with tacky flux, especially for components such as BTC/QFN components, with the need for solder joint standoff height for reliability of the BTC/QFN solder joint.

3.4.2 Soldering Temperatures and Times

Soldering temperatures and times for lead-free non-standard component rework are similar to lead-free BGA/CSP rework and depend on the size of the component to rework and thermal mass of the board being soldered to. The maximum component package and solder joint temperatures and times during rework are indicated for tin-lead and lead-free soldering in Tables 3.3 and 3.4.

For the rework of BTC/QFN components there are not very well standardized rework techniques compared with BGA/CSP and hand soldering. Issues involve how to apply the solder paste during BTC/QFN rework. Processes for reworking new/non-standard component types are being established with common industry procedures for rework for these types of components, which include leadless device rework referred to in the IPC 7711 standard [12].

3.4.3   Non-standard Component Temperatures in Relation to IPC JEDEC J-STD-020 Standard and Component Warpage Standards

These would be similar to BGA/CSP rework in terms of component temperatures and component warpage during rework. There is also a need to consider temperatures used for temperature-sensitive components. IPC/JEDEC J-STD-075 standard [13] is also used to help classify these types of components. This standard discusses assembly classification for chip components, TH connectors, aluminum capacitors, crystals, oscillators, fuses, light-emitting diodes (LEDs), relays, and inductors. The purpose of this specification has been to establish an agreed set of worst-case solder process limits (tin-lead and lead-free) which can safely be used for assembling non-semiconductor electronic components on common substrates. The standard discusses Process Sensitivity Level (PSL), which is a rating used to identify a component that is solder process sensitive because the component cannot be used in one or more of the base solder process conditions.

3.4.4   Equipment and Tooling Updates for Lead-Free Non-standard Component Rework

As already indicated, issues related to BTC/QFN components also include paste deposition. The ability to ensure proper deposition of paste on, for example, the internal, staggered rowed 0.4 mm pitch terminals can be challenging, especially when pitches decrease down to 0.3 mm. Mini-stencil paste deposition methods will have to be improved. In some cases the stencil and component fixturing has been integrated into the rework system itself to help to eliminate operator handling further and increase yield. Areas to address for BTC/QFN rework include: removal and replacement of the parts, stencil design, solder paste selection, voiding reduction, and package body temperature monitoring. Challenges for these types of components involve rework of these components with thermal vias in the center of the component, best practices for rework for these components, how to reduce voids in the center ground connections, use of appropriate flux type/quantity, and deactivation processes to ensure product reliability.

With respect to non-standard SMT area array connector rework, the main issue is related to the height and thermal mass of the connector, on a large, thermally massive PCB; with the connector body being damaged at temperatures required to reflow the solder joints. Development work into new nozzle designs, and increased local bottom side heating to assist in reworking this type of connector is required.

For LED rework, the LED components have sensitive plastic/silicone bodies. LED assemblies have challenges during rework with thermally sensitive optical components populated along a substrate's edge, or near to other devices with minimal component spacing which requires total system and process control during rework [14].

3.4.5 Adjacent Component Temperatures

For non-standard component rework such as LED rework, partial reflow of adjacent devices or discoloration of critical surface finishes, or delamination of the PCB may occur during rework. Minimizing thermal exposure is a key to successful rework [14].

3.4.6 Non-standard Component Rework Solder Joint Reliability

For non-standard component rework, there is a trend to use tacky flux from the electronic manufacturing service (EMS) company perspective due to ease of use and speed of rework, but from an original equipment manufacturer (OEM) perspective based on solder joint reliability concerns, solder paste on the part is needed for applications such as BTC/QFN rework. The use of solder paste is required as the replacement component and the cleaned pads have no solder on them which would be required for an electrical connection and good solder joint reliability.

3.5 PTH (Pin-Through-Hole) Wave Rework

3.5.1 Alloy and Flux Choices

3.5.1.1 Alloys

In terms of alloy choices for lead-free PTH rework for mini-pot/solder fountain rework, these are based on the alloys used for the assembly operations. The lead-free rework alloys used would be chosen from Sn3-4Ag0.5Cu (mp: 217 °C) and Sn0.7Cu (mp: 227 °C) based alloys. Sn0.7Cu based alloys are lower in cost but are higher in melting temperature than Sn3Ag0.5Cu. There is also a consideration to reduce copper dissolution of the board during the lead-free PTH rework operation.

Types of alloys used for good holefill and low copper dissolution include Sn0.7Cu+Ni and Sn0.7Cu0.3Ag0.03Ni (JEITA) alloys. Although higher board and component temperatures are a concern during lead-free PTH rework due to copper dissolution concerns, SnCuNi alloys are generally used.

3.5.1.2 Flux

In terms of lead-free PTH rework flux trends, there are choices in terms of the use of liquid flux versus tacky flux. The use would depend on the type of component being reworked and the ease of use in rework operations as well as the ability of the liquid flux or tacky flux to ensure good rework soldering. There would need to be care taken that the liquid flux or tacky flux was thermally activated during the PTH rework operation to avoid any reliability issue if the liquid or tacky flux was not exposed to heat.

3.5.2 Soldering Temperatures and Times

Soldering temperatures and times for lead-free PTH wave rework depend on the size of the component to rework and thermal mass of the board being soldered to. The maximum component package and solder joint temperatures and times are indicated for lead-free PTH rework in Table 3.5.

Table 3.5 shows the maximum mini-pot temperatures and contact times for lead-free Sn3Ag0.7Cu (mp: 217 °C) and Sn0.7CuNi (mp: 227 °C) PTH rework. The maximum pot temperatures and contact times follow the maximum component package temperature and time rating using by component and board suppliers. This helps to ensure that there is minimal thermal damage to the components or board during PTH rework. It should be noted in Table 3.5 that there is a trend for some components and boards to exceed the current maximum component and board temperatures and contact times during lead-free PTH rework.

As a comparison between tin-lead (Sn37Pb; mp 183 °C) and lead-free (Sn0.7CuNi: mp 227 °C) PTH rework soldering for tin-lead PTH rework, the use of a board pre-heater prior to solder fountain/mini-pot rework would be optional and only for large thicker boards. Tin-lead PTH component part removal would use a pot temperature of 260 °C for up to 10 seconds of contact time. Tin-lead PTH component part replacement would use a pot temperature of 260 °C for up to 15 seconds of contact time.

For lead-free Sn0.7CuNi PTH rework the use of a board pre-heater prior to solder fountain rework would typically be needed. Lead-free Sn0.7CuNi PTH component part removal would use a pot temperature of 270–275 °C for up to 15 seconds of contact time. Lead-free Sn0.7CuNi PTH component part replacement would use a pot temperature of 270–275 °C for up to 30 seconds of contact time. For lead-free PTH rework there would need to be awareness of the board copper barrel/pad dissolution issues with the higher tin containing lead-free solder and the increased pot temperatures and contact times used, which would tend to increase copper dissolution with lead-free PTH rework.

Along with lead-free rework, surface mount assembly and wave soldering operations can lead to dissolution of component and board terminations which can affect solderability and reliability. Tables 3.6–3.11 and Figures 3.7–3.12 show

Table 3.5 Lead-free PTH rework [2, 3].

Soldering process	Parameter	Units	2015	2017	2019	2025	2027
	Alloy type used	—	SAC 305/SnCuNi (low copper barrel dissolution)	SAC 305/SnCuNi (low copper barrel dissolution)	SAC305/SnCuNi (low copper barrel dissolution)	SAC305/SnCuNi (low copper barrel dissolution)	SAC 305/SnCuNi (low copper barrel dissolution)
	Type of preheat used	—	Convection/IR	Convection/IR	Convection/IR	Convection/IR	Convection/IR
	Board preheat temperature if preheat is used (depends on board thickness)	°C	125–155	125–155	125–155	125–155	125–175
Pb-free	Max. pot temperature	°C	260–275	260–275	260–275	260–275	260–280
	Total component contact time (total time to remove and replace component)	s	15–30	15–30	15–30	15–30	15–45
	Minimum remaining board copper thickness required after rework for reliability	Um	12.7	12.7	12.7	12.7	12.7

Table 3.6 Gold dissolution rate in lead-free and tin-lead solder [15–17].

Gold (Au) dissolution rate/(μm s^{-1})

Solder type	Solder pot	Sample	Pot temperature (°C)											Ref.
			203	233	241	242	247	250	255	271	272	277	280	
Sn3Ag0.5Cu	Large static	Wire						6.7					20	Hillman
Sn3.5Ag	Small static	Wire			3.5				4.5	10.2				Bath
Sn2.5Ag0.8Cu0.5Sb	Small static	Wire				3.2			4.1		8.1			Bath
Sn0.7Cu	Small static	Wire					3.7		5.4					Bath
Sn40Pb	Small static	Wire	0.9	2.3					3.3			8.4		Bath
Sn40Pb	Small static	Wire						4.1					8.5	Bader

Table 3.7 Silver dissolution rate in lead-free and tin-lead solder [15–17].

Silver (Ag) dissolution rate/($\mu m\,s^{-1}$)			Pot temperature (°C)											
Solder type	Solder pot	Sample	203	233	241	242	247	250	255	271	272	277	280	Ref.
Sn3Ag0.5Cu	Large static	Wire						0.8					2.2	Hillman
Sn3.5Ag	Small static	Wire			0.4				0.6	1.38				Bath
Sn2.5Ag0.8Cu0.5Sb	Small static	Wire				0.6			1.2		1.7			Bath
Sn0.7Cu	Small static	Wire					2		2.4			4.4		Bath
Sn40Pb	Small static	Wire	0.26	0.7					1.3					Bath
Sn36Pb2Ag	Small static	Wire	0.02	0.1					0.3					Bath
Sn40Pb	Small static	Wire						1.3					2.1	Bader

Table 3.8 Copper dissolution rate in lead-free and tin-lead solder [15–17].

Copper (Cu) dissolution rate/(μm s^{-1})

Solder type	Solder pot	Sample	Pot temperature (°C)											Ref.
			203	233	241	242	247	250	255	271	272	277	280	
Sn3Ag0.5Cu	Large static	Wire						0.29					0.56	Hillman
Tin-lead	Large static	Wire						0.16					0.29	Hillman
Sn3.5Ag	Small static	Wire			0.62				0.73	0.69				Bath
Sn2.5Ag0.8Cu0.5Sb	Small static	Wire				0.13			0.29		0.22			Bath
Sn0.7Cu	Small static	Wire					0.27		0.49			0.67		Bath
Sn40Pb	Small static	Wire	0.18	0.4					0.43					Bath
Sn40Pb	Small static	Wire						0.23					0.41	Bader

Table 3.9 Palladium dissolution rate in lead-free and tin-lead solder [15–17].

Palladium (Pd) dissolution rate/($\mu m\,s^{-1}$)

Solder type	Solder pot	Sample	Pot temperature (°C)											Ref.
			203	233	241	242	247	250	255	271	272	277	280	
Sn3Ag0.5Cu	Large static	Wire						0.071					0.14	Hillman
Sn3.5Ag	Small static	Wire			0.11				0.12	0.19				Bath
Sn2.5Ag0.8Cu0.5Sb	Small static	Wire				0.2			0.18		0.17			Bath
Sn0.7Cu	Small static	Wire					0.17		0.16			0.29		Bath
Sn40Pb	Small static	Wire	0.3	0.33					0.33					Bath
Sn40Pb	Small static	Wire						0.16					0.3	Bader

Table 3.10 Nickel dissolution rate in lead-free and tin-lead solder [15–17].

Nickel (Ni) dissolution rate/(μm s^{-1})			Pot temperature (°C)											
Solder type	Solder pot	Sample	203	233	241	242	247	250	255	271	272	277	280	Ref.
Sn3Ag0.5Cu	Large static	Wire						0.018					0.033	Hillman
Sn3.5Ag	Small static	Wire			0.01					0.02				Bath
Sn2.5Ag0.8Cu0.5Sb	Small static	Wire				0.005					0.04			Bath
Sn0.7Cu	Small static	Wire					0.01		0.01			0.03		Bath
Sn40Pb	Small static	Wire	0.0001	0.0002					0.0002					Bath
Sn40Pb	Small static	Wire						0.002					0.0054	Bader

Table 3.11 Iron dissolution rate in lead-free and tin-lead solder [16].

Iron (Fe) dissolution rate/(μm s^{-1})			Pot temperature (°C)									
Solder type	Solder pot	Sample	203	233	241	242	247	255	271	272	277	Ref.
Sn3.5Ag	Small static	Wire			0.004			0.01	0.02			Bath
Sn2.5Ag0.8Cu0.5Sb	Small static	Wire				0.01		0.009		0.02		Bath
Sn0.7Cu	Small static	Wire					0.017	0.015			0.04	Bath
Sn40Pb	Small static	Wire	0.01					0.01				Bath

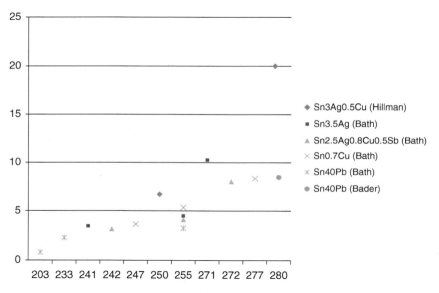

Figure 3.7 Gold dissolution rate (μm/s) in lead-free and tin-lead solder in relation to solder pot temperature (°C) [15–17].

results from evaluations in the industry to understand the dissolution of various metals (gold, silver, copper, palladium, nickel, iron) in tin-lead and lead-free solders at different solder pot temperatures [15–17]. Gold dissolution is generally higher with the high-tin lead-free solders than tin-lead solder. Silver dissolution depends on whether the solder alloy itself contains silver and the tin content of the alloy, which is the same case for copper dissolution. Palladium dissolution was slightly higher with tin-lead solder than with lead-free high tin solder. Nickel dissolution is higher with high-tin lead-free solder than with tin-lead solder. Iron dissolution is similar for high-tin lead-free solders and tin-lead solder.

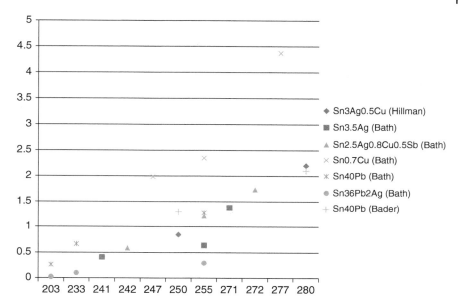

Figure 3.8 Silver dissolution rate (μm/s) in lead-free and tin-lead solder in relation to solder pot temperature (°C) [15–17].

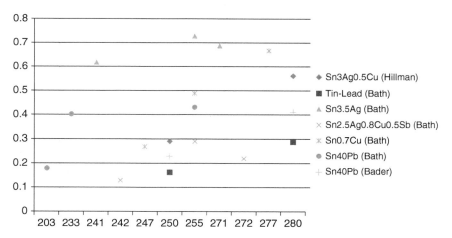

Figure 3.9 Copper dissolution rate (μm/s) in lead-free and tin-lead solder in relation to solder pot temperature (°C) [15–17].

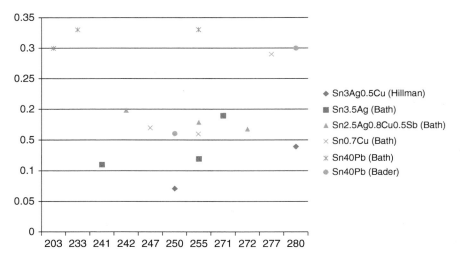

Figure 3.10 Palladium dissolution rate (μm/s) in lead-free and tin-lead solder in relation to solder pot temperature (°C) [15–17].

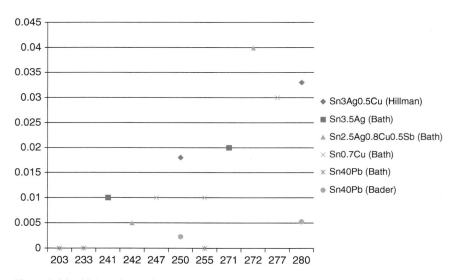

Figure 3.11 Nickel dissolution rate (μm/s) in lead-free and tin-lead solder in relation to solder pot temperature (°C) [15–17].

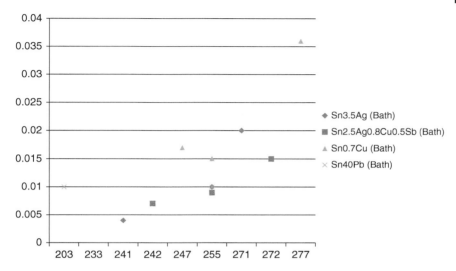

Figure 3.12 Iron dissolution rate (μm/s) in lead-free and tin-lead solder in relation to solder pot temperature (°C) [16].

3.5.3 Component Temperatures in Relation to Industry and Board Standards During PTH Rework

3.5.3.1 Component Temperature Rating Standards

As already indicated, during lead-free PTH rework, one of the challenges is to ensure the maximum component temperatures are not exceeded during lead-free PTH rework. The standard that the industry uses to define the component temperature and time limits during lead-free wave soldering are JEDEC JESD-B106-E [18] and JEDEC JESD22-A111B [19]. For JEDEC JESD-B106 standard for tin-lead (Sn37Pb) first pass wave soldering the maximum pot temperature is 260 °C for 10 seconds. For lead-free (Sn3Ag0.5Cu) first pass wave soldering the maximum pot temperature is 270 °C for 7 seconds. There is an optional lead-free PTH rework pot temperature for rating components at a pot temperature of 270 °C for 15 seconds. JEDEC JESD22-A111B standard evaluates wave soldered immersed bottom-side small body SMT components to 260 °C pot temperature for 10 seconds. It should be noted that a 270 °C pot temperature would cause the bottom side of a board at the solder joint to reach 260 °C during a wave or wave rework soldering operation.

3.5.3.2 Bare Board Testing Standards and Methods for PTH Rework

In addition to the component temperatures during lead-free PTH rework there is also the challenge to ensure the maximum board temperature is not exceeded during lead-free PTH rework. IPC 4101 [20] includes Glass transition temperature

(Tg), Time to delamination at 288 °C (T288) and Laminate decomposition temperature (Td). There are also thermal cycling test methods conducted on boards which include IST (Interconnect Stress Test) and conditioning of boards prior to IST testing, typically involving a reflow process conditioning of 3X or 6X at 245 or 260 °C. Therefore the maximum board temperature would be up to 260 °C, similar to components which restrict the solder pot temperature used for wave soldering. As already indicated, a 270 °C pot temperature would cause the bottom side of a board at the solder joint to reach 260 °C during a wave or wave rework soldering operation.

3.5.4 Equipment Updates for PTH Component Rework

With tin-lead PTH rework, most of the mini-pots have no preheat system and board copper dissolution was not a concern or was not fully understood, so solder pot dwell time was increased when needed.

As products such as telecommunications equipment are moving or have moved to lead-free, the requirement to rework larger, more thermally massive PCBs have increased, with an increase in soldering temperatures and times used for lead-free compared with tin-lead rework. Further development is required on establishing a robust process to remove and replace PTH component types assembled on large-mass PCBs. Work is required to increase the process window to allow the removal and replacement of a PTH part in a single step to reduce solder pot dwell times and copper board dissolution concerns.

Automated PTH rework technology has been developed to rework these challenging applications that are beyond the capability of the traditional solder fountain/mini-pot. This technology features automated component alignment, large board handing and preheat capability, non-contact barrel cleaning as well as focused convective top side and bottom side heating [21].

3.5.5 Adjacent Component Temperatures During PTH Rework

During PTH rework, care should be taken to ensure adjacent components are not reflowed or partially reflowed during rework. This would involve good design of rework nozzles and masking of areas close to the areas being reworked.

3.5.6 PTH Component Rework Solder Joint Reliability

3.5.6.1 Copper Dissolution
As already indicated, further development efforts will be required for lead-free PTH rework process optimization using existing equipment sets, new equipment and nozzle designs with flux development and evaluations and further alloy

development such as nickel added to the lead-free high tin solder alloy to reduce board copper dissolution [11].

In addition, IPC standards such as IPC 6012 [9] will need to be updated to increase the amount of copper in the barrels for incoming as-received boards to 1.3 mils (33 μm) so that there will be sufficient copper knee barrel thickness remaining to survive the wave soldering and wave rework soldering processes on large thermal mass network board type product to provide a reliable solder joint. Data available in the industry indicates that if the copper barrel knee thickness after PTH rework is below 0.5 mils (12.7 μm) copper knee thickness, there would be cracking on the copper knee during reliability testing [22, 23].

Other alternative methods of removal and replacement of PTH components during PTH rework using BGA rework equipment are being explored. In a specific case solder paste with solder preforms was used on thicker boards during PTH rework, using BGA rework equipment to obtain good hole fill after rework while reducing the amount of copper dissolution in the board barrel [24]. Questions still remain in terms of the survivability/temperature rating of the PTH component undergoing the rework as well as the cycle time of the BGA rework equipment process compared with the mini-pot solder fountain rework process.

3.5.6.2 Holefill

For lead-free PTH rework, the challenge is to improve board solder holefill but not increase the solder pot dwell times too much, which would increase the incidence of copper board barrel dissolution and thermal damage to the board or component. By optimization of board preheat, flux, solder alloy selection and equipment development, good holefill has been achieved, but the process window is narrow. More development of equipment and standardized training for the lead-free PTH rework process is required to ensure successful rework of product.

3.6 Conclusions

This chapter reviewed the rework processes which can be affected in lead-free soldering with discussion of hand soldering, BGA/CSP rework, and PTH rework.

For hand soldering rework, soldering temperatures should not increase with lead-free soldering versus tin-lead soldering, but there is an increased tendency for soldering iron tip erosion, with different lead-free alloys being assessed to reduce this effect.

For BGA/CSP rework, lead-free soldering temperatures are higher than tin-lead soldering, which creates challenges for the boards and components being reworked. The lead-free BGA/CSP rework process window is narrowed, which requires optimization of the equipment and process and training of personnel.

For lead-free PTH rework, lead-free soldering temperatures are higher than tin-lead soldering, with challenges related to board copper dissolution concerns for product boards and potentially reduced lead-free holefill with the reduced process window. There should be optimization in terms of equipment and process and training of personnel.

With the introduction of newer components which have both smaller and larger sizes, reduction in component pitch and potential for board and component warpage during rework and temperature-sensitive components the challenges for assembly and rework are increasing. Equipment updates, process development and personnel training will all play a role in successful lead-free assembly and rework.

References

1 Joshi, S., Bath, J., Mori, K., et al. (2017) An investigation into lead-free low silver cored solder wire for electronics manufacturing applications. IPC APEX Conference.
2 INEMI 2017 Rework Roadmap.
3 JISSO meeting presentation: May 2018.
4 IPC/JEDEC J-STD-020 *Moisture/reflow sensitivity classification for non-hermetic solid state surface mount devices.* IPC International.
5 JEDEC Publication 95 (JEP 95). *JEDEC registered and standard outlines for solid state and related devices.* JEDEC.
6 JEITA-ED-7306 (2007) *Measurement methods of package warpage at elevated temperature and the maximum permissible warpage.* JEITA.
7 JEDEC JEP 95 SPP-024A (2009) *Reflow flatness requirements for ball grid array packages.* JEDEC.
8 IPC (2015) Technology Roadmap. IPC International.
9 IPC-6012. *Qualification and performance specification for rigid printed boards.* IPC International.
10 IPC-6016. *Qualification and performance specification for high density interconnect (HDI) layers or boards.* IPC International.
11 INEMI rework studies (2014) IPC APEX Conference Forum.
12 IPC 7711. *Rework modification and repair of electronic assemblies.* IPC International.
13 IPC/JEDEC J-STD-075. *Classification of non-IC electronic components for assembly processes.* IPC International.
14 Lilie, D. and Cabral, A. (2016) Advanced rework applications in a shrinking world. IPC APEX 2016.

15 Hillman, D., Wilcoxon, R., Pearson, T., and McKenna, P. (2018) Dissolution rate of specific elements in SAC305 solder. SMTAI Conference.

16 Bath, J., Ahluwalia, H., Chaggar, K., and Nimmo, K. (1998) Dissolution and Intermetallic Growth Rates of Gold, Silver, Palladium, Copper, Iron and Nickel in Molten Tin-Lead and Lead-free Solders. ITRI (International Tin Research Institute) Research Report, LFS 984. ITRI.

17 Bader, W.G. (1969). Dissolution of Au, Ag, Pd, Pt, Cu and Ni in a molten tin-lead solder. *Welding Research Supplement* 28: 551S–557S.

18 JEDEC JESD22-B106-E (2008) *Resistance to solder shock for through-hole mounted devices*. JEDEC.

19 JEDEC JESD22-A111B (2010) *Evaluation procedure for determining capability to bottom side board attach by full body solder immersion of small surface mount solid state devices*. JEDEC.

20 IPC 4101. *Specification for base materials for rigid and multilayer printed boards*. IPC International.

21 Czaplicki, B. (2013) Advanced through-hole rework of thermally challenging components/assemblies: an evolutionary process. IPC APEX 2013 Conference.

22 Hamilton, C. Snugovsky, P., Moreno, M., et al. (2009) Does copper dissolution impact through-hole solder joint reliability? SMTAI Conference.

23 Ma, L., Donaldson, A., Walwadkar, S., and Hsu, I. (2007) Reliability challenges of lead-free (LF) plated-through-hole (PTH) mini-pot rework. *Proceedings of IPC/JEDEC Lead-free Reliability Conference*, Boston, MA (April 2007).

24 Subbarayan, G., Anderson, L., and Raut, R. (2011) Reworking of pin through hole components using 'hot air' – paste in hole process. SMTAI Conference.

4

Solder Paste and Flux Technology

Shantanu Joshi[1] and Peter Borgesen[2]

[1]*Koki Solder America, Cincinnati, OH, USA*
[2]*Integrated Electronics Engineering Center, Binghamton University, State University of New York, NY, USA*

4.1   Introduction

In the world of electronics materials, solder paste occupies a relatively small volume, yet it plays a critical role in electronics manufacturing/packaging [1–5]. Solder as a joining material provides electrical, thermal, and mechanical functions in electronics assembly. The performance and quality of the solder paste are crucial to the integrity of a solder joint, which in turn is vital to the overall function of the assembly.

Solder is one of several interconnecting materials that can serve as a bonding agent between metallic surfaces of an assembly, under proper conditions. Solder paste, in the "deformable" viscoelastic form, can be applied in a selected shape and size and can be readily adapted to automation. Its "tacky" characteristic provides the capability of holding parts in position without additional adhesives before forming permanent bonds. The metallic nature of solder provides relatively high electrical and thermal conductivity. Solder paste is the primary means of applying solder for high-speed, high-volume, and precision-soldering applications. The creamy characteristics of this material allows it to be manipulated with automated precision-deposition technology, such as printing, dispensing, or pin-transfer. The tacky nature of solder paste enables it to be used as a temporary glue during component placement and soldering steps. The mass reflow soldering technologies such as forced air convection, infrared, conduction, or vapor phase reflow generally deliver a consistent heating and a well-controlled soldering process.

The solder paste technology using lead-containing solder alloy has matured after continuous development and evolution for three decades. The design rules are well established, the processes are fine tuned, and the reliability and failure

Lead-free Soldering Process Development and Reliability, First Edition. Edited by Jasbir Bath.
© 2020 John Wiley & Sons, Inc. Published 2020 by John Wiley & Sons, Inc.

modes are well studied. However, the global move toward lead-free soldering has posed challenges to this well-understood solder paste technology. The primary challenges are poorer solder wetting/spreading and poorer component or substrate performance caused by the higher soldering temperatures. Other challenges are also of concern, such as insufficient lead-free solder alloy reliability data, lack of alternatives for high-lead solder alloys, and problems caused by lead-contamination during the transition period and mixed solder alloys in general. This chapter will introduce the state of the art of lead-free solder paste technology, including the materials, the process, the reliability, and the challenges encountered.

Five common ingredients constitute a standard solder paste: a solder alloy, a flux, an activator, a viscosity modifier, and a solvent. Solder paste is a creamy mixture of solder powder and flux. This creamy nature of solder paste allows it to be maneuvered by automated deposition equipment, such as the stencil printer or dispenser, thus enabling the implementation of high-speed, high-volume throughput production. The particle size of the solder is dictated by the end application, with smaller particle sizes used for smaller paste volume deposition.

As to the flux, it serves two functions in the solder paste. The first and primary function of the flux is as a soldering aid. During soldering, the flux removes metal oxides as well as other surface materials such as grease or metal carbonates, hence allowing the coalescence of the solder powder and the wetting of parts by the molten solder. The second function of the flux is to serve as a vehicle for the solder powder. The rheology of the flux vehicle is required to provide not only a stable suspension of solder powder during storage and handling, but also a solder paste which can be easily handled by paste deposition equipment. In addition, the rheology of the solder paste needs to sustain the subsequent reflow process without slumping and bridging issues. With properly formulated solder paste, the material can be homogeneous, thus allowing the composition of the mixture to be consistent from dot to dot during paste deposition.

Solder pastes are complex mixtures containing leaded solder powder such as Sn37Pb or lead-free solder powder, such as Sn3Ag0.5Cu, and flux chemistry composed of various rosins, activators, solvents, etc.

The flux system, which is being used as a chemical for the removal of oxides from the solder particles, later also removes oxides formed on component or board pad surfaces. Due to its nature, once it is mixed with solder powder it starts reacting with oxides even at room temperature at a relatively slow rate. This chemical reaction is accelerated with the increase of environmental temperature as its molecular activity becomes greater.

4.2 Solder Paste

There are various types of solder paste available today. These products are generally categorized by the type of flux medium in the paste. The fundamental key to good solderability lies in ensuring that the surfaces to be joined are "scientifically" clean. Cleanliness must then be maintained during soldering so that a metallic continuity at the interface can be achieved. This cleaning process is called fluxing, and the material used is the flux. Customarily, the flux is classified based on its activity and chemical nature, namely rosin-based such as water-soluble, and no-clean. Fluxes are applied to the surface to react with metal oxides or non-metallic compounds, thus "cleaning" them from the metal surfaces. The flux activity can be determined by the combined measurements of water extract resistivity, copper mirror (IPC TM-650 Ref: 2.3.32), halide, and surface insulation resistance (SIR) tests (Figure 4.1).

4.2.1 Water-Soluble Solder Paste

Water-soluble flux is designed so that its residue after soldering can be removed by using either pure water or a water medium with addition of saponifier or another benign additive. Considering performance, process, reliability, and cost, a flux chemistry that requires only water for cleaning is the preferred choice. When using a water-soluble paste, to avoid flux entrapment and incomplete residue removal, ultrasonic cleaning is an effective aid. Also, for water-soluble pastes controlling the temperature profile, particularly the peak temperature and dwell time at peak temperature, is important to avoid overheating the flux.

Water-soluble solder paste must be cleaned with pure deionized water. The main component, which tends to remain on the substrate after reflow before cleaning, is normally a thixotropic material. Because easy-to-clean thixotropic materials tend to be deficient in effectively controlling slump resistance of the solder paste, not

Figure 4.1 Classification of paste according to cleaning type. Source: Courtesy Koki Solder.

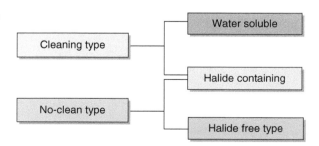

much attention is paid to the cleanability when selecting thixotropic materials, as more importance is attached to the slump resistance feature. Increased package density combined with a complete ban of CFCs (chlorofluorocarbons) greatly influenced the industry into the implementation of no-clean processes. As requirements of the process for both workability and reliability are becoming stricter and more precise, no-clean solder pastes have been developed accordingly.

4.2.2 No-Clean Solder Paste

From the user's point of view, a no-clean flux (incorporated into the solder paste) requires the following:

- Minimal amount of residue
- Residue that is translucent and aesthetically acceptable
- Residue that will not interfere with bed-of-nails ICT (in-circuit test) testing
- Residue that will not interfere with conformal coating where applicable
- Residue that is non-tacky
- Residue that stays inert under exposure to temperature, humidity, and voltage bias
- Ability to flux effectively without solder ball formation

Common solder paste tests of chemical and physical characteristics continue to apply to no-clean systems, so the industry's established test parameters and methods can be used to assess the quality and properties of the assemblies. These include ionic contaminant test and visual examination. However, the tests for no-clean systems have one difference. These tests should be conducted after reflow or soldering. The solder paste chemical makeup measurement in terms of ionic mobility must also be taken after exposure to a specified reflow condition, not before exposure. This procedure is designed to target the characteristics of the residue left on the board, not the as-is paste chemistry.

It is often required that a no-clean solder paste be halide-free. However, halide-containing solder pastes may also end up being used in a no-clean process, because to clean or not is determined in accordance with the reliability requirements of the finished products. The materials specific concerns with respect to using a water-soluble paste without cleaning do include the fact that many of these leave very active residues behind. On a side note, at the other extreme some companies particularly concerned about residues have been known to use a no-clean paste and still clean. This is, however, not trivial as no-clean residues can often only be removed by aggressive cleaning that may damage components or the board. As a general comparison, a halide-containing paste tends to have better workability, such as solderability, but may often exhibit inferior reliability compared to a halide-free type in terms of SIR. This can be attributed mainly to its

Figure 4.2 Relationship between solderability and surface insulation resistance reliability. Source: Courtesy Koki Solder.

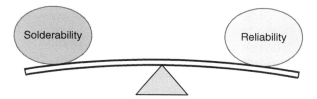

higher activation strength. Figure 4.2 illustrates the need for a balance between solderability and SIR reliability.

With proper cleaning processes and reflow parameters, a water-soluble process can produce clean assemblies in both function and appearance. Importantly, the nature of its chemistry imparts wider fluxing latitude, better accommodating the inherent variations in solderability of components and boards. Cleaning requires initial capital equipment investment, added operating costs in energy and water consumption, and expenditures on consumables for a closed-loop recycle system. No-clean systems eliminate one process step, which is clearly an economic advantage. It should be noted that the cleaning process has been perceived as a step to remove residues from solder flux or paste, yet it actually has also provided the cleaning function for components and boards for many operations without being noticed. It is not unusual for boards, before fluxing and soldering, to contain higher amounts of ionic contaminants than after soldering and cleaning. The level of as-received contamination may exceed the acceptable level, because most steps in board fabrication and component plating involve use of highly ionic chemicals.

For a no-clean system that requires reflow soldering under a protective atmosphere such as nitrogen (N_2), the cost of N_2 may offset or exceed the savings gained from no-clean operations, depending on N_2 consumption and the unit cost of N_2, which varies with the location. Other factors that may also complicate the assessment of a no-clean system do, however, include solder balling and the acceptability of residue appearance. In general, both water-clean and no-clean routes are viable application systems under appropriate conditions. A basic understanding of the principles behind each practice and the compliance with application requirements is essential to the success of implementing either manufacturing system.

4.3 Flux Technology

4.3.1 Halide-Free and Halide-Containing

As far as a no-clean soldering process is concerned, with high reliability in mind, it is desirable for the paste to be halide-free. However, we would prefer as high a solderability as offered by halide-containing pastes, to ensure as wide a process window as possible. The main challenge for no-clean halide-free solder pastes is thus to improve activation. Most halide-free solder pastes contain organic acids as an activator instead of halides in attempts to retain a certain activation strength.

Although a variety of organic acids are available, generally, the lower the molecular weight of the organic acid, the more activation it has. Since the activation strength of organic acids themselves is much weaker than halide, relatively active and large amounts of organic acids tend to be formulated in the flux system. However, such high activation organic acids tend to absorb moisture. Thus, there is a danger that flux residues left on the substrate could become ionized, reacting with moisture/water, leading to deterioration of electrical properties such as SIR and electromigration performance. Activation systems in a solder paste therefore employ a combination of less hygroscopic organic acids and specially developed non-ionic activators. These special activator systems do not become ionized as they do not have the property of dissociation and are electrically very stable and safe, and some exhibit as high an activation strength as halogens. Since the activation temperature of a non-ionic activator is relatively high, its combination with carefully selected organic acids provides a longer activation time at the reflow stage further enhancing solderability and ensuring high reliability at the same time.

In general, the behavior and properties of a solder paste and its residues are of concern in each of the following production stages: Storage → Printing → Placement → Reflow → Inspection → Cleaning

At the various stages, good performance is required with respect to the associated characteristics, as shown in Table 4.1.

It is very difficult to develop a solder paste that will perform perfectly in all these respects. All solder pastes have advantages and disadvantages, so it is critical to verify exact requirements and applications in order to select the best-suited solder paste product.

Table 4.1 Characteristics for good solder paste performance.

Stage	Characteristics to be checked
Storage	• Stability of properties (viscosity, solderability, etc.)
Printing	• Fine pitch printing … fine (0.5 mm)/super fine pitch (0.4 mm) • Stencil life • Stencil idle time (abandon time) • Rolling property • Squeegee separation • Print speed (normal speed 20~40 mm s^{-1} → high speed 200 mm s^{-1}.) • Viscosity change • Clogging in stencil apertures • Smearing
Mounting	• Tack time • Tack force • Slump resistance

Table 4.1 (Continued)

Stage	Characteristics to be checked
Reflow	• Bridge (short circuit) • Solder bead • Micro solder ball • Tombstoning • Wetting • Voiding • Head-on-pillow
Inspection	• Visual cleanliness (flux residue) • In-circuit test probe testability
Cleanability	• Visual cleanliness (flux residue) • Ionic contamination

Source: Courtesy Koki Solder.

4.4 Composition of Solder Paste

No-clean solder pastes are basically composed of an alloy and a flux as shown in Figure 4.3.

4.4.1 Alloy

There are various methods of manufacturing solder powder, such as gas atomization and centrifugal atomization. One of the main methods to manufacture solder powders is by the atomization method as shown in Figure 4.4.

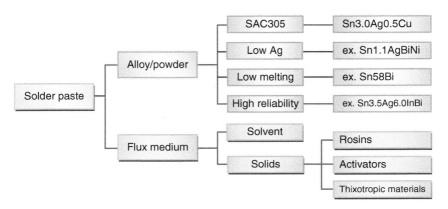

Figure 4.3 Solder paste composition. Source: Courtesy Koki Solder.

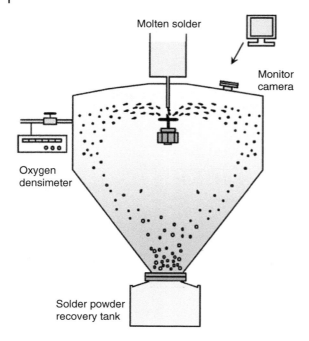

Figure 4.4 Atomization method for solder powder manufacture. Source: Courtesy Koki Solder.

Features of this method are:

- Suitable for the manufacture of fine solder particles
- Easy to control the oxide film formation on the solder particle surface
- Low oxidation level of solder particles

The powder manufacturing chamber is purged with nitrogen, resulting in a very low oxygen density. Ingot solder is melted in the solder pot situated at the top of the tank and molten solder is dripped directly on to a high-speed spinning spindle. When solder drops hit the spindle, it splashes toward the wall of the tank and before it reaches the wall the solder becomes spherical and solid. Solder powder diameters obtained at this production stage range from about 1 to 100 μm.

After this stage, solder powder is brought to the classification stages. There are two different powder classification processes. The first classification takes place by using a nitrogen blower. At this stage, because of the different weight of each particle, particles smaller than those required are removed. The rest of the powder is then moved to mesh sieving to eliminate particles larger than those required.

There are two key issues for solder powder. Firstly, the particle size distribution directly affects the rheology/printability of the solder paste, such as the rolling behavior, separation from the stencil, slump resistance, etc. The sizes must first of all be small enough to fit through the apertures in the stencil or screen used in printing or the nozzle used in dispensing so the distribution must be selected with the minimum pitch and population density of the fine pitch components in mind.

Secondly, the overall oxide content relative to the metal should be kept as low as possible. In order to minimize the oxide content, the inclusion of particles smaller than 20 μm must be as low as possible, because when the particle size decreases the relative surface area increases and as indicated in Figure 4.5, particles smaller than 20 μm have a high oxidation level. In general, a large number of high oxide content particles in the solder powder can result in micro solder balls and poor wetting when reflowed.

A variety of solder pastes are available today, and it is important that the correct paste is chosen for the intended application. Fluxes vary widely but in general careful attention needs to be paid to the particle size distribution as shown in Figure 4.6, and to the rigidity of the classification, as these affect both the printing performance and the solderability.

Figure 4.5 Solder particle size versus oxide content. Source: Courtesy Koki Solder.

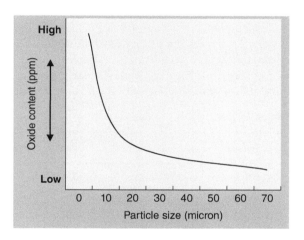

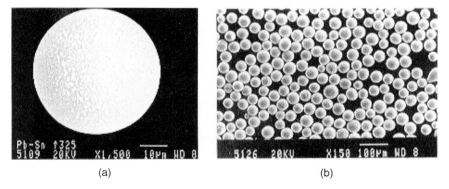

Figure 4.6 (a) and (b) Solder powder particle (left) and solder powder size distribution (right). Source: Courtesy Koki Solder.

4.4.2 Flux

In addition to the solder particles, the flux medium is another critical factor that has a large influence on the properties and performance of the solder paste. Keeping in mind that it may also need to facilitate the reflow of pre-existing solder on substrate pads (HASL – hot air solder leveling) or area array components, the fundamental roles of the flux medium are very similar to those of a wave soldering flux, i.e.

1. Elimination of oxide films: The flux chemically melts and removes the oxide films formed on the surface of electrical components, substrates, and solder.
2. Prevention of re-oxidation: The substrates, components and solder powder, when exposed to the heating environment in the reflow oven, tend to rapidly oxidize. Solids in the flux soften into a liquid state, effectively covering these surfaces and preventing them from re-oxidation.
3. Reduction of surface tension of the solder to enhance wetting: The flux temporarily reduces the surface tension of the molten solder and helps increase the contact area (wetting) with the substrate and components.
4. Providing the rheology and viscosity needed to make the solder powder printable. The components of the flux are shown in Table 4.2.

Table 4.2 Components of a flux.

	Component	Responsible for:	Characteristics
Rosin	• Rosin	• Printability • Solderability • Slump resistance • Tackiness • Color of residue • ICT (in-circuit test) testability	The rosin softens during the preheat stage (softening point is around 80~130 °C) and spreads to the surface of the solder particles and substrate. Rosins normally used are all from natural rosin. Depending on how it is processed, color, activation strength and softening point vary. In order to control workability (slumping, tackiness, etc.) properties, and residue (color, flow, ICT testability) properties, a mixture of different rosins are normally formulated together.
Activator	• Organic acids • etc.	• Activation strength (solderability) • Reliability (SIR (Surface Insulation Resistance), electromigration, corrosion) • Shelf life	These are especially determinative elements in the strength of oxide removal. Along with softening and liquefying of rosins, these activators wet to metal surfaces and react with oxide substances. Activator influences directly the electrical and chemical reliability.

Table 4.2 (Continued)

Component	Responsible for:	Characteristics
Thixotropic agent	• Printability • Viscosity • Thixotropic index • Slump resistance • Odor • Cleanability	These agents help make the solder paste resistant to shear stress during printing and recover viscosity after the solder paste is deposited on the board pad. These agents are employed to improve smooth release of paste from stencil apertures for better paste printability.
Solvent	• Stencil life/tack time • Slump resistance • Odor	Depending on boiling point, evaporation rate varies and determines open time and slump properties of solder pastes. Boiling point of the solvents that are normally used are typically around 220~290 °C.

Source: Courtesy Koki Solder.

4.4.3 Solder Powder Type

There are various solder powder sizes used in solder pastes in the electronics manufacturing industry. The three main powder sizes used are Types 3, 4, and 5. Type 3 has 80 wt% minimum of the powder sizes between 25 and 45 µm. It is the more standard/traditional solder powder size used for general-purpose soldering (for 0.5 mm pitch components, etc.). Type 4 has 80 wt% minimum of the powder sizes between 20 and 38 µm. It is being increasingly used for finer pitch components and small components (0.4 mm pitch chip scale package [CSP], 0201 [0603 metric] components). Type 5 has 80 wt% minimum of the powder size between 15–25 µm. It is typically used for very fine pitch and very small components (0.3 mm pitch CSP, 01005 [0402 metric] chip components). Type 6 solder pastes with powder size between 5 and 20 µm are being considered for paste printing of 0201 metric chip components.

The sizes must first of all be small enough to fit through the apertures in the stencil or screen used in printing or the nozzle used in dispensing. A semi-empirical "rule of thumb" intended to reduce the risk of particles clogging a stencil aperture to an acceptable statistical level is to keep the aperture width above five mean particle sizes. This should be used in conjunction with two recommendations with respect to the stencil aspect ratio greater than 1.5 and stencil area ratio greater than 0.66. In the case of dispensing the corresponding "rule of thumb" is for the needle inner diameter to be larger than 10 mean particle sizes.

Studies have also been done to examine the effect of powder size on solder paste transfer for Package on Package (PoP) assembly where the solder balls on the top component are dipped into the solder paste. Here tests with Type 4 (20–38 µm)

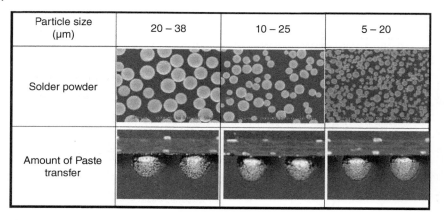

Particle size (µm)	20 – 38	10 – 25	5 – 20
Solder powder			
Amount of Paste transfer			

Figure 4.7 Different powders sizes for PoP [6].

solder powder size versus 10–25 µm versus 5–20 µm showed the best paste transfer results with 5–20 µm (Figure 4.7). This is important because a better paste transfer helped reduce head-on-pillow component soldering issues with the PoP component during reflow.

4.4.3.1 Oxide Layer

One of the main drawbacks of using finer particle sizes is that the surface area to solder volume ratio is higher so there is more oxide to be removed during soldering (Figure 4.6). This means that there is more need for nitrogen during reflow, for example with Type 5 or Type 6 powder size solder pastes, to reduce the potential for solder graping issues.

4.5 Characteristics of a Solder Paste

4.5.1 Printing

Studies show that up to 70% of all surface mount technology (SMT) defects have their root cause in solder paste printing [7]. An ideal paste would be pseudoplastic but most pastes are thixotropic with a viscosity that varies during the printing process to ensure effective filling of small apertures in relatively thick stencils or screens but also effective release and retention of tackiness. Assuming an otherwise optimized paste, general rules apply for the transfer through stencil apertures. The width of an aperture such as those used for leaded devices divided by the stencil thickness (stencil aspect ratio) should be greater than 1.5 to ensure effective transfer. When it comes to near-circular or square apertures, the ratio of the aperture opening area to the area of the aperture wall surfaces must be at least 0.66 for optimum paste transfer efficiency with increasing challenges in transferring sufficient paste volume when area ratios are below this value. Not only are transfer ratios reduced at lower values, the relative scatter in deposit volumes increases rapidly. Area ratios below 0.4 are prohibitive for most purposes in manufacturing.

As component sizes become smaller with smaller board pad sizes, thinner stencils are required to meet these guidelines, and for a mix of large and small components on the same board, there are more challenges to print both large and small paste deposits. Work is ongoing to use ultrasonic squeegees or pressurized printing of solder paste on a board with a mix of coarse and very fine pitch components (01005 chip components, 0.3 mm CSP) for applications such as mobile phones. However, the vast majority of printing is still based on the conventional approach. When necessary, a step stencil may also be an option.

In order to help define the printing parameters for a solder paste, it is a good suggestion to develop a Design of Experiment (DOE) to determine the process window of the paste using the following parameters

- Squeegee pressure
- Squeegee speed
- Stencil separation speed

The printing tests should be conducted on the most challenging component locations such as the finest pitch components and the smallest chip components, i.e. those with the smallest stencil apertures. These could for example include 0.4 mm pitch CSP and QFP components and 0402 metric (01005 Imperial) and 0603 metric (0201 Imperial) chip components.

4.5.1.1 Printing Parameters

As already indicated, the three main printing parameters to consider when printing a solder paste with a stencil are squeegee speed, squeegee pressure and stencil separation speed. For squeegee speed this is dependent on the product being assembled and solder paste supplier recommendations. For squeegee pressure this is typically 1–1.5 lb per linear inch of blade length. For example a 16 in. (41 cm) length of squeegee blade would use a 16–24 lb pressure (7–11 kg pressure). For stencil separation speed this would be dependent on solder paste supplier recommendations.

One other parameter which would typically be defined would be stencil cleaning frequency during paste printing. The printer machine cleaning function would be used to clean solder paste in the stencil apertures after printing paste. There is typically dry and wet stencil wipe/cleaning. The dry machine wipe on the bottom side of the stencil would typically be employed after a few paste prints. The wet machine wipe with a solvent could also be done after an increased number of paste prints. The actual under stencil wipe clean frequency would be dependent on the product type, stencil apertures and solder paste type used.

4.5.2 Reflow

The reflow profile employed for lead-free soldering would be dependent on the solder paste supplier recommendations for the specific paste. The typical reflow and cooling stages for lead-free Sn3Ag0.5Cu solder paste would involve a preheat/soak temperature and time between 150 and 217 °C for 60–150 seconds. The peak soldering temperature would be 235–250 °C with time over 217 °C of 30–90 seconds and ramp up rate of 1–3 °C s^{-1}. The cooling rate from the peak reflow temperature down to 217 °C would be 1–4 °C s^{-1}.

4.5.2.1 Wetting/Spreadability of Lead-Free Solder Paste

Lead-free Sn3Ag0.5Cu solder typically has reduced spreadability compared with tin-lead Sn37Pb solder paste due to the increased surface tension of the lead-free Sn3Ag0.5Cu solder. The reduced wetting/spreading of the lead-free paste can be an advantage in certain situations. For example, there would be less tendency for the lead-free SnAgCu solder paste to cause solder bridging on fine pitch components such as QFPs compared with the better wetting/spreading Sn37Pb solder paste. Also, the risk of tombstoning of 0603 metric (0201 Imperial) and 0402 metric (01005 Imperial) chip components may be reduced.

4.5.2.2 Bridging

As already indicated in the previous section, lead-free Sn3Ag0.5Cu paste has reduced wetting/spreading with less tendency to bridge on finer pitch components versus tin-lead solder paste. In order to reduce solder bridging, should it occur, some areas to consider include whether the stencil apertures are too large, or there is excess component placement force causing solder paste squeezing on the board pad or if the solder paste is wicking on the component termination during reflow.

4.5.2.3 Micro Solder Balls

Along with solder bridging, micro solder balls can form during reflow. These solder beads or mid-chip solder balls can be due to incorrect stencil aperture/thickness, incorrect reflow profile, or ineffective understencil cleaning [8].

4.5.2.4 Voiding

For lead-free or tin-lead solder pastes, one area of focus is to reduce voiding during reflow soldering. It is typical to develop a suitable lead-free solder paste reflow profile to reduce voiding. This would involve assessment of the solder paste being used and any guidelines to reduce voiding from the solder paste supplier for the specific solder paste in question. Consideration would be given to solderability/void exit issues under QFN/MLF/BTC (Bottom Termination Component) components.

In terms of voiding categories a ball grid array (BGA)/CSP component solder joint with more than 30% voiding is termed a defect according to IPC-A-610 standard [9]. This does not mean that a full correlation exists between the level of voiding and thermal cycling or mechanical reliability issues. If anything, the location and distribution of voids is known to be more critical than the percentage. In fact, the voids of greatest concern may be too small to detect by X-ray inspection and distributed as a dense network along a contact pad. There could be potential void issues in terms of electrical/thermal requirements for QFN/MLF/BTC components but there is limited guidance on what void % is an issue for this type of component [10]. From a practical perspective the best recommendation would seem to be to minimize all observable voiding.

Work has been done by many solder paste companies to develop solder pastes with reduced voiding. In one study a SnAgCu solder paste was developed to help reduce voiding in QFN/BTC/MLF components. Tests were conducted on different board surface finishes on soldered power transistor components, which showed different levels of voiding % as indicated in Figure 4.8.

In this work voiding was reduced by improving the wetting property of lead-free Paste C, which extended the duration of the fluid/active state of the flux. This helped to push out more of the entrapped gas during reflow and also enabled the component to come closer to the board pad, lowering the component solder joint standoff height as shown in Figure 4.9.

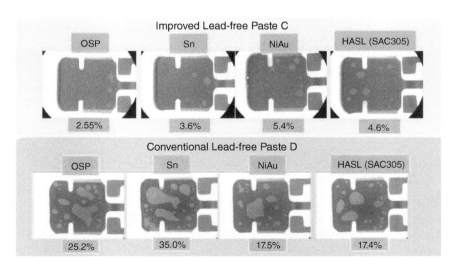

Figure 4.8 Voiding comparison on a power transistor BTC component using conventional lead-free SnAgCu Paste D and specially designed lead-free SnAgCu Paste C to reduce solder voiding on different board surface finishes [11].

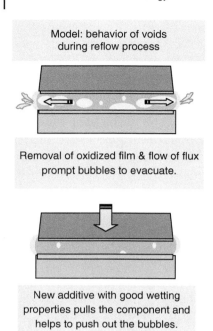

Model: behavior of voids
during reflow process

Removal of oxidized film & flow of flux
prompt bubbles to evacuate.

New additive with good wetting
properties pulls the component and
helps to push out the bubbles.

Figure 4.9 Model showing behavior of developed lead-free SnAgCu solder paste to remove voiding under a power transistor/BTC component [11].

4.5.2.5 Head-on-Pillow Component Soldering Defect

In addition to other challenges and defects occurring during soldering there is also the potential for warpage of the component or board during reflow which can lead to the Head-on-Pillow component soldering defect.

There has been an increase in the rate of Head-on-Pillow component soldering defects which interrupts the merger of the BGA/CSP component solder spheres with the molten solder paste during reflow. The potential reasons for Head-on-Pillow include:

- Warpage of the component or board
- Ball coplanarity issues for BGA/CSP components
- Non-wetting of the component based on contamination or excessive oxidation of the component coating.

The issue occurs not only in lead-free soldered assemblies where the increased soldering temperatures may give rise to increase component/board warpage but also in tin-lead soldered assemblies. An example of a Head-on-Pillow component soldering defect is shown in Figure 4.10.

One of the challenges for Head-on-Pillow is that it is difficult to detect in production during inspection and functional level testing as there is partial contact between the reflowed solder paste and the ball sphere, providing electrical

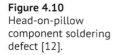

Figure 4.10
Head-on-pillow
component soldering
defect [12].

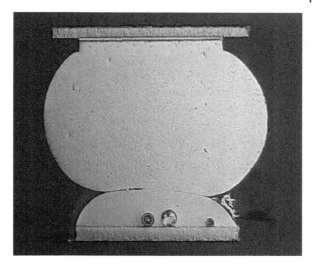

contact, even though there is no real metallurgical bond. The problem can occur in first-pass assembly or during BGA/CSP rework and standard 2D X-ray inspection cannot typically identify it. Often X-ray imaging of a board by tilting can identify the issue but that requires a skilled operator and time on the machine so only sampling is compatible with regular production. After assembly and shipment of the product to the field, however, thermal or mechanical cycling could cause an open solder joint and thus a field failure.

There have been many head-on-pillow case studies done in the industry. One involved head-on-pillow issues on 0.8 mm pitch 473 I/O PBGA, 84 I/O FBGA and 64 I/O PBGA components [13]. Head-on-pillow defects were reduced by monitoring solder paste out of jar exposure time, increasing stencil printer automatic paste inspection, manual inspection of all BGA paste printed board pads, use of a head-in-pillow developed solder paste, and use of laser cut/electroformed stencils. Increasing stencil apertures increased paste volume, which helped to reduce head-on-pillow, but increasing the stencil aperture openings also increased the potential for solder bridging.

In another lead-free head-on-pillow investigation [14] there was found to be a 10% head-on-pillow defect rate on a 0.5 mm pitch BGA component (Sn1Ag0.5Cu spheres) soldered with Sn3Ag0.5Cu paste. The defect rate was reduced down to zero with the use of a developed head-in-pillow paste, more balanced copper in the printed circuit board (PCB) design (less board warpage), a new BGA mold cap (giving less component warpage), and nitrogen atmosphere in the reflow oven (which maintained flux activity in the solder paste during reflow). In addition, the BGA component spheres were dipped in flux using tacky flux (allowing for more flux to maintain flux activity for wetting during reflow).

If a Head-on-Pillow component soldering defect occurs, the typical areas to investigate would be:

- Warpage issue of the component or board
- Ball coplanarity issue for BGA/CSP components
- Non-wetting of the component or board which may be due to contamination or excessive oxidation of the coating
- Insufficient solder paste: which would need checking of stencil aperture and stencil area/aspect ratios
- Reflow profile optimization
- Paste-related issues

4.5.2.6 Non-Wet Open

Two different defects can occur when a BGA ball is lifted from the solder paste by warpage during reflow, Head-on-Pillow and Non-Wet Open.

Non-Wet Opens occur when the solder paste from the board adheres to the BGA component sphere during component or board warpage. During the reflow the solder paste fully melts around the component ball sphere with little or no paste remaining on the board pad to create a solder joint.

In terms of general reflow problems on the board leading to solder opens the areas to investigate would be insufficient solder paste (check stencil apertures and stencil area/aspect ratios), non-wetting of the component or board due to contamination or excessive oxidation of coating, ball coplanarity issue for BGA/CSP components, reflow profile (requiring reflow profile optimization), and any specific issues related to the solder paste in terms of shelf life, solder paste dry-out, and type of solder paste used.

4.5.2.7 Tombstoning

Component misalignment/tombstoning during reflow is usually due to incorrect board pad footprint or stencil design. It can also be due to component misplacement or insufficient component placement pressure. The reflow profile would also need to be checked to see if this may have an effect on tombstoning occurrence or amount [8].

4.5.3 In-Circuit Test (ICT) Probe Testability

For no-clean lead-free solder paste, there is a potential issue with probing because of flux residues after reflow at the higher soldering temperatures interfering with ICT probes.

It has generally been found that chisel-type probe tips are less sensitive to flux residues than crown probe tips, but there are limits on the amount of probe force which can be used to penetrate the no-clean lead-free flux residue on the board.

There has been found to be differences in probeability of the no-clean flux residue in air versus nitrogen reflow atmosphere with a nitrogen atmosphere helping to make the flux residue more probeable. The time after reflow assembly before ICT probe testing of the no-clean flux residue can also have an effect on probeability with flux residue generally hardening over time.

The interaction of the lead-free SnAgCu solder paste with different lead-free board surface finishes can also have an effect on ICT. There are potential issues with OSP (organic solderability preservative) and immersion tin board surface finishes. A lead-free SnAgCu paste will tend to have a more non-uniform spread during reflow on OSP test pads and vias compared with tin-lead solder spreading. This would affect ICT probing. There may be a need to not print on test vias, which would create exposed/oxidized OSP pads during reflow and would affect ICT probing. Also, for an immersion tin surface finish which did not have lead-free SnAgCu paste printed and reflowed on the test vias or test pads, excessive aging of the exposed tin coated board surface finish could lead to the complete transformation to a copper-tin intermetallic compound. This would be difficult to solder to and rework and it may potentially be difficult to probe the test vias and test pads during ICT testing.

4.5.4 Flux Reliability Issues

The flux used in the solder paste helps to clean the board and component areas to be wetted by the solder. It also helps to remove oxide on the solder surface aiding wetting. There needs to be a certain level of flux activity in order to ensure good soldering/wetting, but if the flux activity is too high this could cause corrosion and reliability issues. For no-clean flux chemistries, the flux residues are typically designed to be left on the board surface after soldering.

Flux reliability testing such as SIR can help to validate the reliability of the no-clean flux residue on the board surface. In certain cases, the appearance of the reflowed no-clean flux residue may be a cause for concern. If a flux residue appears wet and tacky, it may have not been exposed to an adequate heat reflow profile to fully cure the no-clean flux residue. IPC-A-610 standard [9] indicates that flux residue appearance may vary depending upon flux characteristics and solder processes.

A flux residue may appear soft and tacky based on the nature of the flux ingredients. Some flux residues are softer for ease of ICT pin probeability. If the flux residue areas on the board appear wet and tacky and other areas appear fully cured, dry, and hard, then there may be an issue with improper or too low soldering temperatures and times on certain parts of the board during reflow. The reflow profile should be checked to see if it is optimized according to the solder paste supplier recommendations. Some components which shield flux residue from reflow heat such as under QFN/BTC/MLF components may have soft or tacky flux residue under the component which may be an electrical reliability concern.

In these cases, the solder paste supplier could be contacted to provide typical flux residue images of the fully cured reflowed solder paste which passed electrical SIR reliability tests and compared with the flux residue appearance that are seen on the product board.

4.6 Conclusions

This chapter reviewed solder paste and flux and the characteristics needed for these materials to ensure good performance. Different solder powder particle sizes/types used in solder pastes were reviewed in relation to their suitability for different component pitches and sizes of component being assembled. The relation between powder size and oxidation level of the solder powder was explored.

Printing and reflow parameters for the lead-free SnAgCu solder pastes was reviewed. There are many defects which can occur during electronics manufacturing, including micro solder balls, voiding, tombstoning, bridging, opens, head-on-pillow, and non-wet opens. These defects and their causes were reviewed.

ICT pin testability of lead-free SnAgCu solder pastes and the interaction of the lead-free SnAgCu paste with different board surface finishes as well as electrical reliability of lead-free no-clean SnAgCu solder pastes were also discussed.

Development of lead-free solder pastes with the specific solder materials and fluxes used is a complex issue with many trade-offs for the different electronic products built. Development of lead-free SnAgCu pastes is ongoing to address the assembly challenges in electronics manufacturing.

References

1 Hwang, J.S. (1989). Introduction. In: *Solder Paste in Electronics Packaging* (ed. J.S. Hwang), 3–21. Dordrecht: Springer.

2 Lee, N. (2007). Lead-free solder paste technology. In: *Lead-Free Electronics* (eds. E. Bradley, C.A. Handwerker, J. Bath, et al.), 125–182. Hoboken, NJ: Wiley.

3 Fang, T. (1999). Environmentally sound assembly processes. In: *Wiley Encyclopedia of Electrical and Electronics Engineering* (ed. J.G. Webster). Hoboken, NJ: Wiley.

4 Lee, N. (2002). Solder paste technology. In: *Reflow Soldering Processes and Troubleshooting: SMT, BGA, CSP and Flip Chip Technologies* (ed. N. Lee), 37, 37–56, 56. Boston, MA: Newnes.

5 Prasad, R.P. (1997). Solder paste and its application. In: *Surface Mount Technology* (ed. R.P. Prasad), 383–443. Boston, MA: Springer.

6 Bath, J., Itoh, M., Clark, G., et al. (2010) An investigation into the development of lead-free solder paste for Package on Package (PoP) component manufacturing applications. SMTAI Conference.

7 Gershenson, M. (2008) SPI for yield improvement. IPC International Test and Inspection Technology Conference.

8 IPC 9111 (2016) *Process Effects Handbook Draft for Industry Consensus*. IPC International.

9 IPC-A-610 *Acceptability of Electronic Assemblies*. IPC International.

10 Bath, J. (2011) A review of the challenges and development of SMT, wave and rework assembly processes in the electronics industry. SMTA Los Angeles/Orange County chapter EXPO presentation, November.

11 Bath, J., Itoh, M., Clark, G., et al. (2010) An investigation into the development of tin-lead and lead-free solder pastes to reduce voiding in large contact area power transistor/QFN type components. SMTAI Conference.

12 Bath, J. (2011) Head-in-pillow review and reduction. SMTA Los Angeles/Orange County Chapter presentation, February.

13 Nowland, R., Coyle, R., Read, P., and Wenger, G. (2010) Telecommunications case studies address head-in-pillow (HnP) defects and mitigation through assembly process modifications and control. IPC APEX Conference.

14 Oliphant, C., Christian, B., Subba-Rao, K., et al. (2010) Head-on-pillow defect – a pain in the neck or head-on-pillow BGA solder defect. IPC APEX Conference.

5

Low Temperature Lead-Free Alloys and Solder Pastes

Raiyo Aspandiar, Nilesh Badwe, and Kevin Byrd

Intel Corporation, Hillsboro, OR, USA

5.1 Introduction

5.1.1 Definition of Low Temperature Solders

Solders for electronics product assembly can be classified based on their melting point and soldering temperatures. Table 5.1 depicts the solder classifications, which are similar to those proposed in the Printed Circuits Assembler's Guide to Low Temperature Soldering [1]. The classes include: high temperature solders that typically contain Au and hence are expensive; mainstream lead-free solders mainly in the SnAgCu (SAC) family; medium temperature solders, which entail the addition of Bi, In, and Zn to lower the melting point of SAC solders; low temperature solders, which are typically in the Sn-Bi and Sn-In system, and ultra-low temperature solders which have a majority of Bi or In in their compositions. This table does not contain Pb or Cd containing solders since these are banned for environmental concerns. The classical Sn-Pb solders would fall into the medium temperature solders category.

Figure 5.1 graphically illustrates this classification of solders based on their melting temperature ranges and soldering temperature ranges. The boxes with dashed line boundary represent the melting temperature range and the shaded boxes represent the soldering temperature range. The soldering temperature, which if reflow soldering in an in-line oven is the peak reflow temperature (PRT), is higher than the solder melting temperature because the solder needs to flow, wet, and form the solder joint between the board lands and the component terminations.

The focus here is on low temperature solders. The soldering temperature range for low temperature solders is low enough to achieve various significant benefits for the board assembly processes employed in the assembly of specific electronics

Lead-free Soldering Process Development and Reliability, First Edition. Edited by Jasbir Bath.
© 2020 John Wiley & Sons, Inc. Published 2020 by John Wiley & Sons, Inc.

Table 5.1 Classification, elemental composition, melting range and metallurgical type of available lead- and cadmium-free solders.

Classification	Sn	Ag	Au	Bi	Cu	Ga	In	Sb	Si	Zn	Other	Melting range	Type
High temperature solders			96.8					3.2				363	Eutectic
			88.0								Ge(12)	356	Eutectic
	22.0		78.0									280–300	Non-eutectic
	20.0		80.0									280	Eutectic
				100.0								271	Elemental
	95.0							5.0				237–240	Non-eutectic
Mainstream lead-free solders	95.0	5.0										221–240	Non-eutectic
	65.0	25.0					10.0					233	Eutectic
	99.3				0.7						Ni	227	Eutectic
	89.3	3.8			0.9		0.5	5.5				221–228	Non-eutectic
	98.5	1.0			0.5						Mn	217–226	Non-eutectic
	90.0	3.8		1.5	1.2		3.5					216–226	Non-eutectic
	95.5	4.0			0.5							217–225	Non-eutectic
	89.7	3.4		3.2	0.7		3.0				Ni	210–221	Non-eutectic
	96.5	3.0			0.5							217–221	Near-eutetic
	91.3	3.5		3.0	0.7		1.5				Ni	208–218	Non-eutectic
Medium temperature solders	89.2	3.5		0.5	0.8							202–206	Non-eutectic
	91.8	3.4		4.8								211–213	Non-eutectic
	86.9	3.1					10.0					204–205	Near-eutectic
	91.0									9.0		199	Eutectic
	77.2	2.8					20.0					175–187	Non-eutectic
	86.5			3.5			4.5			5.5		174–186	Non-eutectic
Low temperature solders	60.0			40.0								138–170	Non-eutectic
							100.0					156.7	Elemental
						0.7	99.5					150	Eutectic
		3.0					97.0					143.3	Eutectic
	42.0	1.0		57.0								139–140	Near-eutectic
	42.0			58.0								138.3	Eutectic
	52.0						48.0					118–131	Eutectic
	50.0						50.0					118–125	Non-eutectic
	48.0						52.0					118	Eutectic
Ultra-low temperature solders				67.0			33.0					109	Eutectic
	46.0						52.2			1.8		108	Eutectic
	16.3			54.0			29.7					81	Eutectic
	17.0			57.0			26.0					79	Eutectic
				33.7			66.3					72	Eutectic
	16.5			32.5			51.0					60	Eutectic
						100.0						29.8	Elemental

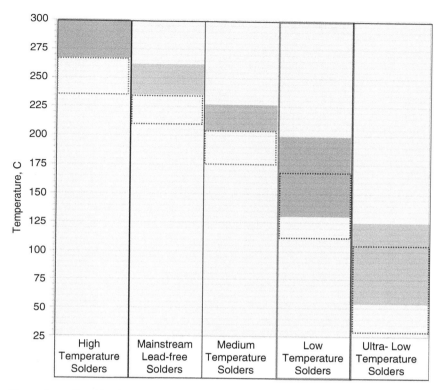

Figure 5.1 Classification of solders used in electronics assembly based on their melting temperature ranges and soldering temperature ranges. The boxes with dashed line boundary represent the melting temperature range and the shaded boxes represent the soldering temperature range.

products such as mobile, desktop, and server computers. These benefits are reduced manufacturing costs, reduced environmental impact, and improved manufacturing yields. These benefits are covered in detail in the subsequent section.

5.1.2 Benefits of Low Temperature Soldering

The two main drivers for recent adoption of low temperature solders have been economic and environmental. The economic driver has been related to the reduction in manufacturing assembly costs, particularly the energy costs for operating the soldering equipment. The environmental driver is related to the electronic products' life cycles. These two are considered below.

5.1.2.1 Reduced Manufacturing Cost

Reduced manufacturing cost when using low temperature solders in comparison to the current mainstream SnAgCu (SAC) solders are realized due to lower material costs for certain low temperature solders and lower electricity costs due to power use savings in reflow ovens.

The BiSnAg-based low temperature solders have lower amounts of Ag, in the 0.4–1 wt% range, compared to the mainstream SnAgCu solders, which have Ag in the 3.0–4.0 wt% range. This reduction in Ag will directly result in savings of the material cost of the solder. Currently this is estimated to be about 10% savings on the SnAgCu solder cost.

Bismuth is a very rare element in the earth's crust, with estimates of its occurrence varying from 0.085 to 0.480 ppm by mass% [2]. The spare capacity for Global Bismuth Metal usage per year is 4000 t currently [3]. The Global Solder Market for paste per year is approximated to 20 000 t. Assuming the computer and office equipment market segment which comprises of 33% of the electronics manufacturing industry [4] and for which the BiSnAg solder paste is of primary interest, converts entirely to BiSnAg solder paste, the amount of bismuth in the solder paste metal needed would be about 3400 t. This would all but use up the spare capacity, but would not significantly affect the price. If, however, other electronics manufacturing industry segments such as telecom and automotive would convert to this low temperature solder, then the bismuth metal costs would increase and the metal cost savings could be negligible.

5.1.2.2 Power Use Savings

Using lower temperature solders lowers the peak temperatures necessary during the reflow soldering process step on an electronics manufacturing SMT line. The primary source of manufacturing cost savings when using low temperature solders is the reduced cost of electrical power due to a smaller current draw being required to operate the ovens during reflow. This is a direct consequence of the set temperatures of the zones in the oven being at a lower value than those used for higher melting lead- free solders.

This lower power usage when soldering at lower peak temperatures has been confirmed in a few independent studies within the industry. To determine the cost savings accrued when wave soldering is replaced by a pin-in-paste (PIP) process employing low temperature solders, it was shown that energy consumption can be decreased from the 20–24 kW h^{-1} range when reflow soldering SnAgCu solder paste at 245 °C PRTs to the 15–17 kW h^{-1} range when reflow soldering BiSn solder paste at 190 °C PRTs [5].

A similar study was undertaken at the company SMT manufacturing facility in Oregon, USA to assess the extent of power use savings realized by using BiSn-based solder pastes instead of the standard lead-free Sn3Ag0.5Cu solder

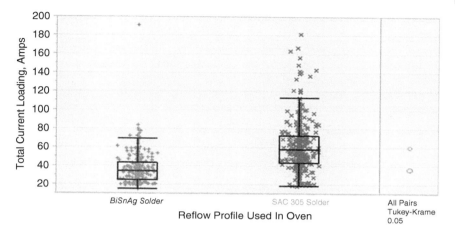

Figure 5.2 Comparison of current loading for a reflow oven when running a low temperature BiSnAg soldering versus a standard SAC soldering reflow profile.

pastes [6]. Figure 5.2 depicts the current readings for low (BiSnAg) and high (SnAgCu) temperature reflow processes. The amperage readings were measured at one-minute intervals for the full duration of a typical six-hour operation. Each data point in the plot denoted each amperage reading taken. The average current draw during SnAgCu reflow was measured to be 60.4A, while the average low temperature BiSnAg reflow current was measured to be 36.7 A. This difference represents a 39% reduction in current flow and is both statistically and technically significant.

To evaluate the economic benefits of this power reduction, a cost estimate, on a per oven basis, was calculated assuming an 80% utilization of the oven and an energy cost of USD $0.11 per kWh. This energy cost was on the low end of US energy estimates [7] and the high end of China energy estimates [8]. Based on these assumptions, the cost savings of low temperature soldering operations is estimated to be $168 per oven per week or $8749 per oven per year as shown in Table 5.2.

A similar evaluation was repeated at a high-volume facility of an ODM and results consistent with the internal company study were obtained.

5.1.2.3 Environmental Benefits
As listed in Table 5.2, low temperature solder use generates savings in KWh consumed by reflow ovens on the SMT manufacturing lines. This reduction in KWh in itself generates environmental benefits by reducing the generation of CO_2 greenhouse gases. Based on US EPA (Environmental Protection Agency) estimates [9] 0.0007 metric tons of CO_2 are produced for every kWh consumed, an 11.7 kW power reduction due to low temperature reflow equates to 1.1 metric tons of CO_2

Table 5.2 Estimated cost-saving comparison for SnAgCu versus BiSnAg soldering reflow processes.

SAC 305 Paste		Sn/Bi/Ag Paste	
Oven energy consumption (kWh)	**29.5**	**Oven energy consumption (kWh)**	**17.8**
80% utilization (h wk^{-1})	134.4	80% utilization (h wk^{-1})	134.4
Energy cost-$/kWh (PRC 2013)	$0.11	Energy cost/kWh (PRC 2013)	$0.11
Oven cost per week	$424	Oven cost per week	$256
	BiSnAg savings (per oven per week)		$168
	BiSnAg savings (per oven per year)		$8749

that will not be produced per oven each week relative to high temperature reflow, assuming 80% oven utilization. This is approximately 57.2 metric tons of CO_2 per oven per year. For reference, 57 metric tons of CO_2 is the greenhouse gas equivalent of burning nearly 6000 gal of gasoline, or the average monthly CO_2 production of 60 US households [9]. The company reported that single oven energy savings are estimated to scale to a CO_2 reduction in excess of 25 000 metric tons per year based on a modest 20% share of the worldwide mobile laptop market by one of their OEMs for board assemblies manufactured at a partner ODM [6].

5.1.2.4 Manufacturing Yield Improvements

The need for low temperature solders has also arisen recently due to technical reasons. Over the past six decades, Moore's Law has been the driving force for improved performance, lower power, and reduced cost per transistor for silicon technology. Figure 5.3 plots the decreasing trend of the node size of a silicon transistor with time starting from the 1960s. To keep pace with this decreasing silicon transistor size, the component packages in which the silicon die are housed have become thinner with finer termination sizes at increasingly tighter pitches.

Reduction in package thickness creates new challenges for their reflow soldering assembly. Due to various mismatches in the coefficient of thermal expansion (CTE) of materials comprising these electronic packages, their resultant warpage increases markedly at the current SnAgCu reflow peak temperatures. Figure 5.4 illustrates the trends in the company's FCBGA Package Platforms with time for the Package Z-height and the package coplanarity at 260 °C, which is the upper limit for PRTs when using SnAgCu solders.

Though the package coplanarity at the PRTs is depicted in Figure 5.4, while heating up to the PRT the warpage of the FCBGA package changes dynamically. Dynamic warpage occurs due to the differential expansion of materials. Essentially, during the heating of the package in the reflow oven, the silicon

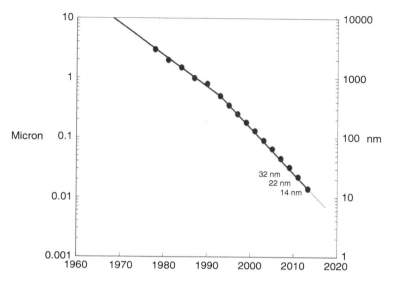

Figure 5.3 Moore's law graph of year versus node size for silicon transistors.

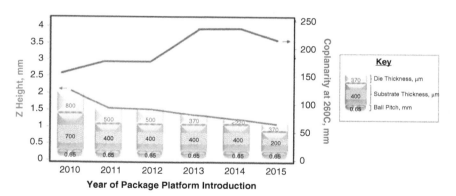

Figure 5.4 Trends for package Z-height (left y-axis) and coplanarity at 260 °C (right y-axis) for the company's recent FCBGA Package Platforms.

die expands much less than the package substrate laminate. This results in the warpage configurations for room temperature and reflow shown in Figure 5.5a,b. FCBGA packages typically have a convex (positive) warpage at room temperature and a concave (negative) warpage at the reflow temperatures when using SnAgCu solder pastes. Similarly, the shape of the PCB warpage can vary based on the PCB layer construction and whether pallets are used or not, and the design of the pallets, if used, during the reflow soldering process.

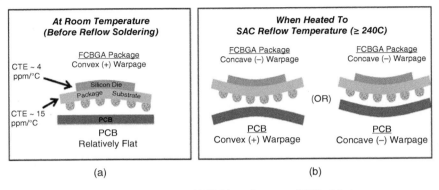

Figure 5.5 Typical warpage shapes of FCBGA packages and PCBs (a) at room temperature and (b) at SnAgCu solder paste reflow temperatures.

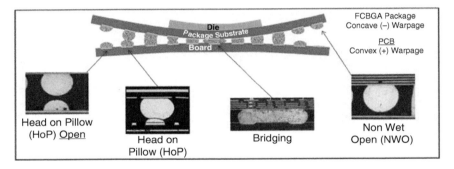

Figure 5.6 Description of defects that can be caused by dynamic warpage of FCBGA components and/or PCBs during the reflow soldering process.

These dynamic warpage characteristics will lead to a gap being created between the BGA solder ball and the solder paste on the board land. This in turn can lead to solder joint defects that affect the yield of the board after reflow soldering. The types of solder joints defects, generated due to the dynamic warpage of such packages, are shown in Figure 5.6. They include Head-on-Pillow (HoP), where, though there is physical contact, there is no coalescence of the ball with the solder mass from the solder paste [10], HoP Open, where no physical contact occurs between the solder ball and the post-reflow solder paste mass, Non-Wet Open (NWO), where there is no contact between the solder ball and the printed board land with little or no evidence of solder wetting on the board land

[11, 12] and solder bridging, where two or more neighboring solder joints are connected together.

Reducing the reflow peak temperature improves the SMT margin by reducing the dynamic warpage and keeping the ball and paste in contact during reflow. This becomes obvious when the dynamic warpage versus temperature plot for FCBGA packages is observed. Figure 5.7 typifies such a plot, with the data points in the plot showing the measured warpage of (a) FCBGA package and (b) PCB with and without pallets, by the Shadow Moiré technique. Note that the temperature scale is specific and not continuous. If the peak reflow soldering temperature is lowered to the 160–180 °C range, the warpage of the FCBGA component at the PRT is reduced by 30–50%.

The warpage data for a 0.8 mm thick notebook motherboard is shown in Figure 5.7b [13]. Shadow Moiré data was collected at 260 °C versus 200 °C. This data shows a decrease in panel warpage of between 19% and 23% depending on the use of a reflow pallet. However, there is no consistent shape across the gamut of board designs. Depending on the board layer stack-up design, materials of construction, designs of support pallets during reflow, the board profiles can be flat, convex, or concave.

5.1.3 Drawbacks

5.1.3.1 Brittleness
Most of the low temperature solders are based on the Sn-Bi alloy system with anywhere from 35 to 58 wt% bismuth. One of the major drawbacks of this system is the brittle nature of bismuth. The presence of the brittle bismuth phase in a solder joint can create brittleness to the joint leading to a significant degradation in the reliability performance. However, the addition of alternate alloying elements in a small quantity can improve the ductility of the Sn-Bi alloys to minimize the brittleness of bismuth through grain refinement.

5.1.4 Other Low Temperature Metallurgical Systems

One alternative to the Sn-Bi system in the low temperature category is the Sn-In system as listed in Table 5.1. The eutectic temperature of this system is at 118 °C, which would require an even lower reflow temperature and hence result in higher energy savings when compared to the Bi-Sn system. However, one of the biggest disadvantages of this system is the cost. The typical price of indium metal is at a par with silver (Ag) and with ~50% presence in the alloy makes it very expensive.

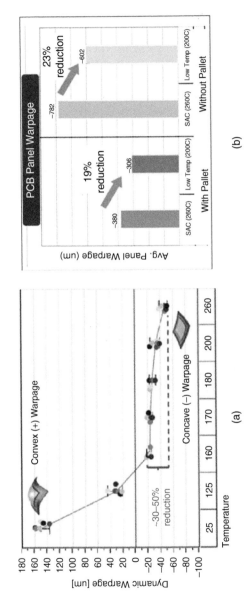

Figure 5.7 Dynamic warpage plots with temperature for (a) a FCBGA package and (b) PCB, with and without reflow pallets. The amount of dynamic warpage reduction for soldering at lower temperatures is indicated.

5.2 Development of Robust Bismuth-Based Low Temperature Solder Alloys

5.2.1 Bismuth-Tin (Bi-Sn) Phase Diagram

Figure 5.8 shows the phase diagram for the Sn-Bi alloy system. This system has a eutectic point at 58 wt% of bismuth where there is a single melting point for the alloy at 139 °C. Once the bismuth composition goes above or below this point, the alloy exhibits a pasty range from 139 °C to a corresponding point on the liquidus line. Here the alloy starts melting at 139 °C and continues to melt until it reaches the liquidus temperature for the given composition.

Given the limitations of the conventional Sn-Bi solder alloys including Sn-Bi eutectic and SnBi+ 0.4–1% Ag owing to the brittle nature of the bismuth phase, there has been significant efforts in the industry in the last three to five years toward improvement of mechanical properties of this alloy system through additional alloying/dopant elements. Some of the techniques involve:

1. Reduction in the bismuth content: One of the solutions involves reduction in the brittle bismuth phase in the alloy from the eutectic composition of 58%. Typical PRTs for low temperature soldering are limited to 190 °C. In order to provide at least 15 °C of overheating, the complete melting of the solder alloy is expected to occur at 175 °C or below. Based on the phase diagram, the reduction in the bismuth content of the solder alloy has been limited to 40%.
2. Bismuth phase refinement: This approach involves addition of different alloying elements which during the solidification process can provide nucleation sites for the bismuth phase leading to a refinement in the bismuth particle size. This approach also focused on improvement in the mechanical properties, mainly the ductility, through reduction of the large-sized bismuth particles in the microstructure which could easily crack under stress leading to poor reliability performance.
3. Tin grain refinement: Similar to the above approach, several different alloying elements were explored to provide a refinement in the bulk tin grain microstructure as well. Usually for SnAgCu lead-free solder alloys, the entire solder joint contains very few grains (<10) depending on the size of the joint. The alloying elements helped refine the microstructure of the bulk tin leading to much smaller grains. These small grains can help to increase the yield and tensile strength of the material through grain boundary strengthening where the grain boundaries in the material impede dislocation motion during any deformation.

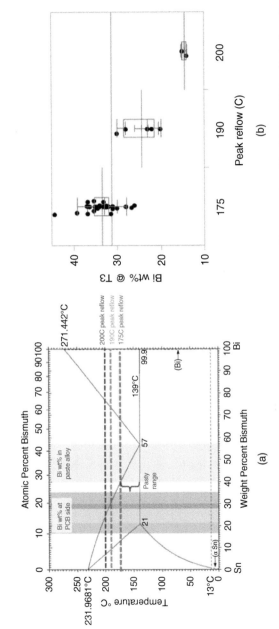

Figure 5.8 (a) Sn-Bi phase diagram along with distinct regions with bismuth composition near the solder joint-to-PCB land interface at different peak reflow temperatures. (b) Bismuth composition measured using energy dispersive X-Ray spectroscopy (EDX) at the solder joint-to-PCB land interface versus peak reflow temperature for multiple paste materials from several corner and die shadow BGA solder joints [14].

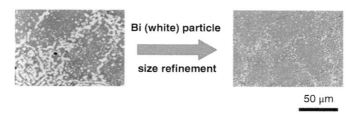

Bi (white) particle

size refinement

50 µm

Figure 5.9 Different alloying elements can provide nucleation sites for bismuth particles during solidification leading to refinement in the bismuth particle size.

4. Solid solution strengthening: This approach involves additions of extra alloying elements. These alloying elements and their compositions were chosen based on their solubility in the Sn-Bi alloy system where they would not precipitate out but stay inside the bulk of the material in solid solution. Typically such elements having different atomic sizes either reside at an interstitial position or at a substitutional position. During deformation of the material, these atoms provide a resistance to any dislocation motion resulting in an improvement in the yield and tensile strength of the material.

5. Precipitation hardening: Addition of an alloying element beyond its solubility can lead to precipitation of the element in the microstructure. Certain alloying elements can also form intermetallic compounds (IMCs) with tin or bismuth which precipitate throughout the microstructure. These precipitates can provide resistance to the dislocation motion during deformation. They can also provide nucleation sites for bismuth during solidification leading to finer bismuth phase particles. All of these can help improve the yield and tensile strength of the alloy as well as improve ductility through bismuth particle size refinement as seen in Figure 5.9.

The potential alloying elements which can provide different benefits can be obtained through studying phase diagrams. The choice of dopants can be further narrowed down by using Hume-Rothery rules for metals/alloys. The alloying can have a significant impact on the solidus and liquidus of the material. Current efforts are focused on keeping the liquidus at or below ~150 °C while minimizing the pasty range (liquidus – solidus) of the alloy. This helps enable reflow with peak temperature 165–190 °C for both hybrid as well as homogeneous solder joint formation.

5.2.2 Mechanical Properties

Typical Sn-Bi based alloys usually show significantly reduced elongation at high strain rates which makes them high-risk alloys if used for drop-shock performance. The alloys with dopants targeted at bismuth phase refinement to

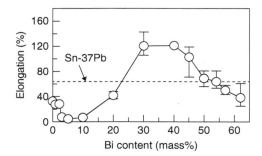

Figure 5.10 Elongation of Sn-Bi alloy maximized between 25 and 45 wt% of bismuth at slow strain rates [15].

achieve high ductility help reduce this risk. Figure 5.10 shows elongation versus bismuth composition for Sn-Bi alloys. It indicates that the slow strain rate ductility maximizes at ~25–45 wt% of bismuth for Sn-Bi alloys. That along with the creep and fatigue resistance of the alloys will dictate its performance during thermal cycle fatigue whereas overall tensile strength, fracture toughness and high strain rate ductility will affect its drop-shock performance. Several consumer electronic manufacturers use different adhesive materials to improve the drop-shock performance. Hence the industry focus is more concentrated toward the improvement for mechanical properties of the Sn-Bi alloys targeted at maximizing the thermal cycle reliability.

5.2.3 Physical Properties

Electrical conductivity of Sn-Bi is ~50% of SnAgCu alloys. However, this can be increased with the choice of high conductivity alloying elements. Thermal conductivity of typical Sn-Bi alloys is ~25–60% of that of the SnAgCu alloys depending on the alloying elements and bismuth concentration. This reduces any heat transfer through conduction in the solder during operation and may affect overall performance of the components.

Another important physical property is the CTE. The CTE of Sn-Bi based alloys is ~18–20 ppm °C^{-1}, lower than typical SnAgCu alloys which is ~26 ppm °C^{-1}. This is expected to reduce overall stress on the solder joints as it is closer to the typical CTE of the substrate and PCB (~15 ppm °C^{-1}) as well as the silicon die (~3.5 ppm °C^{-1}). Small amounts of alloying elements have minimal effect on the CTE.

5.2.4 Alloy Development Progress

Currently all of the major solder paste suppliers are engaged in the development of next generation Low Temperature Solder (LTS) alloys beyond the conventional Sn-Bi eutectic and Sn-Bi-Ag alloys. Some of the recent evaluations have pointed toward thermal cycle and drop-shock reliability performance of SnAgCu-LTS hybrid BGA component joints meeting the user defined requirements. While technical challenges of low temperature solders are being addressed, overall adoption of the technology in the mobile, laptop and desktop PC industry is also progressing fairly rapidly.

5.2.5 Fluxes for Low Temperature Solders

Any soldering typically requires fluxes. The fluxes usually consist of rosins, organic acids, amines, halides, solvents, and thixotropic agents. The solder alloys in the BGA form or powder form in a paste have an oxide layer on the surface. Here the activators (rosins, organic acids, amines, halides) help with wetting and cleaning the oxide on the solder surface during a reflow process so that the molten solder can coagulate to form a single joint. The function of solvents and thixotropic agents is to control rheological properties of the solder paste or flux in order to facilitate processing. Given the lower PRTs for the Sn-Bi based low temperature solder alloys, the fluxes used for soldering also need reformulation to ensure sufficient activity during the low temperature reflow. This usually warrants for high activity fluxes to be used for solders as the oxide to be cleaned mainly remains Sn oxide – the same as SnAgCu solders, for the Sn-Bi based low temperature solders.

To assess the impact of the new fluxes of the Sn-Bi based low temperature solder alloy solder pastes on the surface insulation resistance (SIR) of boards after reflow soldering, the iNEMI Low Temperature Solder Process and Reliability (LTSPR) project team evaluated 13 different solder pastes from 4 distinct solder paste categories [16]. The SIR measurements were performed according to IPC-650 Method 2.6.3.7 [17]. IPC-B-24 comb pattern coupons, with 4 nets per coupon, were the test vehicle. Except for one of the 13 solder pastes, 3 coupons were prepared for each solder paste, by the paste supplier. Control coupons, with no flux or solder paste applied, were also tested from each supplier. The iNEMI LTSPR project team ran the test and collected the data at an independent testing house.

The results from this test on all 13 solder pastes are shown in Figure 5.11. Each solder paste is grouped with its similar category: either the SnAgCu standard, the Sn-Bi baseline which included three different SnBiAg solder pastes, the ductile SnBiX$_1$X$_2$ category which included the solder paste whose alloys were strengthened by the addition of various dopants, and the resin-reinforced Sn-Bi solder pastes. The results show that all solder pastes had SIR values above the 1×10^8 Ω minimum requirement. Most of the solder paste SIR data (plotted in red) was very similar to that for the control coupons (plotted in blue) which had no solder flux paste applied to them. However, three solder pastes, two in the Sn-Bi category and one in the resin-reinforced solder paste category, had the SIR values either much

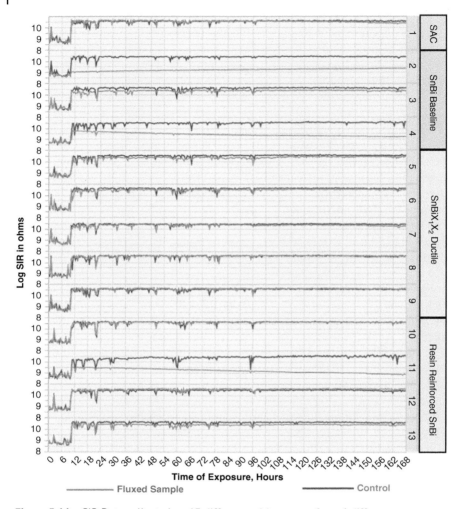

Figure 5.11 SIR Data collected on 13 different solder pastes from 4 different paste categories [16].

lower than the control values or SIR values that exhibited a decreasing trend with time. Though decreasing trends in SIR with time indicate potential ion migration none of the trends are sharp and do not fall below the minimum requirement. Hence, the iNEMI LTSPR project team concluded that all 13 solder pastes evaluated met the SIR requirements [16]. After the SIR tests were completed, each sample was inspected under 30X magnification for the presence of dendrites with no evidence of dendrites reported.

5.3 SMT Process Characterization of Sn-Bi Based Solder Pastes

5.3.1 Printability

Solder paste print performance is influenced significantly by the rheological properties of the solder material [18]. In general, Sn-Bi based low temperature solders have material properties (viscosity, thixotropic index, etc.) comparable to SnAgCu based paste materials. A typical Sn3Ag0.5Cu paste material is compared with a selection of low temperature solder pastes formulated with a range of Bi% in Table 5.3.

As can be seen from the data, there are only minor distinctions between the key properties for Sn-Bi based materials and their SnAgCu counterparts.

Based on the materials properties, it would be expected that a low temperature solder paste would provide very similar solder print performance to that demonstrated by a SnAgCu solder paste. Figure 5.12 provides a comparison of solder print volume performance for a SnAgCu control paste versus two low temperature solder materials.

Table 5.3 Rheological properties of two SnBi low temperature solder pastes compared with a SnAgCu paste material.

Paste	Viscosity at 10 RPM (Pa s)	Thixotropic index
Sn3Ag0.5Cu	207	6.4
LTS_A	195	7.5
LTS_B	206	5.9

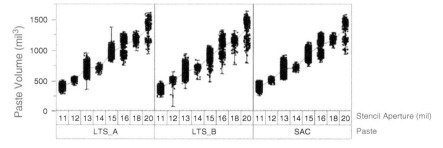

Figure 5.12 Solder paste printed volume versus aperture size (round) for SnAgCu and two Sn-Bi based low temperature solders.

Both low temperature solders exhibit a solder paste volume virtually identical to the SnAgCu control paste. This performance is maintained over a wide range of stencil apertures with Area Ratios ranging from 0.69 to 1.25, with a stencil thickness of 4 mils (100 μm).

Due to the similarities in materials performance, no paste print equipment modifications are required when setting up a low temperature solder print process. While setup-specific updates in print speed, print pressure, stencil separation speed, etc. will be necessary, the overall development of a low temperature solder paste print process will look much the same as a SnAgCu process. Stencil material and aperture design can be directly leveraged from an existing SnAgCu process. If eliminating wave solder in favor of PIP, through-hole barrel fill may be lower than observed with the wave solder process. This result can be compensated for by modifying the aperture design to provide an increase in solder paste volume.

5.3.2 Reflow Profiles

A key distinction between a SnAgCu-based reflow profile and a low temperature profile is the PRT that must be reached in order to complete the soldering process. The ability to utilize a PRT below 190 °C rather than the 235–245 °C peak temperature required by a SnAgCu solder paste is a significant benefit in terms of reducing package and board warpage as well as board thermal stress. Figure 5.13 compares a typical SnAgCu-based reflow profile to a low temperature solder profile.

Depending on the selected solder paste supplier and low temperature solder product, a traditional soak-ramp to peak or a ramp to peak profile may be suggested by the paste supplier. Typical reflow profile parameters are listed in Table 5.4. Even though the time above liquidus (TAL) is slightly higher than a typical SAC profile (60–90 seconds), overall reflow time is similar due to lower

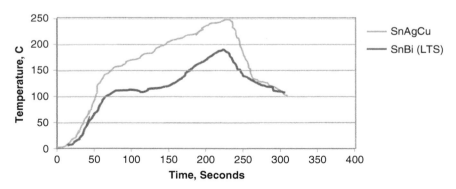

Figure 5.13 Soak-ramp-peak profile for low temperature solder compared with the SnAgCu solder.

Table 5.4 Reflow profile critical success criteria used to establish the common lead-free low temperature reflow profile.

Attributes	Value
Peak ref low temp	180–185 °C
Initial ramp rate ($°C\,s^{-1}$)	1–3
Soak time (100–120 °C)	60–90 s
Time above liquidus	100–120 s
Max rising slope ($°C\,s^{-1}$)	1–3

PRT. The higher TAL helps provide the flux sufficient time to clean the oxide and the solder to form a joint.

5.3.3 Rework

When developing a low temperature solder rework process, several factors must be considered. For BGA packages the primary consideration is BGA ball metallurgy. If the component to be reworked uses SnAgCu BGA spheres, then a mini-stencil will be required if a low temperature reflow profile is to be used. The design of the mini-stencil thickness and aperture sizes should be such that good-quality solder joints result along with no bismuth present at the package side interface. Studies have shown that solder joint reliability issues can occur, if Bi diffuses to the package land-to-solder interface [13]. Rework profiling can be accomplished in the same way as a SnAgCu rework profile while making the necessary reductions in PRT. When reworking a SnAgCu BGA component, a flux only process is not possible when using low temperature solder PRTs. With no low temperature solder paste to create the metallurgical bond, the SnAgCu BGA ball will not melt and a solder joint will not result.

In the case of a BGA component with low temperature solder BGA spheres or any board component that will have a homogenous low temperature solder joint (passives, lead-frame components, most connector types, etc.), rework is accomplished in exactly the same manner as a SnAgCu rework process, but at the reduced PRT with the low temperature soldering material. Either a mini-stencil or flux-only process can be utilized for the BGA component. For non-BGA components, a mini-stencil can be used for larger components or low temperature solder wire can be employed for smaller components or for touch-up processes. Due to the difficulty in extruding Sn-Bi based wire, many solder suppliers do not support flux-cored low temperature solder wire. Most solder suppliers would offer a low temperature solder wire and can provide a low temperature solder optimized flux material. It is also advised to purchase solder iron tips with a lower operating temperature. A soldering iron temperature rating of 500–600 °F (260–316 °C) has been shown to produce acceptable results for low temperature solder versus the typical temperature rating of 800–900 °F (427–482 °C) for SnAgCu hand solder processes.

5.4 Polymeric Reinforcement of Sn-Bi Based Low Temperature Alloys

The presence of the brittle bismuth phase in the microstructure of the Sn-Bi based solder joint [19] increases their risk of brittle fracture during application of mechanical forces, particular under high strain rates [20]. The inherent brittle nature of bismuth is largely attributed to its rhombohedral crystal structure [21], which has very few of the slip planes that are necessary for material ductility. In fact, bismuth has only 1/3 the slip planes found in Sn and 1/6 the slip planes found in Cu, Ni, Al, and Pb [22]. The brittleness of BiSn-based solder alloys can limit their use in cell phones, tablets and other mobile devices, which can be subjected to multiple drops during use. Mixed alloy BGA solder joints formed by soldering SnAgCu solder balls with BiSnAg solder paste have been shown to exhibit significant reduction in mechanical drop reliability when compared with solder joints formed using SnAgCu-based solder pastes [23–25]. The failure interface is at the PCB land-to-solder joint interface where the bismuth phase is located for mixed solder joints.

5.4.1 Current Polymeric Reinforcement Strategies

In electronic assemblies today, strengthening of solder joints at risk for brittle or even fatigue failure is achieved by use of extrinsic polymeric reinforcement. Figure 5.14 depicts diagrammatically various polymeric reinforcement strategies in use today for Area Array Packages, such as FCBGAs. Figure 5.15 illustrates the

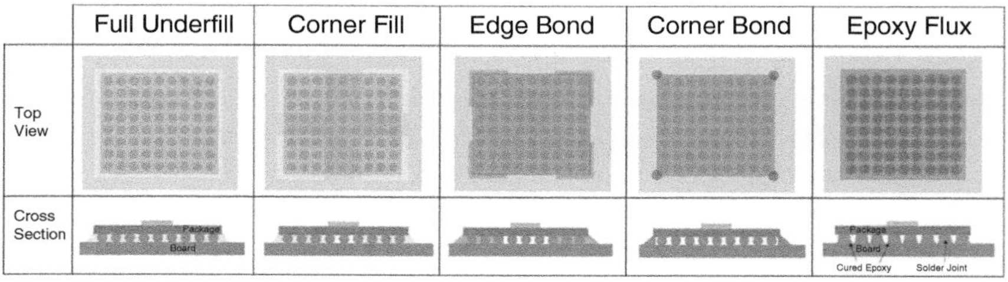

Figure 5.14 Various polymeric reinforcement strategies for area array packages in use today.

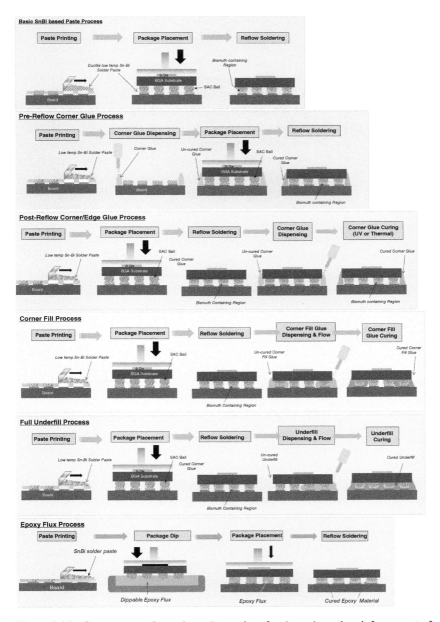

Figure 5.15 Process steps for various alternatives for the polymeric reinforcement of area array packages.

process steps for each of these polymeric reinforcement strategies. These include the following.

- *Full underfill*: This board-level underfill process [26–31] is similar to the chip level underfill process. After the component is soldered using the SMT process, underfill is dispensed on the edges of the package body and flows under the package driven by capillary forces. Once the underfill has flowed under the entire area of the package, the board assembly containing the area array package is heated up to a set temperature for a set time to cure the underfill. This alternative provides excellent shock resistance and very good temperature cycle resistance if the underfill material is chosen properly based on its relevant properties. However, this full underfill process is not conducive to a high-volume manufacturing (HVM) environment since it is an off-line process, and requires an additional curing step. As area array package body sizes have become larger, the ball diameter smaller, the ball pitch tighter and less uniform, the time required for the underfill to flow and cover the entire package area has increased significantly [32–37], thereby leading to a slowdown in the throughput of assemblies on the HVM manufacturing lines. There is also a material incompatibility concern between the underfill and the flux residue left from the solder paste after reflow. This incompatibility creates voids in the underfill and leads to subsequent delamination and solder extrusion [38, 39]. Another compounding concern is that the area array components with full underfill are nearly impossible to rework, requiring a highly skilled operator to avoid damage to the PCB lands when removing the remnants of the underfill after the defective BGA has been de-soldered [30, 40].
- *Corner fill*: This process evolved from the full underfill process to diminish the time required for the underfill to flow under the large package body. Underfill is dispensed near each of the four corners of the package instead of just in one location. The amount dispensed is also just sufficient to cover the corner of the package about three to six rows deep of solder joints, depending on the package body size. The process steps are identical to that of the full underfill, but with a shorter time for underfill flow. Since in many cases, to attain the required reliability levels for shock and temperature cycling resistance, full underfill was an over-requirement, corner fill is adequate in this regard.
- *Edge bond*: This process is used when enhancement in shock resistance is the primary goal to meet the product reliability requirement. Glue is dispensed on all four corners of the area array package to about three to six rows of solder joints in on each side, depending on the body size of the component. The glue's rheological properties are markedly different from that of the underfill glue. Once dispensed it does not flow under the package body. Depending on its formulation, this glue can be either thermally cured or cured under ultraviolet (UV)

light, since most of it is applied outside the package body edges. Since UV curing takes much less time than thermal curing for underfill, this process evolved from the underfill process to significantly reduce the curing time step and thereby increase the board throughput in the HVM environment. This edge bond process also reduces the amount of glue material used when compared to the underfill processes. However, it does not enhance the temperature cycling resistance of the package solder joints to any appreciable degree.

- *Corner bond*: This process is very similar to the edge bond process, but only a dot of glue is used at the corners of the package, and hence uses the least volume of glue material of all the polymeric reinforcement processes. But the level of enhancement for shock reliability is also least of the other polymeric enhancement processes. One variant of this corner bond process is to dispense the glue at the corners of the area array land patterns on the board after printing of the solder paste, but before placement of the component on the board. The glue then cures during reflow. This process is also shown in Figure 5.15. However, the material and process control requirements for this pre-reflow corner bond process are very stringent since the glue should not flow on to the corner lands before the solder joints are formed on those lands. Further, the glue material chosen should have a low coefficient of expansion (CTE) and high glass transition temperature (Tg) to avoid deleterious impact on the temperature cycling reliability of the corner solder joints.

- *Epoxy flux*: This process was developed to mitigate the drawbacks of underfills, edge bond and corner glues [41–49]. The epoxy flux process entails dipping the area array component into a tray containing the epoxy flux material, which enables the flux to coat the solder balls of the component. The component is then placed on the board, after solder paste has been printed on the board lands. The board with the components is then reflow soldered. The epoxy flux cures and this cured resin then functions as a polymeric reinforcement material that adheres to the solder joint and the surface of the board, including the solder-mask and laminate, in the close vicinity of the solder joint. There is significantly more flexibility in how the epoxy fluxes are applied to the area array component solder balls than there is for underfills or corner glues. Besides dipping the components into a reservoir of the material before being placed on the board the epoxy fluxes can be screen printed, dispensed, or jetted on the PCB lands before component placement. The rheology of the epoxy fluxes, however, will need to be optimized by these different application methods.

Curing of the epoxy fluxes is achieved during the reflow soldering process and hence an additional post-reflow step for cure is not required. This characteristic imparts the main processing advantage to epoxy fluxes when compared to underfills or corner glues. Nevertheless, the formulation of the epoxy flux has to be optimally engineered to ensure that it will cure during the reflow soldering process step without impeding the BGA solder joint formation processes. The resin within the epoxy flux should not begin gelling (i.e. no longer being a liquid and having lost its ability to flow) until the solder in the paste has melted, wetted the PCB land and the solder ball and the solder joint has collapsed fully. If the epoxy flux is used in conjunction with solder paste printed on the PCB lands (for instance, if the need arises to accommodate substantial dynamic warpage of FCBGA components during reflow), it should have sufficient miscibility with the solder paste.

This combination of epoxy flux and solder paste requires "venting channels" within the solder joint array for the volatiles evolving from the SJEM (Solder Joint Encapsulant Material) and solder paste to escape during the reflow soldering process. But, generally, this is not an issue since, unlike underfills, the cured resin from the epoxy fluxes encapsulate individual solder joints only rather than the entire area under the BGA component substrate, thereby leaving interconnected channels for the emitted gases to escape [24, 25]. In this regard, epoxy fluxes provide polymeric reinforcement at the individual solder joint level whereas underfills, and corner glues provide polymeric reinforcement at the entire package level.

All these current polymeric reinforcement strategies can be applied to electronic assemblies soldered using low temperature Sn-Bi based solder paste, but with a caveat related to the cure schedule for thermally cured glues. The time and temperature used for the glue curing schedule should meet a minimum requirement to attain at least 90% cure for the cured glue to have the desired mechanical properties. Some underfills and glues have a cure temperature above the liquidus temperature of the Bi-Sn alloys which is in the 139–150 °C range. This cure temperature should be avoided for low temperature solder assemblies. Instead a longer time at a lower temperature is recommended. A temperature of 100 °C for 30 minutes is the most appropriate.

5.4.2 Joint Reinforced Pastes (JRP)

A better polymeric reinforcement option for low temperature soldered assemblies to strengthen the Bi-Sn solder joints is to use low temperature resin-reinforced solder pastes, also known as Joint Reinforced Pastes (JRP). These are solder pastes that contain an uncured resin as part of their flux component. During the reflow process, when the solder paste melts, this resin is displaced away from the molten solder and coats the molten solder externally. As the reflow process proceeds further, the resin starts to cure. Eventually, after the reflow process is completed, the cured resin forms a fillet around the solder joints, providing the necessary mechanical reinforcement. Figure 5.16 diagrammatically illustrates this process.

This concept of the JRP material is obviously similar to that of epoxy fluxes, but with two important differences. One is that epoxy fluxes require an additional

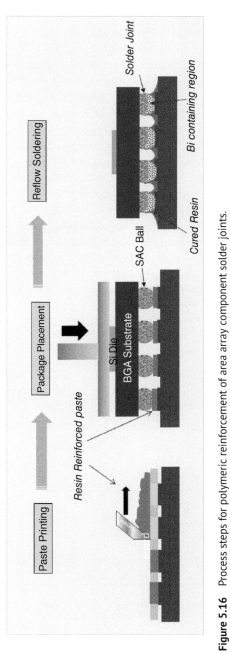

Figure 5.16 Process steps for polymeric reinforcement of area array component solder joints.

process step to apply the material to the solder ball or termination of the component, whereas the standard SMT process step of solder paste printing through a stencil is all that is required for JRP material application. No additional process step is required. Second, epoxy fluxes are applied to area array devices and hence only strengthen part of the solder joints on a typical product board, albeit those at the highest risk for failure in the field. But, for the brittle Sn-Bi solder joints, many of the solder joints of other components on an electronic product board, such as leaded devices (QFPs, SOICs) and bottom termination components (BTCs) also need strengthening. The JRP materials are applied to *all* the lands on the board and hence the polymeric resin strengthening is imparted to all these solder joints.

The resin in JRP materials cures is designed to gel and cure during the reflow process. This fact, in itself, necessitates a curing phase in the reflow profile for JRPs. Figure 5.17 shows two distinct reflow profiles for Sn-Bi JRPs and compares it to a typical reflow profile for a standard (non-resin) Sn-Bi solder paste. The red reflow profile in Figure 5.17 is the long JRP profile during which the resin cures after the solder has melted and is still molten when the curing occurs. This profile is called a trapezoidal profile, and is almost twice as long as the standard Sn-Bi reflow profile and has a negative effect on the board throughput on the SMT manufacturing line. To increase the throughput, some JRP material suppliers have

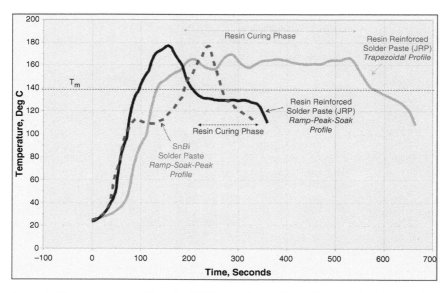

Figure 5.17 Comparison of two Sn-Bi JRP material profiles with a standard (non-resin) Sn-Bi solder paste reflow profile. The solid red line denotes a trapezoidal JRP profile, the solid green line denotes the Ramp-Peak-Soak JRP profile and the dashed blue line denotes the Ramp-Soak-Peak standard Sn-Bi solder paste profile.

developed the shorter green profile, during which the solder paste is swiftly taken above the solder liquidus temperature to form the solder joint and then brought back down just below the liquidus and kept there to cure the resin. This profile is called the Ramp-Peak-Soak profile and is as long as the standard Ramp-Soak-Peak profile for the standard Sn-Bi solder pastes.

The appearance of a typical BGA solder joint formed when using Bi-Sn JRPs is shown in Figure 5.18. This is a dark field contrast optical microscopy image. A dark field setting is needed to reveal the presence of the cured resin in the image. This particular BGA component had a SnAgCu solder ball and the resulting solder joint was a mixed alloy, hybrid solder joint, with the lower part of the joint containing bismuth. The two-phase microstructure is visible. The cured resin is seen as a fillet around the lower half of the solder joint. The resin encompassed the solder joint and adhered to the solder mask and laminate around the periphery of the solder joint. This particular solder joint had a metal defined PCB land and the resin filled the "trench" around the PCB land up to the edge of the soldermask opening. Wetting of the solder on the BGA land was excellent. This shows that when the solder paste melted the resin, which had yet to gel it was pushed away to the periphery of the molten solder paste as it wetted the PCB land and the solder

Figure 5.18 An optical microscopy cross-section image of a SnAgCu ball BGA solder joint formed using a Sn-Bi JRP solder with a trapezoidal reflow profile.

ball. As discussed later, this is an important requirement for successful formation of solder joints when using JRP solder materials.

Figure 5.19 portrays dark field optical microscopy images of solder joints for various types of components formed using JRP solder materials. The cured resin can be observed on the outside of the solder joints. This encapsulation of the solder joint by the resin and its adherence to the soldermask, and laminate around the vicinity of the solder joint provides the mechanical reinforcement to strengthen these solder joints under external mechanical and thermomechanical stresses during use.

The ramp rate of the JRP paste reflow profile is very critical to ensure good quality solder joint formation. A slower than necessary ramp rate will create poorly wetted solder joints. Figure 5.20 illustrates examples of such poor quality BGA solder joints.

The reason for such poor-quality solder joints can be explained by comparing the ramp rate for two trapezoidal JRP reflow profiles shown in Figure 5.21. The Sn-Bi solder in the JRP has to melt, wet the board lands and the SnAgCu solder ball *before* the resin starts to gel and its viscosity increases as its cure progresses. If ramp rate is slow, the JRP resin will gel and cure before the solder has fully wetted the PCB land and solder ball, and consequently this will lead to partial wetting. This is illustrated by the slower ramp rate line crossing the resin curing phase region in Figure 5.21.

The presence of the resin in JRP solder pastes creates limitations in the stencil printing process step of the SMT manufacturing process. Stencil aperture area ratios are critical to ensuring that sufficient solder paste is printed on the PCB lands to form good quality solder joints. The current industry standard for an acceptable area ratio is 0.66 but technological advances are pushing this limit further down to below 0.60 [50]. In a comparison study done by the iNEMI LTSPR project team, JRP (resin-reinforced Sn-Bi) solder pastes had a more pronounced decrease in transfer efficiency of the solder paste as the stencil apertures went below 0.60. This is shown in Figure 5.22. The JRP pastes' transfer efficiency was equivalent to that of the SnAgCu, and standard BiSn solder paste when the stencil aperture area ratio was 0.75. For a stencil aperture ratio of 0.59, though statistically the JRP paste's transfer efficiency was equivalent to that of SAC and standard BiSn solder paste, there were more outliers at both the upper and lower ends of the data distribution. For 0.5 aperture ratio, the JRP pastes had lower transfer efficiency than the other two category pastes.

One more drawback of JRP pastes is the impact that the cured resin, encapsulating the solder joints, has on the rework process. Though not as extensive as underfills, the presence of cured resin when using JRP resin pastes tends to leave more material on the board after part removal. This necessitates extra time and effort by the rework operator to remove this resin before re-soldering of a new

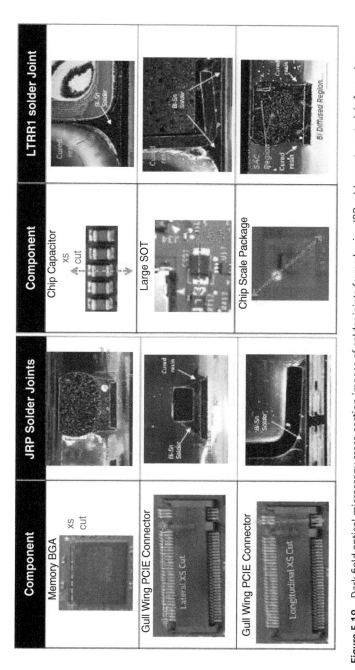

Figure 5.19 Dark field optical microscopy cross-section images of solder joints formed using JRP solder paste materials for various electronic component packages [6].

Cured resin

Figure 5.20 Partial wetting defects for BGA solder joints caused when using incorrect reflow profiles for JRP solder pastes [16].

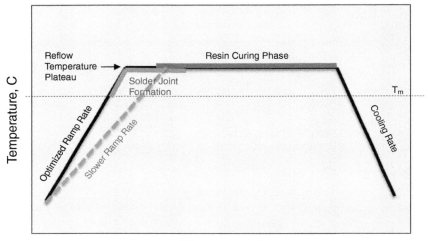

Figure 5.21 Effect of a slower ramp rate for JRP solder pastes on the solder joint formation mechanism.

component on that particular PCB land site. However, the presence of the resin was instrumental in limiting damage to the land during the site redress process step in rework [16].

To assess the enhancement of solder joint reliability when using JRP paste to form Sn-Bi solder joints encapsulated by cured resins, the iNEMI LTSPR project team subjected two different types of BGA packages, assembled on specially designed boards, to mechanical shock drop testing and monitored the solder joint resistances in situ [51, 52]. The data was used to generate Weibull plots. The characteristic life, η, which represents the shock drop level at which 63.2% of the samples in each category would have failed, was then plotted. Data from three types of solder pastes were compared: (i) standard Sn-Bi (no resin) solder paste, (ii) Sn-Bi JRP (resin containing), and (iii) SnAgCu paste as a control.

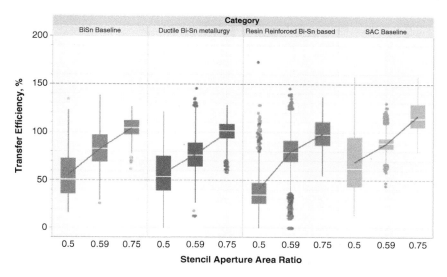

Figure 5.22 Transfer Efficiency of BiSn, JRP and SnAgCu solder pastes through stencil apertures for various aperture area ratios [16].

Two packages were evaluated. One was a small System on Chip (SoC) BGA package, 14 mm × 14 mm body size. The other was a larger 42 mm × 24 mm FCBGA package. Figures 5.23 and 5.24 show the box plots of the Characteristic Life, η(eta), extracted from Weibull plots drawn using results of mechanical shock drop tests on the small SoC component and the larger FCBGA component, respectively. Details of the package and board design, as well as the mechanical shock test parameters are also given above the plots in each figure.

From these two plots the following results can be ascertained:

- For both components, the Sn-Bi solder joints are significantly less robust under mechanical shock drop tests than SnAgCu solder joints. This was expected due to the presence of the brittle bismuth phase in the Sn-Bi solder microstructure.
- For both components, the box plot of the characteristic lives of the solder joints formed by the JRP pastes are higher than that for the standard Sn-Bi solder pastes. The solder microstructure of the solder joints for both these cases is expected to be the same, with the difference being that the JRP paste solder joints have a cured resin encapsulating the lower part of the solder joints. These results therefore confirm that the cured resin around the solder joints significantly enhances the reliability of the BGA solder joints during mechanical shock drop tests.

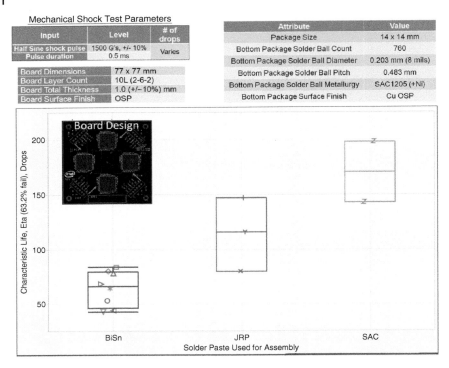

Mechanical Shock Test Parameters

Input	Level	# of drops
Half Sine shock pulse	1500 G's, +/- 10%	Varies
Pulse duration	0.5 ms	

Board Dimensions	77 x 77 mm
Board Layer Count	10L (2-6-2)
Board Total Thickness	1.0 (+/- 10%) mm
Board Surface Finish	OSP

Attribute	Value
Package Size	14 x 14 mm
Bottom Package Solder Ball Count	760
Bottom Package Solder Ball Diameter	0.203 mm (8 mils)
Bottom Package Solder Ball Pitch	0.483 mm
Bottom Package Solder Ball Metallurgy	SAC1205 (+Ni)
Bottom Package Surface Finish	Cu OSP

Figure 5.23 Comparison of the Characteristic Life, η (eta), extracted from Weibull plots drawn using the results of mechanical shock drop tests on 14 × 14 mm SoC packages on a JEDEC B111A board, assembled using different solder pastes. Details of the package and board design, as well as the mechanical shock test parameters are also given.

- For the smaller SoC component, the box plot for the JRP pastes is lower than that for the SnAgCu solder paste, though there is some overlap. Therefore, the enhancement provided by the resin around the Sn-Bi solder joints is not sufficient to bring the solder joints reliability under mechanical shock drop up to the SnAgCu solder joint level. Solder pastes from three different JRP paste manufacturers were used for this investigation, and the data point for each manufacturer is represented by a different marker symbol for its data point. A closer examination of the plot in Figure 5.23 reveals that the highest data point in the JRP category is on a par with one of the data points for the SnAgCu category.
- For the larger FCBGA component, the characteristic lives of five of the JRP solder pastes is higher than that for the SnAgCu solder paste and one of the JRP pastes is on a par with it. In fact, there is a large variability in the characteristic life of solder joints for the components based on which JRP manufacturer's paste was used to assemble the board and form the solder joints.

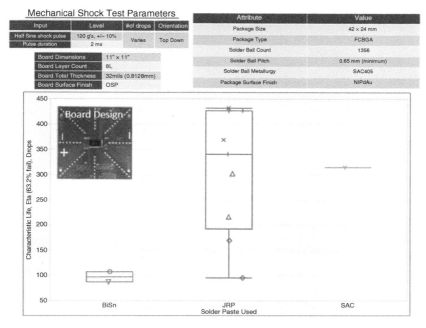

Mechanical Shock Test Parameters			
Input	Level	#of drops	Orientation
Half Sine shock pulse	120 g's, +/- 10%	Varies	Top Down
Pulse duration	2 ms		

Board Dimensions	11" x 11"
Board Layer Count	8L
Board Total Thickness	32mils (0.8128mm)
Board Surface Finish	OSP

Attribute	Value
Package Size	42 x 24 mm
Package Type	FCBGA
Solder Ball Count	1356
Solder Ball Pitch	0.65 mm (minimum)
Solder Ball Metallurgy	SAC405
Package Surface Finish	NIPdAu

Figure 5.24 Comparison of the Characteristic Life, η (eta), extracted from Weibull plots drawn using the results of mechanical shock drop tests on 42 × 24 mm FCBGA packages on a special designed shock test board, assembled using different solder pastes. Details of the package and board design, as well as the mechanical shock test parameters are also given.

The reason for this large variability in the characteristic lives of the solder joints under mechanical shock drop tests when assembled with JRP pastes from different manufacturers was ascertained when the height to which the solder joints were encapsulated by the resin were measured from the cross-sections done on the solder joint [53]. Figure 5.25 shows the plot of characteristic life from the Weibull plots under mechanical shock versus the resin height that the cured resin encapsulates the solder joints at, as a % of the total solder joint height. This is for the 14 mm × 14 mm SoC component, whose reliability results are shown in Figure 5.23. The box plots represent the distribution of 50 data points for each of the JRP paste manufacturers. The means of these distributions are indicated by the red square within each of the box plots. It is easily apparent that as the resin height % increases the characteristic life for the solder joint reliability under mechanical shock also increases. From Figure 5.23 it can be deduced that due to the higher resin height encapsulation of the BiSn solder joints resulting when using JRP manufacturer C's solder paste, the resin-reinforced solder joints were

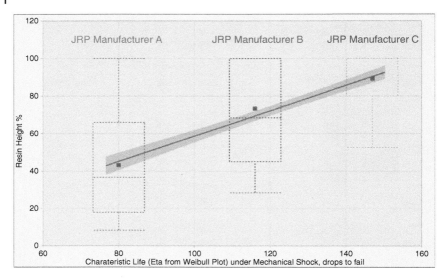

Figure 5.25 Plot of Characteristic Life, η, under mechanical shock for SoC BGA solder joints versus the resin height % that these solder joints were encapsulated to when using JRP solder pastes from three different JRP paste manufacturers.

on a par with SnAgCu solder joints for their characteristic life in mechanical shock reliability tests.

The variation in the resin height to which the solder joints are encapsulated between the JRP pastes from different manufacturers is due to the volume of resin incorporated in the solder paste being different for each supplier. The amount of resin also modulates the pastes viscosity and therefore needs to be optimized based on each formulation of the JRP solder paste flux.

5.4.3 Polymeric Reinforcement Summary

To overcome the brittleness of Sn-Bi solders caused by the presence of the brittle bismuth phase in the solder microstructure, a wide range of resin reinforcement alternatives are available. Many of these alternatives have their advantages and disadvantages. The choice of which to use depends on the application for which the low temperature solders are required for.

5.5 Mixed SnAgCu-BiSn BGA Solder Joints

5.5.1 Formation Mechanism

Low temperature solders can be effectively used to solder BGA components with SnAgCu balls. Unlike the normal melting based joining that occurs in a

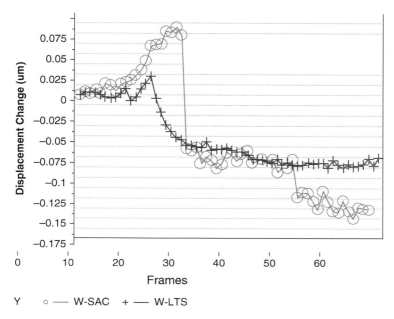

Figure 5.26 Solder joint displacement behavior for a fully melted solder joint (SnAgCu) versus a SnAgCu-Sn-Bi (LTS) solder joint.

homogenous SnAgCu SMT process, the full SnAgCu solder ball will not melt during the low temperature reflow. Instead, the metallurgical bond is created by a diffusion-driven process. As the low temperature solder paste melts and contacts the SnAgCu solder ball, tin from the SnAgCu solder sphere begins to dissolve and inter-diffuse with the low temperature solder paste, creating a coherent solder joint. The difference between melting and diffusion-driven solder joint formation is illustrated in the plot of solder joint displacement versus time, Figure 5.26. The melting solder joint (SnAgCu) exhibits rapid reduction in solder joint height when the BGA ball reaches liquidus. The mixed SnAgCu-BiSn (LTS-Low Temperature Solder) joint shows a more gradual reduction in the solder joint height as diffusion progresses through the reflow cycle.

The resulting mixed SnAgCu-SnBi solder joint is referred to as a "hybrid" joint. The characteristics of the hybrid solder joint are strongly dependent on the PRT. As can be seen in Figure 5.27, the extent of the interdiffusion increases with increased PRT.

Once the PRT exceeds 200 °C, the joint may approach complete homogenization, depending on the size of the BGA sphere and the solder paste volume. The resulting solder joint height will also be reduced as the reflow temperature is increased (Figure 5.28).

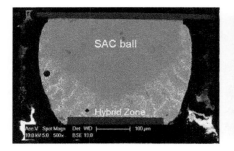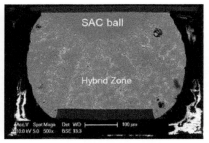

Figure 5.27 Hybrid SnAgCu – Sn-Bi solder joint reflowed at 190 °C (left) and 200 °C (right). Note the increased extent of the hybrid zone with the increase in peak reflow temperature.

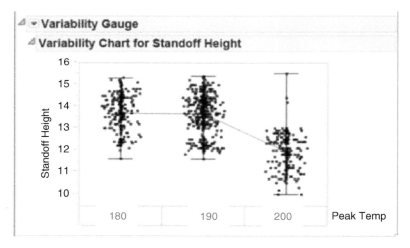

Figure 5.28 Plot of solder joint height versus peak reflow temperature for a hybrid SnAgCu – Sn-Bi solder joint.

The diffusion reaction between the Sn-Bi solder and the SnAgCu solder ball occurs both at the immediate interface between the solder paste deposit and BGA sphere as well as over the surface of the BGA sphere as the molten solder paste wicks up the sides of the solder sphere (Figure 5.29).

The extent to which the molten solder paste extends up the sides of the solder sphere and thus it may ultimately reach the package-ball interface is a function of solder paste volume, reflow temperature, reflow cycle count, and reflow atmosphere. The presence of bismuth at the package-ball interface is a key consideration in the optimization of solder joint reliability for hybrid BiSn-SnAgCu solder joints and will be discussed in section 5.6.

bismuth (<12%) at the package-ball interface will be greater when using nitrogen. The specific risk for a given SMT process should be verified by using post-reflow cross-sections to identify the resulting extent of the hybrid diffusion zone. Bismuth concentrations at the package-ball solder joint interface could be a reliability risk, as discussed in section 5.6.

5.5.2 Microstructural Features and Key Characteristics

At time zero (T0) after SMT reflow, a hybrid SnAgCu-BiSn BGA solder joint exhibits a characteristic microstructure. In the hybrid mixed region, the bismuth will form islands within the tin matrix. The size of the bismuth islands is dependent on the PRT. As the reflow peak temperature is increased, the bismuth regions begin to break down into smaller regions of bismuth and the matrix exhibits an increase in grain size refinement. Figure 5.33 illustrates a typical hybrid SnAgCu-SnBi mixed zone microstructure as reflow temperature is increased.

The initial bismuth concentration of the solder paste does not greatly impact the microstructure of the mixed region for solder pastes with bismuth in the range of 32–58 wt%. Figure 5.34 provides cross-sectional views of time zero solder joints formed using a selection of solder pastes with varying bismuth concentrations. Each paste was reflowed using an identical reflow profile and PRT. For a given set of reflow conditions, the time zero microstructure can be seen to be very similar for each of the bismuth concentrations.

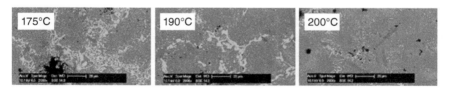

Figure 5.33 Increased hybrid Bi diffusion zone microstructure refinement as reflow peak temperature is increased.

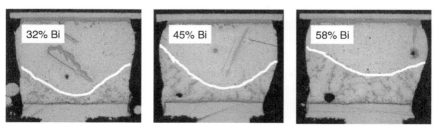

Figure 5.34 Time zero hybrid bismuth zone microstructure versus weight percent bismuth in the solder paste (white line added for emphasis).

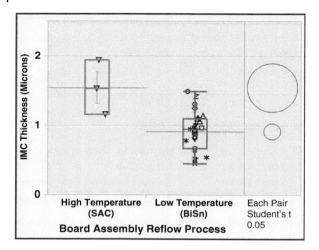

Figure 5.35 Comparison of IMC thickness (on Cu OSP surface finish) for SnAgCu versus BiSn solder pastes.

The impact of the IMCs on the mechanical performance of Sn-Bi based low temperature solder joints has been extensively studied [54–57]. Studies have shown that various IMCs (generally Ni3Sn-based IMCs for NiAu-based surface finishes or Cu3Sn-Cu6Sn based IMCs for Cu OSP surface finishes) can be generated when using Sn-Bi based solders [58]. Due to the potential weak adhesion of the two-phase IMCs [59], they are considered undesirable for solder joint reliability performance. However, under typical low temperature solder reflow conditions (peak temperature < 190 °C, time above liquidus<120 seconds), the risk of the two-phase IMC formation appears to be low [60]. IMC thickness for a low temperature solder based solder joint will typically be thinner than those seen for a SnAgCu-based solder joint [61]. Figure 5.35 provides a comparison of IMC thickness for a SnAgCu solder joint compared with a selection of Sn-Bi based low temperature solder pastes.

5.5.3 Soldering Process Optimization

As the percentage of bismuth in a Sn-Bi solder paste is either increased or decreased away from the eutectic percentage of 58%, the temperature over which the material starts to melt until it is fully liquid begins to increase. This temperature range is often referred to as the "pasty range" of the material. Table 5.5 details the pasty range values for a selection of Sn-Bi solders based on the percentage of bismuth in the material.

When selecting a PRT for the SMT process, it is important to consider the approximate bismuth weight percentage that will result in the hybrid region of the finished solder joint. From Figure 5.10, it can be seen that in the Sn-Bi system, maximum elongation performance is achieved when the bismuth weight percent is in

Table 5.5 Melting start and end temperatures for BiSn solder pastes as a function of bismuth concentration.

Bismuth concentration (%)	Melting start (°C)	Melting peak (°C)	Melting end (°C)	Pasty range (°C)
32	136	174	190	54
45	138	154	166	28
50	139	143	153	14
57	138	141	150	12

the region of 25–45% [15]. Optimizing the elongation performance is desirable for achieving the best solder joint reliability capability.

It is possible to estimate the resulting bismuth weight % for a finished solder joint with knowledge of the starting bismuth % in the chosen solder material and the desired PRT. The starting bismuth percentage in the solder paste and the associated pasty range will influence the required PRT depending on the desired number of degrees over full liquidus ("super heat") to ensure acceptable wetting and solder joint formation across the extent of the board. Typical low temperature solder reflow temperatures of 160–190 °C will result in hybrid regions with bismuth weight % of 20–45%. Actual bismuth concentration will be influenced by the additional dopants present in the chosen solder paste. However, good agreement between theoretical and actual results has been demonstrated. Figure 5.36 illustrates the post-SMT bismuth weight % in the hybrid region for joints formed at 175 and 190 °C PRT and pastes with 50% and 57% starting bismuth concentration and additional supplier proprietary dopants.

5.5.4 Possible Defects

Estimates of process-induced voiding for both BGA component solder joints as well as difficult to control components such as QFN/BTC thermal pads consistently show that low temperature solder performs as well as or better than SnAgCu solder pastes. For hybrid SnAgCu-low temperature solder BGA solder joints, typical levels of process voiding can be expected to be approximately the same as for a homogenous SnAgCu BGA solder joint. Figure 5.37 compares void percentage for a hybrid SnAgCu-low temperature solder BGA case and the same component with a homogenous SnAgCu solder joint.

A recent study [62] comparing a eutectic Sn-Bi solder, two ductile low temperature solders, and SnAgCu solder demonstrated that the low temperature solders significantly outperformed SnAgCu solder for void % in a QFN/BTC component thermal pad solder joint. Table 5.6 summarizes the void % data for the four solder paste examples.

B39 440x

Solder Paste – 50% Bi
Reflow Peak - 175ºC

100 μm

| Weight % | | | | | | |
Spectrum	Ni	Cu	Ag	Sn	Bi	Total
1	0.46	0.77	3.95	92.14	2.69	100
2		0.99	3.13	63.46	32.42	100

B40

Solder Paste – 57% Bi
Reflow Peak - 175ºC

100 μm

| Weight % | | | | | | |
Spectrum	Ni	Cu	Ag	Sn	Bi	Total
3	0.58	1.12	3.86	92.91	1.54	100
4		0.25	1.6	65.73	32.41	100

B47

Solder Paste – 50% Bi
Reflow Peak - 190ºC

100 μm

| Weight % | | | | | | |
Spectrum	Ni	Cu	Ag	Sn	Bi	Total
17	0.59	0.92	3.89	93.79	0.82	100
18		0.48	3.55	70.87	25.1	100

B48

Solder Paste – 57% Bi
Reflow Peak - 190ºC

100 μm

| Weight % | | | | | | |
Spectrum	Ni	Cu	Ag	Sn	Bi	Total
19	0.53	0.54	3.99	92.39	2.55	100
20		1.16	5.31	68.44	25.1	100

Figure 5.36 Resulting hybrid bismuth diffusion region bismuth concentration versus paste bismuth weight % and reflow peak temperature.

While achieving the necessary bismuth weight % to produce optimized elongation performance would be the primary goal when developing the reflow process, often defect control may require trade-offs with the final reflow peak temperature. Due to the pasty range observed with many Sn-Bi based solder pastes, a defect known as "hot tearing" must be considered. Hot tearing occurs when a component is returning to the room temperature shape after reflow. If the component is pulling away from the board, stress will be generated on the solder joints and the material may begin to separate. If the solder is not fully solidified, the joint can tear and there may not be sufficient liquid material available to heal the separation. The result is a solder joint that exhibits a fatigue-like separation but has occurred immediately after reflow. Figure 5.38 illustrates hybrid solder joints with various degrees of hot tearing.

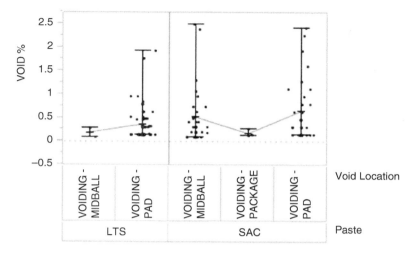

Figure 5.37 Post-SMT void measurements for a 400 μm BGA ball package. Results show comparable performance for LTS and SnAgCu solder pastes.

Table 5.6 QFN/BTC thermal pad void results for a selection of LTS pastes versus the SnAgCu control.

Paste	QFN/BTC void %
SnAgCu (control)	9.9
Ductile LTS (B)	1.8
Eutectic SnBi (C)	5.9
Ductile LTS (D)	8.4

Figure 5.38 Extreme examples of time zero "hot tearing" in a hybrid SnAgCu-LTS solder joint.

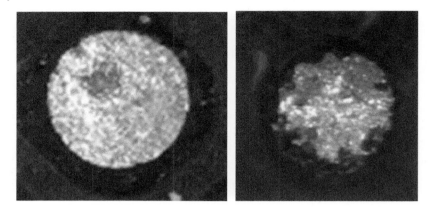

Figure 5.39 Dye and pulled solder joints. No "hot tearing" on left image. Dye stain indicates some level of "hot tearing" on the right image.

Identifying the occurrence of hot tearing requires special attention as the defect is not readily detected with X-ray inspection. As there may only be a partial separation of the solder joint, electrical tests will also not indicate any abnormality. Hot tearing is easily identified in cross-section; however, a more comprehensive screen is to dye and pull the soldered components. At time zero (T0), a solder joint exhibiting hot tearing will show some amount of dye stain when examined optically after pull. Figure 5.39 shows a post-pulled solder joint with dye staining indicating hot tearing.

Elimination of hot tearing has been shown to be most effectively achieved with the reduction of PRT. The amount of temperature reduction required depends on the individual component design as well as the solder paste being used. A secondary method of control is through increased solder paste volume. Due to the risk of small amounts of bismuth reaching the package side interface in hybrid SnAgCu- low temperature solder BGA joints, any increase in solder volume should be attempted with caution. In addition to assessing the hot tearing percentage with dye and pull, cross-sectioning should be used to verify the hybrid mixing profile in the finished solder joints. Other methods of control such as increased TAL, increased cooling rate, and increased soak have not been found to significantly affect hot tearing. Table 5.7 summarizes the effectiveness ranking of hot tearing control strategies.

One of the advantages of a hybrid SnAgCu-low temperature solder when compared to a homogenous SnAgCu BGA solder joint is a significant reduction in the risk of solder ball bridging (SBB). As can be seen from Figure 5.40, the hybrid solder joint height post-SMT assembled is taller than the comparable homogenous SnAgCu BGA solder joint.

Table 5.7 Qualitative ranking of the effectiveness of primary SMT process factors on hot tearing performance.

SMT process factor	Hot tearing control effectiveness	Notes
Reflow peak temperature	High	For large reductions in peak reflow temperature, solder paste non-wet performance should be assessed
Solder paste volume	Medium	Solder paste volume > 0.5 paste:ball ratio should be assessed for risk of bismuth at the package side interface.
Solder paste bismuth concentration	Medium	Lower pasty range has positive effect on hot tearing. Benefit can be modulated by action of additional dopants
Time above liquidus (TAL)	Low	Extended TAL may increase risk of bismuth at package side interface
Cooling rate ($°C\,s^{-1}$)	Low	Increased cooling rates have not been shown to have significant impact on hot tearing performance and may be difficult to achieve in high-volume manufacturing

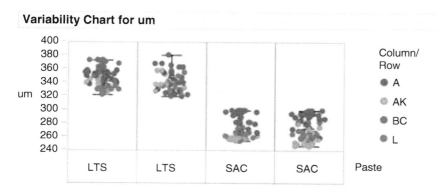

Figure 5.40 Solder joint height post-SMT for a 400 μm BGA ball package – LTS versus SnAgCu.

This increase in solder joint height is due to the incomplete melting of the SnAgCu solder ball. The unmelted SnAgCu solder ball acts as a collapse limiter and as a result reduces the risk of SBB due to solder extrusion created by the joint collapse. This can be a particularly useful feature for fine pitch components where it can be difficult to reduce solder paste volumes sufficiently to prevent SBB during a SnAgCu reflow process.

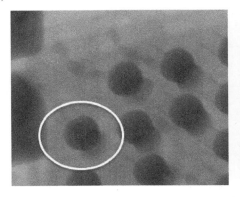 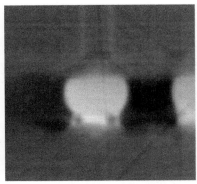

Figure 5.41 Left image shows hybrid solder joint in 2.5D tilt X-ray. Shadow at the joint base would typically indicate a head-on-pillow defect for a SnAgCu solder joint. Right image shows the same joint in a CT scan.

When using X-ray to assess post-SMT assembled solder joint quality, the hybrid SnAgCu-low temperature solder BGA joint presents some unique challenges. Due to the incomplete collapse of the joint, a stretched solder joint may appear as a HoP defect. Figure 5.41 details the view of a hybrid joint in 2.5D tilted X-ray equipment and an associated computed tomography (CT) view. Careful stencil design to ensure the best possible solder joint quality helps reduce the need for this type of judgment call. However, operator training to account for the nuances of hybrid SnAgCu-low temperature solder BGA joint inspection is a necessity.

5.6 Solder Joint Reliability

The solder joint reliability performance of hybrid SnAgCu- low temperature solder BGA joints formed with eutectic Sn-Bi solders has frequently been documented to exhibit significant reductions in capability for both mechanical shock and thermo-mechanical fatigue when compared with a homogenous SnAgCu BGA solder joint baseline [23–25]. The solder paste suppliers have invested significant engineering resources in an effort to identify methods of increasing the ductility of Sn-Bi based solders in order to achieve solder joint reliability performance capable of meeting requirements. The resulting solder paste systems have been referred to as "ductile low temperature solders".

In general for a homogenous SnAgCu system, as long as a solder joint is achieved, a general level of solder joint reliability performance can be assumed. When designing a low temperature solder SMT process for hybrid SnAgCu-low temperature solder joints, solder joint quality and post-reflow morphology very much influence the resulting reliability performance. With a hybrid SnAgCu-low

Table 5.8 Calculation for paste:ball volume ratio square and round stencil apertures.

	Mil	mm
BGA sphere diameter	16	0.4
BGA sphere volume (unit^3)	2144	0.0335
Desired Paste:BGA sphere volume ratio	0.5	0.5
Required solder paste Vol (unit^3)	1072	0.0168
Paste Vol = (BGA Sphere Volume) * Ratio		
Stencil thickness	4	0.102
Aperture size (Round)	18	0.4573
Opening = 2*sqrt(Paste Volume/(Stencil Thickness*π))		
Aperture size (Square)	16	0.4052
Opening = sqrt(Volume/(Stencil Thickness))		

temperature solder BGA component, solder paste volume and reflow peak temperature should be considered critical to function variables.

A convenient way of considering solder volume for a hybrid SnAgCu-low temperature solder BGA joint is in terms of solder paste volume to BGA ball volume ratio (henceforth termed "paste:ball ratio"). Dividing the theoretical (stencil aperture area × stencil thickness) paste volume by the volume of the unmounted BGA solder sphere for a given component results in a paste:ball ratio value. Table 5.8 demonstrates the calculation of the paste:ball ratio for a typical IC component example.

If the paste:ball ratio is too low, often the BGA solder joint quality will be reduced and hot tearing of the solder joints may occur. Figure 5.42 illustrates a hybrid BGA solder joint with poor time zero solder joint quality. For this case, solder joints may demonstrate reduced reliability performance due to poor ball to board pad interface quality.

Using a paste volume that is too high may result in low concentrations of bismuth at the package-ball interface. As can be seen in Figure 5.10, bismuth weight % in the range of 8–12% results in effectively 0% elongation of the solder. This significant reduction in elongation capability paired with the stress riser of the package solder mask – BGA ball interface may lead to brittle crack initiation and early life reliability failures. Figure 5.43 illustrates a hybrid BGA solder joint with a high-risk concentration of bismuth at the package-ball interface.

Testing completed on a variety of BGA packages with BGA ball sizes from 0.009″ (0.22 mm) to 0.016″ (0.4 mm) has shown that a paste:ball ratio of 0.5 results in optimized solder joint performance for both thermomechanical fatigue

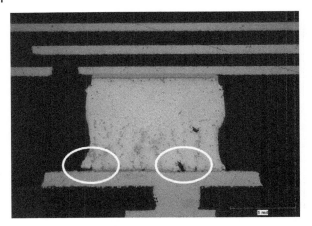

Figure 5.42 Time zero solder joint assembled with a paste:ball volume ratio of 0.3. Evidence of hot tearing highlighted by white circles.

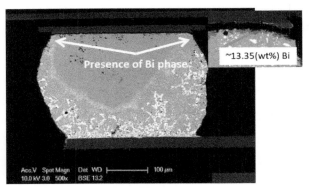

Figure 5.43 Hybrid SnAgCu-SnBi solder joint illustrating bismuth concentration at the package side interface that would exhibit significant reduction in elongation performance.

and mechanical shock. Paste:ball ratios of 0.3 have shown a 50% reduction in temperature cycle (fatigue) capability compared with solder joints built using a 0.5 ratio. Solder joints built with a ratio of 0.8 have shown a further 50% reduction in fatigue capability. Figure 5.44 presents a Weibull plot comparing performance of the three paste:ball ratio cases.

For a hybrid SnAgCu low temperature solder BGA joint, assuming that small concentrations of bismuth at the package side have been avoided, failure in temperature cycling will typically occur just above the IMC on the board side of the solder joint. Figure 5.45 provides a view of a typical hybrid SnAgCu low temperature solder BGA crack signature.

In the case of mechanical shock, a paste:ball ratio of 0.5 has also been demonstrated to provide significant improvements in mechanical shock capability when

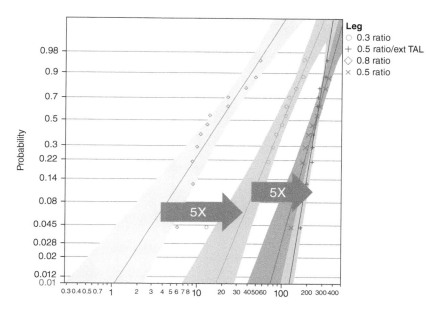

Figure 5.44 Weibull plot of cycles to fail versus paste:ball ratio for a hybrid SnAgCu-SnBi solder joint. Note the limited impact of extended time above liquidus (TAL) on the results for the 0.5 ratio case.

Figure 5.45 Typical fatigue cracking signature in a hybrid SnAgCu-SnBi solder joint. Crack propagates just above the IMC in the bulk solder material.

compared to higher volumes of solder paste. Table 5.9 compares the mechanical shock performance of a BGA component using a homogenous SnAgCu solder joint, and a hybrid SnAgCu-low temperature solder joint built using 0.5 and 0.7 paste:ball ratio.

Table 5.9 Drop/shock capability comparison for SnAgCu solder joints versus hybrid SnAgCu-SnBi solder joints at 0.5 and 0.7 paste:ball volume ratio.

Paste metallurgy_BGA metallurgy	First fail capability	CTF capability (1%, 90% LCL)	Dominating fail mode
SAC_SAC	99.6	135.7	Pad crater
LTS_SAC (Vol ratio 0.7)	63.7	109.9	Board side
LTS_SAC (Vol ratio 0.5)	70.5	120.8	Board side

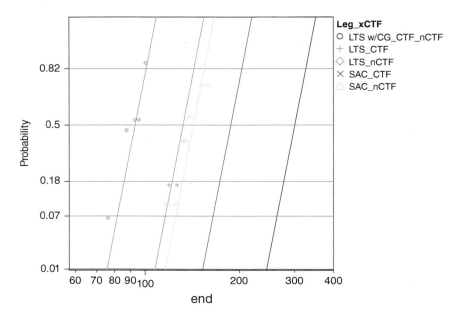

Figure 5.46 Demonstration of improvement in drop/shock performance enabled by the use of a corner glue adhesive (Low Temperature Solder w/CG_CTF_nCTF).

While the hybrid joints do not demonstrate equivalent mechanical shock capability to homogenous SnAgCu joints, the performance gain with the 0.5 paste:ball ratio is clear. It is also useful to note that if additional mechanical performance is required, the use of a corner adhesive can provide significant benefits. Figure 5.46 compares the performance of both hybrid SnAgCu-low temperature solder and homogenous SnAgCu BGA solder joints without reinforcement with hybrid SnAgCu-low temperature solder joints reinforced with a corner glue.

For hybrid SnAgCu-low temperature BGA solder joints in mechanical shock, typical failure location (again assuming there are no low concentrations of bismuth at the package side) is just above the IMC on the board side interface.

Once the correct paste:ball ratio has been established for a given BGA component, the resulting time zero solder joint morphology should be validated before finalizing the SMT process. If the combination of solder paste volume and PRT results in a solder at time zero that shows the presence of bismuth near the package side interface, the process should be reassessed. The risk to reliability performance can be reduced by either reducing the solder paste volume or lowering the PRT. As discussed in the previous section, solder paste volume reductions may increase the risk of hot tearing. However, reductions in PRT help reduce the risk of hot tearing at the same time that migration of bismuth toward the package side is reduced. The minimum acceptable reflow temperature would then need to be determined based on the selected solder paste and its associated pasty range. Within the acceptable temperature working window of a low temperature solder paste, variations in reflow peak temperature have not been shown to create significant deltas in reliability performance assuming bismuth at the package interface is successfully managed.

5.7 Conclusions

In recent years, the economic, environmental, and manufacturing benefits of low temperature solders, as defined by those solders that can be soldered below 200 °C, have driven the development of solder materials with properties that can enable their use to assemble reliable consumer electronic components such as cell phones, mobile and desktop computers.

Predominant low temperature solder alloy development has occurred in the Bi-Sn system, which melt around 140 °C and can be soldered at reflow temperatures in the 160–200 °C temperature ranges. The Bi-Sn metallurgy solder pastes have been in use for niche applications, such as white goods and LCD TVs, for the past 20 years and hence the chemical components of the flux vehicle compatible with the lower temperature during the reflow soldering process have been developed previously. The various SMT assembly equipment and processes for these Bi-Sn solder pastes, such as printing solder paste through stencil apertures on to PCB lands, the soldering profile in the reflow oven, and the component rework process are very similar to those that are currently being used, albeit at lower temperature settings.

However, the inherent brittleness of the bismuth phase in these alloys means that solder joints formed with these alloys are at risk of failure during mechanical shock and drop events, which mobile computer products can be subjected to in the field. To overcome this drawback, the robustness of these alloys has been improved to some extent by reducing their bismuth content below the eutectic composition level, and adding specific alloying elements in quantities ranging from below a tenth of a percent to the 3–4% range. These alloying elements enhance the solder properties by refining the bismuth and tin phases in the solder joint microstructure, and by solid solution hardening and precipitation hardening of the tin matrix. Polymeric reinforcement of these solder joints can also be employed to improve their robustness significantly. This reinforcement can be achieved, both at the package level by use of underfill and corner glue materials after the reflow process step, and at the individual solder joint level by using solder pastes incorporated with resin which subsequently cure to encapsulate the solder joint during the reflow soldering step.

When using Bi-Sn solder alloy pastes for soldering current area array solder sphere termination packages, such as BGAs which have SAC solder metallurgy spheres, special assembly process considerations are required to form robust mixed SAC-BiSn solder joints that meet quality and reliability standards. An optimum solder paste volume to-solder ball volume ratio as well as key parameters of the solder reflow profile, particularly the PRT, are critical to form acceptable mixed SAC-BiSn BGA solder joints. A solder paste volume-to-solder ball ratio that is either too high or too low can give rise to premature fatigue failures at either the solder to PCB land interface (if too low) or the package land-to-solder interface (if too high). Mixed SAC-BiSn solder joints are also prone to hot tearing defects for BGA packages with a high dynamic warpage. But optimum selection of the peak solder reflow temperature, the aforementioned solder paste volume-to-solder ball ratio can mitigate the formation of these defects.

Low temperature soldering has progressed within electronics manufacturing to mobile computer products. It was announced recently that a manufacturer was using a low temperature process for certain laptops where the soldering heat was applied at maximum temperatures of 180 °C, a reduction of 70 °C from the previous method [63]. Since then other computer companies have initiated development of low temperature manufacturing processes for their products. The development of low temperature soldering is an ongoing process with further progress in the coming years.

5.8 Future Development and Trends

Many challenges in the realm of low temperature solders still persist. Some of these challenges are presented below.

Though much development to improve the fundamental solder alloy characteristics in the Bi-Sn solder system has been done in recent years, with multiple generations of solder alloys being developed, further enhancement of the ductility of these low temperature solders is needed. The shock resistance and reliability of these newly developed ductile BiSn solder alloys is still not up to par, without the use of any polymeric reinforcement, compared with that of SAC solders [51, 52].

The benefits presented by the use of the low temperature solders should be maximized. This includes use of lower-cost PCB laminate materials for rigid, rigid-flex, and flexible substrates, packaging materials such as molding compounds, connector, and socket housing materials, where resistance to higher temperatures greater than 200 °C during reflow soldering is not required.

Stacked microvias are used in High Density Interconnect (HDI) designs. These stacked microvias are known to have reliability issues when exposed to reflow soldering temperatures [64]. The effect of using low temperature soldering in mitigating these deleterious reliability effects needs to be ascertained. Further, the potential of low temperature solders possibly enabling higher-density interconnects than those existing today by reducing the thermomechanical stresses on the thin, small surface area interconnect interfaces should be explored.

More-complex package designs with multiple die, thinner substrates, smaller termination sizes, and finer pitches are driving the need to lower package warpage at reflow soldering temperatures [65]. This will require two distinct approaches to lower the CTE mismatch between the die and the package substrate. One is to use new packaging materials and the second is to reduce the soldering process temperature. This second approach is the driver for even lower temperatures than those enabled by current low temperature solders.

Phase change memory (PCM) is an information storage technology that uses the unique properties of phase change materials to achieve a noteworthy combination of features such as fast access time, large electrical contrast, non-volatility, and high scalability [66]. PCM materials store information in their amorphous and crystalline phases, which can be reversibly switched by the application of an external voltage. One potential drawback of the PCM materials is that there is no guarantee of data retention after reflow soldering due to the strong sensitivity on temperature of the amorphous to crystalline transition [67]. The data retention failure rate varies exponentially with temperature and lognormally with time and is well described with the Arrhenius equation [68, 69]. Hence, if soldering temperatures could be dropped down to levels lower than 150 °C, new PCM materials could be developed with the possibility of PCM device switching speeds reaching the levels of DRAM volatile memory switching speeds. This achievement would open up the creation of new computer architectures [66].

The next logical step for low temperature soldering is to go even lower and develop ultra-low temperature solders. However, progressing further into the ultra-low temperature solder range, shown in Table 5.1 and Figure 5.1, has some drawbacks. The homologous temperature, T_{hom}, is defined by a ratio of the

Table 5.10 Various transient liquid phase soldering systems.

TLPS system	Process temp and time	Resulting melting point (°C)	Interacting phases
Sn-Cu	250 °C/8 min	415	Solid –liquid
	250 °C/40 min	676	Solid-solid
Ag-In	206 °C/7 min	>220	
Au-Sn	310–340 °C/1 h	498	
Sn-Bi-Cu	200 °C/50 min	201	Solid-liquid
Ni-Sn	250 °C/15 min	282	
Au-In	Room temp/36 d	540	
(ULT-TLPS)	≤140 °C/<15 min	>200	

temperature of the material during use to its melting temperature when using the Kelvin scale. At 100 °C, the upper dwell temperature of some widely used temperature cycling test protocols in the industry for computer products, the T_{hom} of Sn-Ag-Cu solders is 0.76 and those of eutectic Sn-Bi and Sn-In solders are even higher at 0.91 and 0.95, respectively. At a very high T_{hom} (≥0.95) an exponentially higher rate of diffusion dependent deformation results. Thus at these high T_{hom}s, solders, in particular and metallic materials, in general undergo strength loss, creep deformation, and microstructural coarsening which, in turn, limit their long-term reliability [70].

To overcome the T_{hom} effect for ultra-low temperature alloys, Transient Liquid Phase Soldering (TLPS), using solder pastes, can be employed. Various current TLPS systems are listed in Table 5.10 with their processing time and temperature, and resulting melting point of the solder joint after the process is completed. None of these systems are, however, in the ultra-low temperature range. The current need is to research and develop an ultra-low temperature TLPS system with the requirements noted in the last row of Table 5.10. The requirement is to identify potential low temperature metallurgical systems compatible with TLPS solder alloys, with the melting point of the lower melting constituents being below 140 °C. This research should entail studies of the thermodynamics and kinetics of the solid-liquid diffusion reaction and reactions with surface finishes on the PCB surfaces during the TLPS process. Subsequent characterization of the microstructure of the fully solidified solder joint and resulting appropriate mechanical properties is necessary.

References

1 Ribas, M., Hunsinger, T., Cucu, T. et al. (2018). *The Printed Circuit Assembler's Guide to…™ Low Temperature Soldering*. BR Publishing http://i-007ebooks .com/my-i-connect007/books/lts (accessed 4 February 2020).

2 Wikipedia (2020) Abundance of elements in Earth's crust. https://en.wikipedia .org/wiki/Abundance_of_elements_in_Earth%27s_crust (accessed 4 February 2020).

3 Bath, J., Itoh, M., Clark, G. et al. (2012) An investigation into low temperature tin-bismuth and tin-bismuth-silver lead-free alloy solder pastes for electronics manufacturing applications. *Proceedings of the IPC APEX Conference, San Diego, CA*.

4 Bath, J. (2013) Global trends in electronics and the impact on tin use. Presentation at the 2013 ITRI World Tin Conference, Kunming, China.

5 Holtzer, M. and Mok, T.W. (2013) Eliminating wave soldering with low melting point solder paste. *Proceedings of the SMTAI Conference* (October 2013).

6 Aspandiar, R., Byrd, K., Tang, K.K. et al. (2015) Investigation of low temperature solders to reduce reflow temperature, improve SMT yields and realize energy savings. *Proceedings of the IPC APEX Expo Conference* (February 2015).

7 EIA-US Energy Information Admin (2012). https://www.eia.gov (accessed 4 February 2020).

8 Shenzhen Government Online (2012). http://english.sz.gov.cn (accessed 4 February 2020).

9 EPA (2012). Clean energy. http://www.epa.gov/cleanenergy/energy-resources/ refs.html (accessed 14 September 2015).

10 Amir, D., Aspandiar, R., Buttars, S. et al. (2009) Head-on-pillow SMT failure modes. *Proceedings of SMTA International Conference* (October 2009).

11 Kondrachova, L., Aravamudhan, S., Sidhu, R. et al. (2012) Fundamentals of the non-wet open BGA solder joint defect formation. *Proceedings of the International Conference on Soldering and Reliability (ICSR)* (May 2012).

12 Amir, D., Walwadkar, S., Aravamudhan, S., and May, L. (2012) The challenges of non wet open BGA solder defect. *Proceedings of 2012 SMTA International Conference* (October 2012).

13 Sahasrabudhe, S., Mokler, S., Renavikar, M. et al. (2018). Low temperature solder – a breakthrough technology for surface mounted devices. In: *Proceedings of the 2018 IEEE 68th Electronic Components and Technology Conference*, 1455–1464.

14 Chen, S.-W., Wang, C.-H., Liu, S.-K., and Chiu, C.-N. (2007). Phase diagrams of Pb-free solders and their related materials systems. *Journal of Materials Science: Materials in Electronics* 18 (1–3): 19–37.

15 Takao, H., Yamada, A., Hasegawa, H., and Matsui, M. (2002). Mechanical properties and solder joint reliability of low-melting Sn-Bi-Cu lead free solder alloy. *Journal of Japan Institute of Electronics Packaging* 5 (2): 152–162.

16 Fu, H., Aspandiar, R., Chen, J. et al. (2017). iNEMI project on process development of BiSn-based low temperature solder pastes. In: *Proceedings of the 2017 SMTA International Conference, Rosemont, IL*, 207–220.

17 IPC TM-650 Test Methods Manual (2007) Number 2.6.3.7, SIR Task Group (5-32b). IPC International.

18 Amalu, E.H., Mallik, S., and Ekere, N.N. (2011). Evaluation of rheological properties of lead-free solder pastes and their relationship with transfer efficiency during stencil printing process. *Materials and Design* 32 (6): 3189–3197.

19 Ye, D., Du, C., Wu, M., and Lai, Z. (2015). Microstructure and mechanical properties of Sn–xBi solder alloy. *Journal of Materials Science: Materials in Electronics* 26: 3629–3637.

20 Glazer, J. (1994). Microstructure and mechanical properties of lead-free alloys for low cost electronics assembly. *Journal of Electronic Materials* 23 (8): 693–700.

21 Skudnov, V.A., Sokolov, L.D., Gladkikh, A.N., and Solenov, V.M. (1969). Mechanical properties of bismuth at different temperature and strain rates. *Metal Science and Heat Treatment* 11: 981–984.

22 McCallister, W.D. (2003). *Material Science and Engineering: An Introduction*, 6e, 168. Wiley.

23 Pandher, R. and Healey, R. (2008) Reliability of Pb-free solder alloys in demanding BGA and CSP applications. *Proceedings of the IEEE 58th Electronic Components and Technology Conference (ECTC)*.

24 Liu, Y., Keck, J., Page, E., and Lee, N.-C. (2013) Voiding and reliability of BGA assemblies with SAC and 57Bi42Sn1Ag alloys. *Proceedings of the SMTA International Conference*, 2013.

25 Liu, Y., Keck, J., Page, E., and Lee, N.-C. (2014) Voiding and drop test performance of lead-free low melting and medium melting mixed alloy BGA assembly. *Proceedings of the IPC APEX Conference*, 2014.

26 Chang, S., Ibe, E.S., and Loh, K.I. (2013) Underfill encapsulants and edgebond adhesives for enhancing of board level reliability. *Proceedings of the SMTA International Conference*, 2013.

27 Yeo, S.M., Tay, C.S., Chong, C.C., and Beh, J.S. (2010) Next generation board level underfill (BLUF) for fine pitch BGA and POP. *SMTA International Conference Proceedings*.

28 Shi, H., Tian, C., Yu, D., and Ueda, T. (2012). A comprehensive analysis of the thermal cycling reliability of lead free chip scale package assemblies with various reworkable board-level polymeric strategies. In: *International Conference on Electronic Packaging Technology and High Density Packaging*, 959–970.

29 Xie, F., Wu, H., Baldwin, D.F., et al. (2015) WLCSP and BGA reworkable underfill evaluation and reliability. *Proceedings of the SMTA International Conference*.

30 Rajarathinam, V., Wade, J., Donaldson, A., et al. (2013) Manufacturability assessments of board level adhesives on fine pitch ball grid array components. *Proceedings of the SMTA International Conference*.

31 Shi, H., Tian, C., Pecht, M., and Ueda, T. (2012). Board-level shear, bend, drop and thermal cycling reliability of lead-free chip scale packages with partial underfill: a low-cost alternative to full underfill. In: *Proceedings of the 2012 IEEE 14th Electronics Packaging Technology Conference*, 774–785.

32 Schwiebert, M.K. and Leong, W.H. (1996). Underfill flow as viscous flow between parallel plates driven by capillary action. *IEEE Transactions On Components, Packaging, and Manufacturing Technology, Part C* 19 (2): 133–137.

33 Frear, D.R. (1999). Material issues in area-array microelectronic packaging. *JOM* 51 (3): 22–27.

34 Khor, C.Y., Abdullah, M.Z., and Abdul Mujeebu, M. (2012). Influence of gap height in flip chip underfill process with non-Newtonian flow between two parallel plates. *Journal of Electronic Packaging* 134 (1): 011003.

35 Wan, J.W., Zhang, W.J., and Bergstrom, D.W. (2007). A theoretical analysis of the concept of critical clearance toward a design methodology for the flip-chip package. *Journal of Electronic Packaging* 129: 473–478.

36 Quinones, H. and Ratledge, T. (2010) Capillary underfill physical limitations for future packages. *Proceedings of the Pan Pacific Conference, 2010*, Hawaii.

37 Gilleo, K. (1998). The chemistry and physics of underfill. In: *Proceedings of NEPCON West*, 280–292.

38 Xie, F., Wu, H., Baldwin, D.F. et al. (2015) Evaluation of high reliability reworkable edge bond adhesives for BGA applications. *Proceedings of the SMTA International Conference*.

39 Poole, N. (2013) Factors impacting solder extrusion in reworkable underfills. *Proceedings of the SMTA International Conference*.

40 Toleno, B.J., Hu, S., Yoo, H., and Zhang, R. (2014) New underfill materials designed for increasing reliability of fine-pitch wafer level devices. *Proceedings of the SMTA South East Asia Conference*, Penang, Malaysia.

41 Liu, M. and Yin, W. (2010) A first individual solder joint encapsulant adhesive. *Proceedings of 2010 IMAPS*, Raleigh, NC.

42 Yeo, Y.C., Huang, M., Che, F.X. et al. (2010) Solder joint encapsulation and reliability using dippable underfill. *Proceedings of the 12th Electronic Packaging Technology Conference.*

43 Yin, W., Beckwith, G., Hwang, H.-S. et al. (2012) Epoxy flux bridges the gap to low cost no-clean flip chip assembly. *Proceedings of the SMTA International Conference.*

44 Kondos, P., Meilunas, M. and Anselm, M. (2012) Epoxy fluxes: DIP assembly process issues and reliability. *Proceedings of the SMTA International Conference.*

45 Lee, N.-C. (2013) A novel epoxy flux on solder paste for assembling thermally warped POP. *Proceedings of 2013 IMAPS*, Orlando, FL.

46 Poole, N., E Vasquez, and Toleno, B. (2015) Epoxy flux material and process for enhancing electrical interconnections. *Proceedings of the SMTA International Conference.*

47 Chan, B., Ji, Q., Currie, M. et al. (2009). Epoxy flux technology – tacky flux with value added benefits. In: *Proceedings of the 59th Electronic Components and Technology Conf (ECTC)*, 188–190.

48 Johnson, R.W., Capote, M.A., Zhou, Z.M. et al. (1997). Reflow-curable polymer fluxes for flip chip assembly. In: *Proceedings of the Surface Mount International (SMI) Conference*, 267–272.

49 Zhao, R., Johnson, R.W., Jones, G. et al. (2003). Processing of fluxing underfills for flip chip-on-laminate assembly. *IEEE Transactions on Electronics Packaging Manufacturing* 26 (1): 75–83.

50 Whitmore, M., Schake, J., and Ashmore, C. (2013) Factors affecting stencil aperture design for next generation ultra fine pitch printing. *Proceedings of the South East Asia Technical Training Conference on Electronics Assembly Technologies*, Penang, Malaysia.

51 Fu, H., Radhakrishnan, J., Ribas, M., et al. (2018) iNEMI Project on process development of BiSn-based low temperature solder pastes – Part IV: comprehensive mechanical shock tests on pop components having mixed BGA BiSn-Sac solder joints, *Proceedings of the 2018 SMTA International Conference*, Rosemont, IL (October 2018).

52 Fu, H., Radhakrishnan, J., Ribas, M., et al. (2019) iNEMI project on process development of BiSn-based low temperature solder pastes – Part VI: mechanical shock results on resin reinforced mixed SnAgCu-BiSn solder joints of FCBGA components, *Proceedings of the 2019 SMTA International Conference*, Rosemont, IL.

53 Aspandiar, R. (2019) Post mechanical shock failure analysis on mixed BiSn – SAC BGA solder joints of POP components. *Pan Pacific Conference*, Hawaii.

54 Kejun, Z., Mike, P., Hiroshi, M., and Becky, H. (2012). Optimization of Pb-free solder joint reliability from a metallurgical perspective. *Journal of Electronic Materials* 41 (2): 253–261.

55 Kumar, G.A. and Liangchi, Z. (2016). Growth mechanism of intermetallic compound and mechanical properties of nickel (Ni) nanoparticle doped low melting temperature tin-bismuth (Sn-Bi) solder. *Journal of Materials Science: Materials in Electronics* 27 (1): 781–794.

56 Tomlinson, W.J. and Collier, I. (1987). The mechanical properties and microstructures of copper and brass joints soldered with eutectic tin-bismuth solder. *Journal of Materials Science* 22: 1835–1839.

57 Kim, J.-H., Lee, Y.-C., Lee, S.-M., and Jung, S.-B. (2014). Effect of surface finishes on electromigration reliability in eutectic Sn–58Bi solder joints. *Microelectronic Engineering* 120: 77–84.

58 Mokhtari, O. and Nishikawa, H. (2013). Coarsening of Bi phase and intermetallic layer thickness in Sn-58Bi-X (X=In and Ni) solder joint. In: *14th International Conference on Electronic Packaging Technology*, 250–253.

59 Lee, C.C., Wang, P.J., and Jong, J.S. (2007). Are intermetallics in solder joints really brittle? In: *Proceedings 57th Electronic Components and Technology Conference*, 648–652.

60 Hirman, M., Rendl, K., Steiner, F., and Wirth, V. (2014). Influence of reflow soldering profiles on creation of IMC at the interface of SnBi/Cu. In: *IEEE 37th International Spring Seminar on Electronics Technology*, 147–151.

61 Fu, H., Aspandiar, R., Chen, J. et al. (2018). iNEMI project on process development of BiSn-based low temperature solder pastes — Part II: characterization of mixed alloy BGA solder joints. In: *Proceedings of the SMTA Pan Pacific Microelectronics Symposium*, 1–17.

62 Osgood, H., Geiger, D., Pennings, R. et al. (2019). Low temperature SMT solder evaluation. *IPC APEX Conference*.

63 Webwire (2017) Lenovo™ announces breakthrough, innovative PC manufacturing process. 7 February. https://www.webwire.com/ViewPressRel.asp?aId=207578 (accessed 4 February 2020).

64 Fritz, D. (2019) Weak interface/stacked microvia reliability. Wisdom Wednesday Webinar, 19 April. IPC International.

65 iNEMI (2017) 2017 iNEMI Roadmap. Packaging and component substrates, 10–12.

66 Raoux, S., Xiong, F., Wuttig, M., and Pop, E. (2014). Phase change materials and phase change memory. *MRS Bulletin* 39: 703–710.

67 Zuliani, P., Varesi, E., Palumbo, E. et al. (2013). Overcoming temperature limitations in phase change memories with optimized GexSbyTez. *IEEE Transactions on Electron Devices* 60 (12): 4020–4026.

68 Gleixner, B., Pirovano, A., Sarkar, J. et al. (2007). Data retention characterization of phase-change memory arrays. In: *45th Annual International Reliability 542 Physics Symposium, Phoenix, AZ*, 542–546.

69 Russo, U., Ielmini, D., and Lacaita, A.L. (2007). Analytical modeling of chalcogenide crystallization for PCM data-retention extrapolation. *IEEE Transactions on Electron Devices* 54 (10): 2769–2777.

70 Tarr, M. (2014). The 'homologous' temperature. www.ami.ac.uk/courses/topics/0164_homt/index.html (accessed 13 August 2014).

6

High Temperature Lead-Free Bonding Materials – The Need, the Potential Candidates and the Challenges

Hongwen Zhang and Ning-Cheng Lee

Indium Corporation, Clinton, NY, USA

6.1 Introduction

The lead-free solder alternatives for eutectic tin-lead solder have been widely studied, and Sn-Ag, Sn-Cu, and Sn-Ag-Cu have become the mainstream alternatives for the electronics industry [1–9]. However, the high temperature lead-free (HTLF) bonding materials, replacing the conventional high-lead solders in die-attachment for power semiconductor components, modules, and devices, are still in the early stages of development and implementation despite intensive investigation [10–17].

Die-attachment for the discrete power components requires the usage of high-temperature solders or bonding materials to maintain the joint integrity between the semiconductor die and the lead-frame/substrate in service. The major requirements of die-attachment in the discrete power components [10–14] are (i) a joint softening/remelting temperature no lower than 260 °C if subsequent board-level reflows are needed; (ii) a service temperature up to 150 °C/165 °C; (iii) a drop-in solution compatible to the current high-lead soldering process; (iv) good thermal fatigue resistance; (v) comparable or even better electrical/thermal performance to high-lead solders; and (v) low cost of ownership.

The increasing demands for power semiconductor modules/devices, especially the high power modules/devices of reduced size through the use of SiC/GaN dies, require the bonding materials to provide better thermal conductivity and survive higher power/current density as well as the associated higher junction temperatures up to or even higher than 200 °C. Depending on the package design and the service condition, a wide variety of the die-attach materials [10–17] are being attempted and investigated, including tin-rich solders, the novel HTLF bonding materials (both solders, transient liquidus phase bonding [TLPB] materials, and sintering materials), against the high-lead solders.

Lead-free Soldering Process Development and Reliability, First Edition. Edited by Jasbir Bath.
© 2020 John Wiley & Sons, Inc. Published 2020 by John Wiley & Sons, Inc.

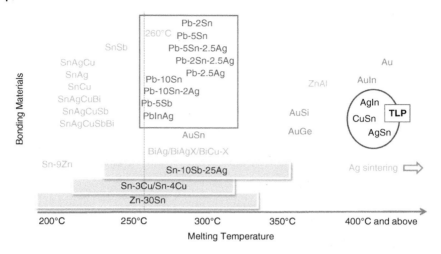

Figure 6.1 Summary of die-attach bonding materials [10, 11].

Solders with melting temperature above 200 °C, as well as TLPB materials and the sintering materials have been summarized in Figure 6.1. High-lead solders, contained in the red box of Figure 6.1, have a melting temperature range from 250 °C for Pb-5Sb to 310 °C for Pb-2Sn, and are being widely used in discrete power components, like diodes, DrMOS (MOSFET/drivers), PQFN, D2PAK and TO, etc. Between 200 and 250 °C, tin-rich lead-free solders, alloyed with one or multiple elements like Ag, Cu, Bi, and Sb, dominate. Tin-rich solders are not suitable for discrete power components because the joint solder would re-melt during subsequent board-level reflow with the traditional peak temperature of 240~260 °C or even higher. Tin-rich solders can be used for die-attachment in the power devices or modules if no subsequent SMT reflow is required. However, it may be a challenge for tin-rich solders to survive the higher junction temperature above 175 °C (Tj > 175 °C) for long-term reliability requirements because it is close to the intrinsic melting temperature of tin-rich solders and solder joint degradation, including both bonding interfaces and solder body, is accelerated at such a high temperature.

Lead-free alternatives, with melting temperatures above 250 °C, are diversified. BiAg/BiAgX, BiCu/BiCuX, and AuSn eutectic exhibit melting temperatures between 260 and 280 °C. Namely, the solidus temperature of BiAg/BiAgX is around 262 °C [18], BiCu/BiCuX around 270 °C [18], and Au80/Sn20 around 278 °C [18], falls in the lower temperature range of the high-lead solders. Tin-antimony based solder alloys, with a ratio of Sb to Sn around 1 to 1, have shown a solidus temperature above 310 °C, which is even higher than Pb-2Sn. Au88/Ge12 [18], Au96.8/Si3.2 [18], and Zn-Al eutectic with aluminum content of 4~6 wt% [18] have melting temperatures above 360 °C, which is much higher than high-lead solders.

In addition to solders, TLPB materials and sintering materials can also be processed around 250~400 °C to form the joint in microelectronics packaging [10–12, 19–42]. TLPB materials normally have two or more alloy constituents, which have different melting temperatures. In a TLPB paste, the low-melting alloy powders melt first during processing to form intermetallic compounds (IMCs) with both the high-melting alloy constituents and the surface metallization to minimize or eliminate the low-melting phases.

Sintering materials (micron-sized silver particles/flakes) have been used for joint formation since the 1980s with the assistance of pressure [28–30]. Different from both solder and TLPB material, sintering materials will not melt during the joint formation. Instead, the bond is formed through solid state atomic interdiffusion, which normally takes a longer time and/or higher temperature to finish and likely needs pressure to assist. When nano-particles were adopted in sintering materials, the processing temperature could be reduced to 250 °C or even lower, and the processing duration can be shortened to a few minutes under pressure. It has also been claimed that nano-Ag sintering materials can possibly be processed under low pressure and even zero pressure (pressure-less sintering) [12, 36, 38–40].

All the potential HTLF bonding materials have shown at least a comparable or much higher bond shear strength than the high-lead solders in Figure 6.2 [39, 43–49]. A variety of testing vehicles were used in the bond shear tests, including (i) SiC die on active metal brazed (AMB) SiN substrate, (ii) Si die on Cu substrate, (iii) Si die on direct bonded copper (DBC) substrate (iv) Cu die on Cu substrate, and (v) chip capacitor on PCB. High-lead solders, including Pb-10Sn-2Ag, Pb-5Sn-1.5Ag, and Pb-5Sn-2.5Ag, have shown similar bond shear strength around 30 MPa at room temperature, regardless of the testing vehicles used. When increasing the temperature from 25 to 250 °C, the bond shear strength for all three high-lead solders drops to around 8 MPa.

Unlike the high-lead solders, BiAgX has exhibited the dependence of bond shear strength on testing vehicles [45-46, 50-59]. The room temperature shear strength of 54 MPa from the testing vehicle (i), SiC/AMB package [45], is almost twice of the strength (28.2 MPa) from the testing vehicle (ii), Si/Cu package [46]. The high coefficient of thermal expansion (CTE) mismatch between Si and Cu may have caused a high internal stress for BiAgX, which in turn resulted in a reduced shear strength compared with SiC/AMB. The bond shear strength for both packages drops linearly at different rates with an increasing temperature. Finally, the bond shear strength of both packages converges at 16 MPa at 250 °C, equivalent to the homologous temperature of 0.98 for BiAgX.

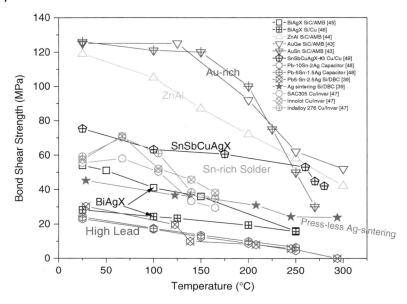

Figure 6.2 The bond shear strength of HTLF candidates [39, 43–49].

Tin-antimony-based solders are strong and stiff, which causes the silicon die to shatter under the bond shear test instead of bonding material rupture [49]. Copper die and copper substrate packages are used in the bond shear tests for the two SnSbCuAgX alloys. These two alloys show a bond shear strength of around 50~80 MPa at room temperature. With increasing temperature, the bond shear strength for both alloys decreases to 40 MPa around 260 and 280 °C, equivalent to the homologous temperature of 0.92–0.95 for SnSbCuAgX. Figure 6.2 has shown the temperature dependence of bond shear strength from one SnSbCuAgX solder alloy.

ZnAl eutectic [44] has the bond shear strength (testing vehicle (i)SiC/AMB package) around 120 MPa at room temperature, which is similar to the hard gold-rich solders. The bond shear strength of ZnAl decreases linearly with an increasing temperature to around 42 MPa at 300 °C, equivalent to the homologous temperature of 0.87.

As a comparison, AuSn eutectic and AuGe eutectic (package/testing vehicle (i)SiC/AMB) have shown a similar bond shear strength at 25 °C of above 120 MPa [43]. The shear strength of both gold-rich alloys did not drop substantially until the temperature increased to 150 °C and above. The bond shear strength of these alloys then drops near linearly with increasing temperature to around 50 MPa at 250 °C, which rank at the top of all the lead-free solders tested.

The bond shear strength of a pressure-less sintered silver joint between Si die and DBC has been measured and compared to all solder materials. At 25 °C, the bond strength of a pressure-less sintered Ag joint is around 45 MPa [39]. The bond shear strength drops with increasing temperature and remains around 25 MPa even at 300 °C.

The reflow processing of solder materials have been well recognized and used commonly for mass production. High-lead solders can be found in paste, wire, and preforms while some of the HTLF candidates cannot. TLPB materials are not easily made into wire and are commonly provided in paste form. In some cases, composite metal foils [60], e.g. tin and tin alloys as well as indium and indium alloys laminated onto silver or copper core layers, have also been tested for high temperature interconnection. Silver-sintering material is commonly in paste form. Silver-sintering film, similar to conductive adhesive film, has been disclosed in the literature [41, 42]. The semi-solid bonding process for TLPB material and the solid state interdiffusion bonding process for sintering material necessitate significant process modifications from the standard deposition and reflow process used for the well-known high-lead materials, which includes the elongated processing time, the pressure assistance, and/or even the vacuum or low-pressure processing [11, 12, 19, 20, 36, 39].

The novel HTLF solder candidates are not completely compatible with the traditional high-lead reflow process. ZnAl eutectic is very reactive and not able to be made into the traditional paste form because of the associated poor shelf life and poor stage operation life. The high reactivity of ZnAl solder would require a vacuum/forming gas/formic acid atmosphere during reflow. Even a pressure-assisted bonding process is recommended during joint formation, in which fresh molten solder would be squeezed out of the oxide crust to complete the wetting and the joint formation.

In this chapter, lead-free solder candidates will be reviewed with a focus on the performance, microstructure, and microstructural evolution. Sintering materials and TLPB materials are also briefly reviewed for comparison. None of the bonding materials can meet the needs of the wide variety of die-attachments, satisfying all the challenges found in terms of processing, performance, and cost. However, each of these materials may be able to cover a niche from the broader categorizations in one or two perspectives based on reliability, processing, performance, and cost.

6.2 Solder Materials

6.2.1 Gold-Based Solders

Gold-tin (Au-20Sn 278 °C), gold-silicon (Au-3.2Si 363 °C), and gold-germanium (Au-12Ge 361 °C) are three known gold-based eutectic solders. Gold-tin eutectic is composed of two intermetallic phases (δ-AuSn and ζ-Au5Sn) and has a eutectic temperature of 278 °C, found in the phase diagram [18]. Both gold-germanium and gold-silicon solders have eutectic temperatures of 361 and 363 °C respectively, and are composed of two solid solution phases, (Au) + (Ge) for Au-Ge and (Au) + (Si) for Au-Si [18].

During soldering, tin from gold-tin eutectic solder will dominate the interfacial reaction to form tin-containing IMCs along the interface [43, 61]. Gold-germanium and gold-silicon have shown different interfacial reactions upon soldering [10, 43, 62]. On a copper surface, AuCu or Au$_3$Cu is formed with gold-germanium solder and AuSiCu is formed with gold-silicon solder. On the Electroless Nickel Immersion Gold (ENIG) surface, Ni$_3$Ge$_5$ or NiGe IMC is formed with gold-germanium, and AuNiSi is formed with gold-silicon. Under a temperature of 300 °C and after aging for 300 hours, the interfacial delamination within or near the IMC layer on the copper surface is observed for both Au-Ge and Au-Si eutectic solders [43, 62]. On the ENIG surface, interfacial failure for both gold-germanium and gold-silicon is barely seen under the same aging condition [43, 62]. Gold-tin is a well-known lead-free solder material with good electrical/thermal conductivity and good strength below 200 °C, and has been used in biomedical, LED, and down-hole applications. For all the gold-based solders, the extremely high cost often prevents them from being widely used in the industry.

6.2.2 Bismuth-Rich Solders

6.2.2.1 Design of Bismuth-Rich Solders

Bismuth-rich alloys have been viewed as the potential candidates to replace high-lead solders as a drop-in solution [45, 46, 50–80]. Bismuth-silver shows the eutectic composition of Bi-2.6Ag and a eutectic temperature of 262 °C [18], below which the molten liquid is solidified into two phases: the pure bismuth matrix and the Ag-rich particles. Bismuth-copper (Bi-Cu) [18] has the eutectic temperature of 270 °C, slightly higher than Bi-Ag, while the eutectic composition is Bi-0.3Cu.

Bismuth and bismuth-rich alloys are brittle and exhibit cleavage rupture through lattice sliding under tension and compression. However, the single crystal bismuth wire has shown the dependence of the ductility on the angle between the loading direction and the orientation of the lattice plane (111). The smaller angle can render a ductility of 100% and above while the larger angle drops the ductility down to 2% [81]. To increase the ductility of bismuth-rich alloys, in practice a few ways of implementation have been considered: (i) alloying the soluble elements into a solid bismuth matrix to impede the dislocation movement and the associated lattice sliding; (ii) dispersing the fine particles with enough

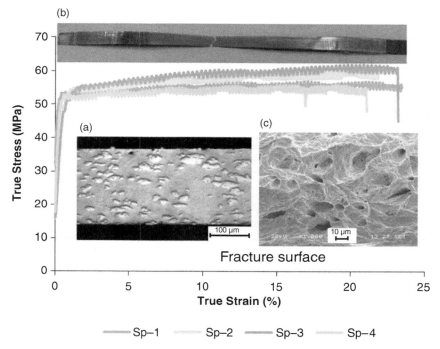

Figure 6.3 Mechanical property, microstructure, and fracture morphology of Bi-11Ag ribbon.

quantity to restrict the dislocation movement and the associated lattice sliding; and/or (iii) refining bismuth grain size to hinder the dislocation movement and the associated lattice sliding through enhanced numbers of grain boundaries.

With fine and well-dispersed silver particles in the cast Bi-11Ag (89wt%Bi–11wt%Ag) alloy, Bi-11Ag can be mechanically extruded into wire and rolled into ribbon under optimized processing conditions. The ribbon and the wire show good strength and ductility. Under testing, Bi-11Ag ribbon [11] shows a true yield strength of about 52 MPa and a true strain above 20% in Figure 6.3. Also, both narrowing and necking have been observed before rupture under tension. The rupture surface of the ribbon shows a vein pattern and dimple voids, which indicates that Bi-11Ag is able to deform in ductile mode if being reinforced with well-dispersed Ag particles and possibly the refined grains. The eutectic composition of Bi-0.3Cu may not make the alloy strong and ductile enough because of the limited copper content to strengthen the bismuth matrix [18]. Increasing copper to 2 wt% or higher would strengthen the alloy significantly if copper precipitate particles are able to be refined, dispersed, and stabilized by alloying the third or even the fourth elements. However, the liquidus temperature increases dramatically when increasing copper content based on the phase diagram [18], and the high pasty range resulting from the high liquidus temperature would increase the reflow temperature accordingly.

Both BiAg and BiCu binary alloys show poor wetting on the common metallization surfaces of Cu, Ni, and Ag [67–72], which limits their application as solder materials. Directly alloying the reactive elements to enhance the interfacial reaction, i.e. Sn or In, shows either little influence on wetting if insufficient additions are used or low-melting phases (BiSn or BiIn) if overdosed. Poor wetting can be improved with the novel design of BiAgX paste [67, 68, 70, 71, 73], in which an additive solder powder (Sn-containing alloy) has been mixed together with Bi-Ag (Ag content ranges from 2.6 to 15 wt%) powder into the paste to improve wetting and promote the interfacial reaction during soldering.

When soldering, the additive solder powder will melt earlier than, or together with the primary solder powder, and react with the surface metallization to promote wetting. The additive alloy powder is designed to dominate the formation of interfacial IMCs and finally, the active elements in the additive solder will be completely converted into IMCs after reflow. The remaining non-active constituents of the additive solders will mix thoroughly with the molten primary solder powder and solidify together to form a homogeneous solder joint.

BiAgX pastes are compatible with current high-lead paste processing. Discrete power components made with BiAgX pastes have passed the preconditioning (3× reflow @ 260 °C peak temperature), moisture sensitivity level (MSL) 1–3, thermal aging test (150 °C), and thermal cycling test (TCT) from −55 to +150 °C. RDS (on) of the MOSFET under various voltages are comparable to the high-lead solder counterparts before and after reliability tests [46]. Under TCT of −40 to +200 °C, BiAgX paste has shown good mechanical performance after 2000 cycles for the SiC/CuMo package [50]. This testing data, together with the reasonable materials cost, and processing compatibility to high-lead solders, has made BiAgX a good lead-free solution for applications up to 200 °C [45, 49–53, 76, 78].

Further increasing the remelting temperature to 270 °C, the mixed alloy powder pastes containing BiCu or BiCu-X(X is the hardening element) and the additives have been designed and tested. Preliminary data has shown the comparable reliability performance to BiAgX [49].

In addition to BiAgX paste, BiAgX wire and preform [49, 69, 74, 75] have successfully been made. Bismuth-silver wire and ribbon are able to be extruded directly from a cast ingot under the optimized processing conditions. Then, the wire and the ribbon are coated with a thin layer of Sn or Sn alloys to improve the wetting performance when soldering. The ribbon is then rolled into the designed thickness and the performs are then mechanically punched from the ribbon with the desired geometry.

6.2.2.2 Mechanical Behavior of BiAgX

6.2.2.2.1 Brittle-to-Ductile Transition

Solder is used to connect two or more objects together and form the interconnection joint. The joint reliability relies not only on the solder material itself but
also the bonding interface. To investigate the joint performance, the appropriate and representative testing vehicles are selected, which includes two types of
die-attachment assemblies/packages: (A) SiC die bonded onto AMB-SiN substrate
(SiC die size: 2 mm/2 mm/0.3 mm; AMB-SiN substrate thickness: 0.62 mm) and
(B) Si die bonded onto Cu substrate (Si die size: 3.175 mm/3.175 mm/0.7 mm; Cu
substrate thickness: 0.56 mm). Package A, mimicking the die-attachment of power
modules, generates the mild thermomechanical stressing condition in the solder
joint because of (i) the small CTE mismatch between SiC die (CTE = 2.4 ppm K^{-1})
and SiN substrate (CTE = 2.1 ppm K^{-1}), and (ii) the smaller and thinner die geometry. The solder joint in Package B, simulating the structure of a semiconductor
power discrete, suffers a harsh stressing condition due to (i) a larger CTE mismatch between Si die (CTE = 2.6 ppm K^{-1}) and Cu substrate (CTE = 17 ppm K^{-1})
and (ii) a bigger and thicker Si die. Die shear tests have been done on both packages from 25 to 250 °C. The bond shear strength relative to temperature has been
summarized in Figure 6.2.

The bond shear force-displacement curves from package B in Figure 6.4 show
a transition from brittle to ductile rupture with increasing temperature [53, 54].
Below 100 °C, the shear force drops abruptly right after reaching the peak load
(~29 kgf or 28.2 MPa at 25 °C and ~25 kgf or 24.3 MPa at 100 °C). The rupture
displacement (the total shear displacement until the rupture) is around 250 μm for
both testing temperatures. Increasing to 125 °C, the peak shear force remains at

Figure 6.4 Shear force
and displacement curve of
BiAgX joint [53, 54].

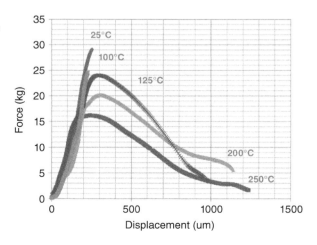

around 24 kgf. However, the shear force drops very slowly after reaching the peak load and the rupture displacement increased dramatically to around 1000 µm, which is four times of that at 100 and 25 °C. Even increasing the temperature to 200 and 250 °C, no abrupt drop of the shear force after the peak load has been observed and the rupture shear displacement increases slightly beyond 1100 µm. The significant change of the rupture displacement between 100 and 125 °C has been recognized as the brittle-to-ductile transition. The transition is actually associated with the recrystallization of the BiAgX joint under the influence of both stress and temperature.

6.2.2.2.2 Creep
The high temperature shear creep behavior of the BiAgX joint (Cu/solder/Cu joint) at 100 °C (373 K), 150 °C (423 K), and 200 °C (473 K) has been studied by Jiang [59]. At 100 °C, the creep shear strain rate of BiAgX is around one to two orders of magnitude lower than that of high-lead under the same stressing condition. The strain rate of Sn3Ag0.5Cu is roughly in the middle between those of BiAgX and a high-lead solder. Raising the temperature to 150 °C and even 200 °C, the shear strain rate of BiAgX is still one to two orders of magnitudes lower than high-lead under the same stressing condition, which indicates the much higher creep resistance for BiAgX (Figure 6.5).

6.2.2.2.3 Thermal Aging
Various BiAgX joint packages were aged at temperatures between 200 and 250 °C to study the bond shear strength degradation. Figure 6.6 has summarized the dependence of the bond shear strength on aging conditions [45, 52, 76–78]. The bond shear strength reduces more quickly at the beginning of the aging time for all aging temperatures and then maintains the bond shear strength or decrease slightly after roughly 1000 hours. The trend of the bond shear strength is similar, regardless of the testing vehicle and testing conditions, while the degradation magnitude may differ.

6.2.2.2.4 Temperature Cycling and Thermal Shock
Degradation of joint bond strength has been monitored under both thermal cycling test (TCT) and thermal shock test (TST). Depending on the testing vehicles being used, the initial bond shear strength varies from around 30–52 MPa. With increasing cycles, the bond shear strength drops in different rates depending on the selected testing vehicle and testing condition. For a mild condition (−55 to +125 °C) [78], the bond strength of the Si/Cu package after 2000 cycles remains around 35 MPa. The combination of low CTE mismatch (Si and CuMo) and harshest cycling condition (−55 to +200 °C) [50] actually maintains well the bond shear strength above 25 MPa even after 2000 cycles. With large CTE mismatch (Si

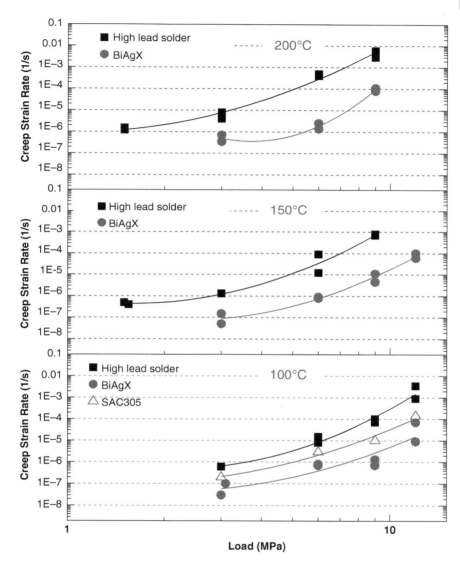

Figure 6.5 Creep behavior of BiAgX joint [59].

and Cu) and harsher condition (−40 to +175 °C), the bond shear strength after 2000 cycles [49] drops to around 6–8 MPa while the original strength is around 30 MPa. Thermal shock from −55 to +150 °C of 2000 cycles [77] maintains the bond shear strength at around 10–16 MPa from the initial strength of around 30–35 MPa (Figure 6.7).

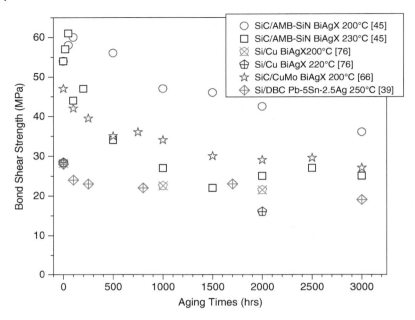

Figure 6.6 BiAgX joint bond shear strength after thermal aging [45, 52, 76–78].

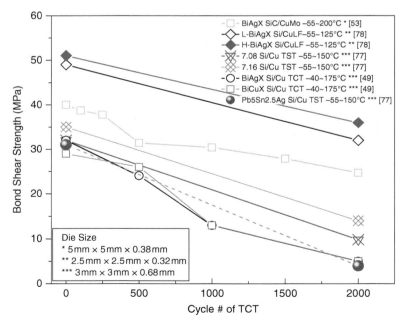

Figure 6.7 BiAgX joint bond shear strength after thermal cycling test [39,45,49,50,66,76–78].

As a comparison, one of the most commonly used high-lead solder alloys (Pb-5Sn-2.5Ag) [77] shows the similar degradation behavior to BiAgX and BiCuX using the same testing vehicle (Si and Cu) under the similar testing environment (TST from −55 to +150 °C).

6.2.2.3 Microstructure and Microstructural Evolution of BiAgX Joint

6.2.2.3.1 Microstructure of As-Reflowed BiAgX Joint

The morphology of a solid joint includes two bonding interfaces and the joint solder body. The factors influencing the microstructure of an as-reflowed joint include (i) the solder composition; (ii) the surface/substrate metallization, and (iii) the reflow profile and reflow atmosphere. The microstructure of the as-reflowed BiAgX joint between Ti/Ni/Ag-plated Si die and Cu substrate has been explored [11], showing two types of AgSn (a mixture of Ag_3Sn and Ag) particles inside the Bi matrix and the joint body (Figure 6.8). Fine AgSn precipitates (submicron) dispersed to form networks and isolated bismuth matrixes into small colonies are shown in Figure 6.8a,b. The scattered micron-sized AgSn particles are enriched along the bonding interface as shown in Figure 6.8a,c.

As a comparison, Bi-11Ag ribbon is reinforced by the dispersed micron-sized Ag particles only (Figure 6.3a) and has no trace of any networked submicron Ag particles. Tin introduced from the additives in BiAgX paste was assumed to assist the formation of the networked submicron AgSn precipitates in addition to improve the wetting during soldering. Tin in BiAgX paste has formed interfacial Cu_3Sn IMCs on the Cu substrate and AgSn at the Si die side shown in Figure 6.8a,c, while BiAg has no interfacial IMCs formed on either Cu or Ag [67, 68]. Tin, introduced from the additive powders, is dominating the interfacial reaction and can be well-controlled through the composition design and the quantity in the paste. All tin elements would be consumed completely in practice to form either interfacial IMCs or IMC precipitates (i.e. networked AgSn and scattered AgSn particles for BiAgX in Figure 6.8) after reflow, and thus no low-melting Bi-Sn phase is present afterwards.

The interfacial Cu_3Sn IMC formed on the Cu substrate during reflow is stable during all the following thermal-mechanical testing because no free tin contributes to the interfacial Cu-Sn IMC growth after soldering [46, 72]. High-lead solders with similar tin content behave similarly, because tin, in both cases, is dominating the interfacial reaction while lead and bismuth are not reacting with either silver or copper to form IMCs.

6.2.2.3.2 Microstructural Evolution

The post-reflow microstructural evolution of a solder joint, driven by free energy reduction, presents typically through solid state atomic diffusion, including lattice diffusion, grain boundary diffusion, surface diffusion, and dislocation core

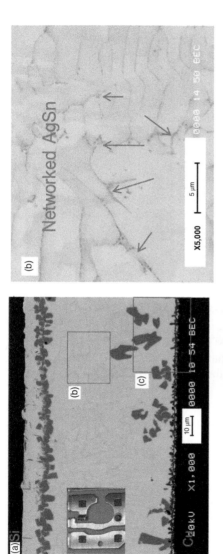

Figure 6.8 Microstructure of BiAgX joint (a) morphology of a BiAgX joint between Si (Ti/Ni/Ag) and Cu lead-frame; (b) the submicron networked AgSn particles isolating Bi colonies; and (c) the micron-sized AgSn particles dispersed close to the IMC interface at the Cu

diffusion, etc. The microstructural evolution in a joint comprises the changes of both the solder and the bonding interface. Inside the solder body, grain growth, phase segregation, precipitate evolution, and even recrystallization are characterized during the microstructural evolution.

On the joint interface, the growth of the interfacial IMC layers, the IMC layer spalling from metallization layer, and even the additional IMC formation between the original IMC layer and the surface metallization layer would dominate the microstructural evolution. Temperature and mechanical stress are two extrinsic factors influencing the solid state atomic diffusion. Higher temperature and higher stress would facilitate atomic diffusion and thus favor the microstructural evolution.

Upon aging, some of the networked submicron AgSn precipitates embedded in the bismuth matrix would continuously grow at the expense of the neighboring AgSn precipitates, which are shrinking or even diminishing. Figure 6.9 has shown the evolution of submicron AgSn precipitates before and after 1000 hours aging at 200 °C. Both secondary electron image (SEI, left) and backscattered electron images (BSEs, right) clearly show the networked precipitates, as marked by the red arrows, before aging. After aging, the networked precipitates, characterized by densely dispersed small precipitates shown in the top images of Figure 6.9, grow into coarse particles with much wider inter-particle spaces. During the growth,

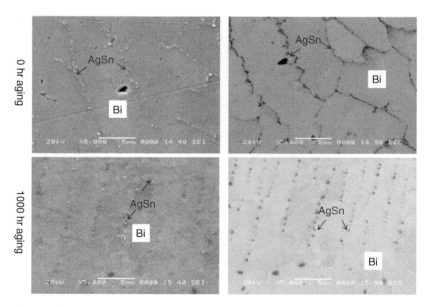

Figure 6.9 Microstructural evolution of networked AgSn precipitates upon aging at 200 °C.

both Ag and Sn elements are diffused through the Bi matrix toward the bigger precipitates from the finer ones. The fine particles shrink and/or even completely disappear, leaving the coarse ones to remain. With reduced particle numbers and wider inter-particle spaces upon aging, AgSn precipitates may not maintain the network continuity anymore.

BiAg alloys are able to be successfully extruded into wire and rolled into ribbon under elevated processing temperatures. The brittle-to-ductile transition during bond shear testing verifies that BiAgX can deform plastically when the transition temperature (between 100 and 125 °C) is reached [54].

To understand the underlying mechanism of the increased ductility under elevated temperature, the interrupted die shear tests at 250 °C have been done on Si/Cu packages and then the microstructural evolution was studied from three regions where the load was intentionally stopped, as shown in Figure 6.10.

Region 1 is around half peak load before reaching the maximum. Region 2 is around the peak load and region 3 is around half peak load passing the maximum. Silicon die has been ground away along the Z-axis to expose the solder beneath for microstructural identification. The polarized optical image of the as-reflowed joint has shown that the columnar grains radiated from an area around the center toward the perimeter of the joint.

Unloading from region 1, there is no significant change of the joint grain morphology while the small and bright spots are scattered along the perimeter. Unloading from region 2, more small and bright spots are observed, dispersed from the fringe of the joint toward the center, and within the columnar grains while the columnar grains are still visible. The morphology of the sample unloaded from region 3 has shown no columnar grains anymore. Instead, the small equiaxed grains, characterized in the inset of the enlarged polarized image, are dominating the whole joint. The observation of morphological transition from

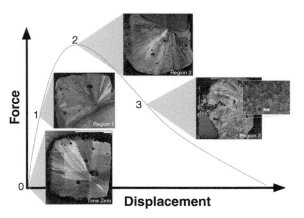

Figure 6.10
Recrystallization of BiAgX joint from the interrupted high temperature shear tests [54].

Increasing Recrystallization Area

Figure 6.11 Evolution of the BiAgX joint under thermal cycling testing: the recrystallization from the perimeter toward the center [11, 80].

the columnar grains to the fine equiaxed grains indicates the recrystallization under high temperature shear. The dispersed bright and small spots within the columnar grains from unloading regions 1 and 2 are likely the new grains recrystallized locally.

In the BiAgX system, the recrystallization is assumed to account for the brittle-to-ductile transition. Recrystallization occurs when the dislocations, which have looped, pinned, and accumulated around the networked AgSn particles, become the grain boundary under thermal stressing.

Solder recrystallization has also occurred under TCTs. Polarized cross-section images from the Z-axial cross-section in Figure 6.11 have shown the microstructural evolution under thermal cycling testing. The time zero joint shows the columnar coarse grains radiated from the center toward the fringe of the joint. After 253 cycles, the fine grains appear along the perimeter of the joint while the columnar grain occupies the majority of the center area. With increasing numbers of thermal cycles, fine grains grow continuously from the perimeter of the joint toward the center and the area occupied by fine grains enlarges substantially. After 2000 cycles, the fine grains occupy around two-thirds of the joint area and surround the remaining columnar grains in the center.

After recrystallization, the micro-crack initiated from the edge (or the corner) of the die boundary grows inwards into the solder joint. After 2000 cycles of the TCT test, the X-ray image has shown the brighter spots within the fringe area of the silicon die in Figure 6.12a. Cross-section images show (i) fine grains along either side and in front of the crack (Figure 6.12c), and (ii) the original coarse grain far away from the crack tip (Figure 6.12b). This indicates that the crack formation follows the recrystallization [54]. The recrystallization may occur when the dislocations are pinned, looped, and accumulated around the networked AgSn particles, and become the grain boundary under stress in TCT. Lead-free SnAgCu solders have

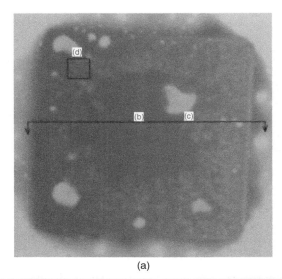

(a)

(b)

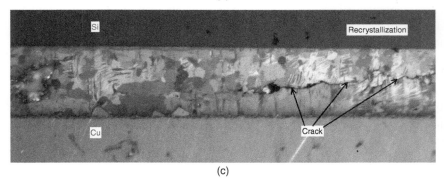

(c)

Figure 6.12 (a) X-ray image of Si/BiAgX/Cu joint after 2000 cycles of TCT tests; (b) the morphology of the coarse grain in area (b) of image (a); (c) the fine grains along the crack banks and in front of the crack tip from area (c) in image (a); (d) the creep fatigue features from area (d) in image (a). [11].

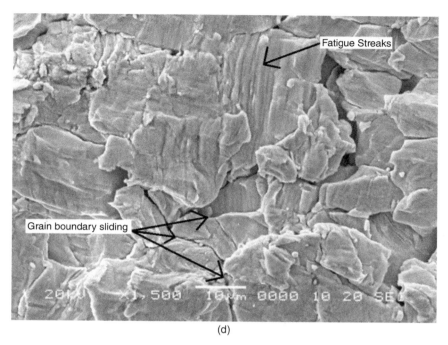

(d)

Figure 6.12 (*Continued*)

also shown similar behavior (localized recrystallization, fatigue crack initiation, and propagation) in TCT tests [82]. After shear testing, the fracture surface morphology of the BiAgX joint, within the fringe area (d) from the X-ray image, shows the fatigue streaks (parallel to each other) within the individual grain, and the grain boundary sliding voids/gaps in Figure 6.12d. This verifies that the solder creep fatigue is the main cause of thermal cycling damage.

High-lead solders and SnAgCu solders have also shown creep-fatigue behavior featured by numerous fatigue streaks inside grains and grain boundary sliding. Solder creep fatigue is normally desired because there is no catastrophic failure to suddenly fail the joint during solder creep fatigue [46, 53].

Beyond BiAgX paste, BiAgX wire and preform have also been produced, although they are still in the research and development stage [49, 69, 74, 75]. BiAgX solder system has shown good compatibility to high-lead solders, including both handling and processing. In addition, good thermal-mechanical performance, and comparable electrical performance make BiAgX solder a drop-in alternative to high-lead in discrete power components, suitable for applications of 200 °C and below while thermal conductivity of BiAgX is inferior to high-lead.

The material cost of BiAgX is higher than that of high-lead solder, but in the same magnitude. In addition to BiAg-based solders, BiCu and BiCu-X alloys have been studied in order to increase the remelting temperature further to 270 °C [15, 49]. So far, BiAgX, including both BiAg and BiCu, are some of the most suitable drop-in solutions to replace high-lead in discrete power components.

6.2.3 Tin-Antimony (Sn-Sb) High Temperature Solders

Increasing antimony (Sb) content in the tin-antimony (Sn-Sb) binary system above 43.6 wt%, Sn_3Sb_2 and β(SnSb) compounds are formed after solidification based on the phase diagram. Sn_3Sb_2 would decompose into two solids β-Sn and β(SnSb) when cooling further to a lower temperature. Copper and silver are alloyed with SnSb to strengthen and convert any remaining β-Sn into Cu_6Sn_5 and Ag_3Sn to eliminate the existence of any tin-rich low-melting phase.

Differential scanning calorimetry (DSC) of a few selected SnSbCuAg alloys (Alloy 1#, 2#, and 3#), in Figure 6.13, has shown that the low-melting phase can be removed. Alloy 1# has a low-melting peak around 227 °C and the onset melting temperature of the main peak is around 302 °C. Alloy 2# also shows a minor trace of melting peak around 226 °C, and the onset melting temperature of the main peak is 304 °C, slightly higher than Alloy 1#. However, Alloy 3# shows

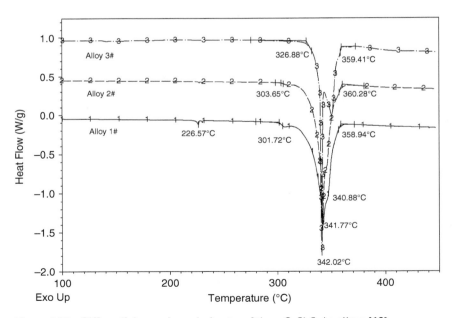

Figure 6.13 Differential scanning calorimetry of three SnSbCuAg alloys [68].

only one melting peak with the onset of 327 °C. All three alloys have a similar peak temperature (~340 °C) and end melting temperature (~360 °C).

Solder paste made of Alloy #2 and Alloy #3 have been used to build two different test vehicles: (i) Si die (0.125″ × 0.125″ × 0.028″ TiNiAu surface finish) onto SiN DBC substrate (copper surface finish), which has little CTE mismatch (silicon: 4 ppm °C⁻¹ and SiN DBC: 4–6 ppm °C⁻¹), and (ii) copper die (0.125″ × 0.125″ × 0.022″) on Invar (Nickel-Iron alloy) substrate, which has a CTE mismatch around 13 ppm (copper: 17 ppm °C⁻¹ and Invar: 4 ppm °C⁻¹). The die shear strength of package (1) for both alloys is much lower than that of package (2), although the CTE mismatch of package (1) is much smaller than package (2). Silicon die shattering during shear tests, shown in Figure 6.14, was attributed to the lower bond shear strength.

After 2000 cycles of TCT (from −40 to +175 °C), the preliminary data has shown the slight changes of the bond shear strength from both packages soldered with Alloy #3 compared with Alloy #2 in Figure 6.15. As a comparison, Alloy #2 shows significant drop of bond strength with increasing numbers of thermal cycles for both packages.

The cross-sectional observation of thin Si die on SiN DBC has not shown the cracks initiated and propagated from the die perimeter toward the joint center as indicated in Figure 6.16. In the joint, SnSb forms the joint matrix, and IMC particles, including both Cu6(SnSb)5 and Ag3(SnSb), are embedded within the joint matrix [15, 49] in Figure 6.16a. After 1000 cycles of TCT, the joint maintains the integrity. AgSn platelets are seen inside the joint although AgSn particles remain

Figure 6.14 Die shattering after bond shear testing for Si/SiN DBC package.

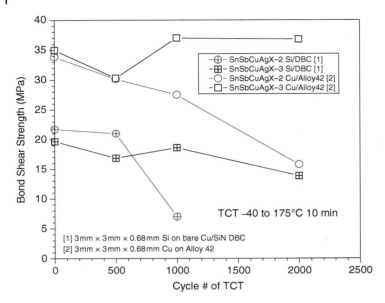

Figure 6.15 Joint shear strength of Alloy #2 and #3 before and after TCT tests.

in certain zones in Figure 6.16b. The thickness of IMC layer on the DBC side increases significantly and both Cu6(SnSb)5 and Cu3(SnSb) are readily identified. There are no horizontal cracks observed from the joint corner or edge while vertical microcracks are found somewhere beside the AgSn platelet and within the thick IMC layer on the DBC side.

Under TSTs from −55 to +155 °C, packages of (i) silicon die (0.125″ × 0.125″ × 0.028″ Ti/Ni/Au surface finish) on Cu and (ii) silicon (0.125″ × 0.125″ × 0.028″ Ti/Ni/Au surface finish) on SiN DBC were used. However, silicon die shattering was the major failure mode for both package types, regardless of both the alloys used. The severe stress condition under TST (involving switching between hot and cold chambers) is assumed to be the major cause of the silicon die shattering. Thin silicon die (2.45 × 2.75 × 0.2 mm) on copper lead-frame packages have been assembled with both alloys and no Si die damage has been observed. However, it is difficult to use die shear testing for monitoring the bond strength degradation. Additional investigations are still ongoing.

6.2.4 Zinc-Aluminum Solders

Zinc-Aluminum eutectic (Zn-(4–6 wt%)Al) has a melting temperature of 382 °C, which is higher than AuSi and AuGe. Zinc-Aluminum eutectic is composed of two phases; η-(Zn) matrix and α-(Al) at room temperature. At 275 °C, there is a solid phase transition of α phase to β phase. Zinc-Aluminum and

Figure 6.16
Microstructure of
SnSbCuAg joint for Si
die/SiN DBC substrate, (a)
before and (b) after
1000 cycles of TCT (−40
to +175 °C).

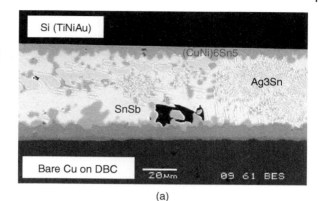

(a)

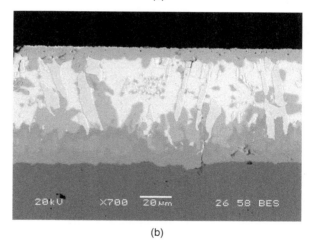

(b)

Zinc-Aluminum-Copper were used as bearing support materials for a few decades, in which the hard, laminar α phases were embedded in the soft zinc matrix [11], as shown in Figure 6.17.

Zinc-Aluminum eutectic alloy has shown yield strength of over 100 MPa, which is around three times that of traditional high-lead solders. Zinc-Aluminum also has good thermal conductivity (100 W mK⁻¹), which is even better than Au-Sn eutectic (60 W mK⁻¹) and high-lead solders (~26 W mK⁻¹) [44, 83]. The cost of Zinc-Aluminum is comparable or even cheaper than that of high-lead solders. However, both zinc and aluminum are reactive metals, and it is difficult to make Zinc-Aluminum paste exhibit acceptable storage life and paste working life during stage operation. Thus, wire and preform are two possible material forms for Zinc-Aluminum for use as a solder.

Soldering with Zinc-Aluminum needs to be done under vacuum, inert gas, forming gas, or even a formic acid atmosphere to minimize the impact from Zinc

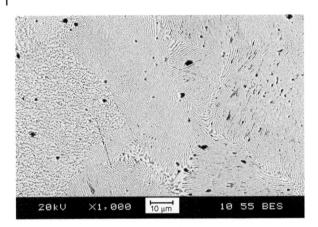

Figure 6.17
Microstructure of cast Zn-4Al alloy.

and Aluminum oxides. Upon soldering, the oxide crust (Zinc and Aluminum oxide) impedes the molten solder flowing and impairs the wetting and interfacial reaction. In order to break the oxide crust [18, 61], pressure is required during soldering. Zinc dominates the interfacial reaction in soldering, and complicated IMCs are formed on various surfaces, like Cu, Ni, and Ag. On the Cu surface, two stacked IMC phases, ε (Cu-Zn4) and γ (Cu5Zn8), are formed when soldering with Zinc-4 wt% aluminum. On the Ag surface, three stacked IMC layers, ε (AgZn3) + ζ (AgZn) + γ (Ag5Zn8), have been observed [44, 83–85].

Tanimoto [44] has reported that the bond shear strength of Zn-Al joints (SiC/ZnAl/AMB-SiN package) drop around one-third from 120 MPa at time zero to 80 MPa after 1500 cycles of TCT (−40 to +200 °C), which is more than 10 times higher than the required bond strength of 6.2 MPa from the IEC 60794-19 standard (Semiconductor devices-mechanical and climatic test methods – Part 19: Die shear strength). However, crack growth into the metallization layer has also been reported after thermal cycling, initiating the interfacial failure.

The high reactivity of zinc-aluminum alloy, prevents it from being made into stable pastes, in addition to the processing requirements of both vacuum and pressure, limits its use in discrete power component manufacturing. But, the high melting temperature, high thermal conductivity, high strength, and low cost may make it a potential candidate to replace Au-Ge/Au-Si for die-attachment in individual power devices under non-corrosive conditions.

6.3 Silver (Ag)-Sintering Materials

Silver-sintering materials have been used for die-attachment in power devices for decades because of the high electrical and thermal conductivities [44, 62, 83].

In the early days, micron-sized silver particles/fillers were used to make pastes, and then they were sintered to form the joint under a pressure of between 9 and 40 MPa and a temperature of 250 °C and above, which was called "pressure sintering". Using nano-silver particles, sintering can be done with a reduced pressure (1–5 MPa) under similar sintering temperatures or even lower to 200 °C [35–37]. It is also desired to eliminate the pressure during sintering altogether, which would allow the silver-sintering paste to be a potential drop-in solution to replace the high-lead solders [13, 31, 36].

Joint formation through sintering is driven by the surface energy reduction [27]. It starts from the contact area (powder-to-powder and powder-to-metallization) through solid state atomic diffusion [86]. The atoms flow toward the contact area and form necking. The necking grows to enlarge the bonding area until the necking curvature radius (r) is comparable to half the diameter (x) of the contact area (Figure 6.18a) in which pores are formed among particles [31]. Afterwards, the decrease in the pore-specific surface area or the spheroidization of the pores will dominate the evolution of the pores' morphology, which can be monitored through the density/porosity change.

Without pressure, the initial compact density (or the metal load of the sintering paste) and the sintering shrinkage are two major factors (in addition to Ag particle sizes) controlling the density/porosity of the sintered joint. Pressure applied during sintering will increase the compact density and enhance the plastic/viscous flow of Ag to grow the necking and minimize porosity, in which the density can reach 95% or higher of the bulk density. Figure 6.18b shows porosity from the strong connection and the weak connection schematically.

Inside silver-sintered joints, the pores act as the precursors of microcracks, which would further grow into cracks and propagate across the whole joint interface. Under shear creep at 325 °C (the homologous temperature < 0.5) and 2 MPa shear pressure, the growth of the pores into microcracks and cracks in a period of 30 hours or less has been identified from the 3-D volume rendered images [32]. Those cracks grew close to each other and interlinked to accelerate the failure in the later stage (from 21 to 29.3 hours). It has also been noted that the higher the shear stress, the greater the porosity and the earlier the failure [32].

Upon thermal cycling, both the pore size and the area/volume percentage of pores increase significantly [31, 32]. The merging and linkage of these pores lead to the initiation and continuous growth of the micro-cracks into macro-cracks, which is similar to the evolution of the pores under shear creep. The original porosity, the size and the shape of the pores, and the evolution of the pores, are believed to dominate the performance of silver-sintered joints [32]. Less porosity, smaller size, and spherical-shaped pores are desired in order to have better reliability. The evolution of both the pores and the bulk silver is inevitably and continuously ongoing to minimize the joint free energy. The evolution of pores includes the spheroidization, the

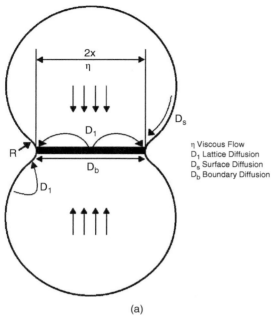

η Viscous Flow
D_1 Lattice Diffusion
D_s Surface Diffusion
D_b Boundary Diffusion

(a)

Figure 6.18 (a) Joining the powders from the contact area during sintering through diffusion and viscous flow and (b) the porosity in weak connection and strong connection [11, 31].

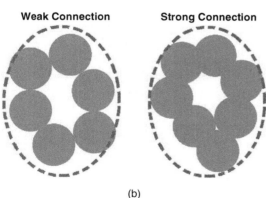

(b)

reduction of the total specific surface area, the sink-in of the vacancies into pores through dislocation movement, and the linkage of the pores. The microstructural evolution of the silver-matrix includes grain coarsening and possibly grain recrystallization under the thermomechanical stress [32].

The joint interface between sintered silver and the bonding surface influences the joint performance too. Normally, silver-sintering paste sinters readily on silver or other noble metal (gold, platinum, and/or palladium) metallization through atomic diffusion. On copper and nickel surfaces, limited success has been reported to date [33]. The natural and tenacious oxides on the surface hinder the interdiffusion, which results in a weak interfacial bond. Additionally, the formation of Kirkendall voids in the interfacial region that originate from the different diffusion rates of the different atom species would weaken the interface.

Pressure-less silver sintering is highly desired as a potential drop-in solution to replace the high-lead solders. The challenges come from the initial porosity and the evolution of the pores inside the joint. The microstructural evolution of bulk silver is also a concern, including grain coarsening and Kirkendall void formation. Interfacial adhesion between silver-sintering and non-noble metal metallization is a major concern too. Pressure-assisted silver sintering improves the joint performance greatly, but limits the adoption into mass production of high-volume discrete power components. In addition to the concern of more than a magnitude increase in material cost (compared to high-lead solders), the influence of nano-silver powder on the environment and human health is still under debate, and many potential users need a solution excluding a nanomaterial.

6.4 Transient Liquid Phase Bonding Materials/Technique

The TLPB process allows for the low-melting constituents to melt first and react with the high-melting one to form high-melting IMCs. When the processing is finished, the low-melting phase will thereafter be completely consumed or minimized, allowing the joint to survive high service temperatures. Figure 6.19 shows the schematic mechanism of the TLPB process, where A is the low-melting phase and B is the solid phase. B (or the elements from B) are dissolved into molten A

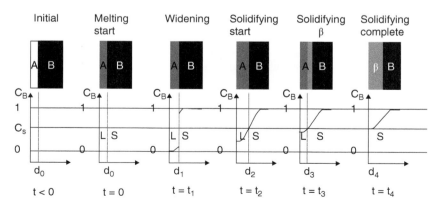

Figure 6.19 The schematic reaction mechanism of TLPB materials to form the high temperature joint [11].

phase and form the high-melting β phase (IMC phases containing both A and B) from the contact interface until low-melting phase A is completely consumed to form β.

In practice, TLPB material can be made into either paste or composite ribbon with a structure having the low temperature metal plating layers on both sides of the high-temperature metal core layer [60]. TLPB paste, i.e. the mixture of copper powder and tin powders, forms Cu_3Sn (adjacent to copper particles) and Cu_6Sn_5 (on top of the Cu_3Sn) around copper particles and on a substrate with a copper metallization surface after reflow as shown in Figure 6.20a. Among IMC phases, the remaining tin phase can still be seen, which results from the incompletion of a low-melting phase transition. The incompletion of the conversion may come from either insufficient reflow time, low reflow temperature, or the slower diffusion rate of atoms through the increasing IMC thickness toward and/or away from the tin-rich region. The pores and voids are dispersed randomly in the joint (Figure 6.20a) instead of consolidating in the center of the joint from the composite ribbon [19, 25, 34]. For a joint formed with the composite ribbon, IMC layers start to form from the contacting interfaces and grow into the molten plating layers. During IMC layer growth, the scallop shape IMCs, instead of flat film growth, evolve from both bonding interfaces and meet in the center of the joint, leaving pores focused in the middle of the joint. Figure 6.20b shows the pore formation from a copper/tin/copper sandwiched assembly. A joint, formed between two copper blocks using a composite preform (15 μm tin layers coated on copper core layer), shows the clustered pores in the middle of the joint in Figure 6.20b.

TLPB can achieve high remelting temperature by converting the joint into IMCs. IMCs are much stronger than high-lead solders, but more brittle. As a result, interfacial delamination is more likely to appear in these rigid TLPB joints [25]. IMC formation is controlled by atomic diffusion kinetics, which may need extended reflow time or higher reflow temperature to allow all the low-melting powders to be converted. Depending on the TLPB materials being designed, the material and processing costs may vary significantly. Also, the rigid IMC joint may lead to the catastrophic interfacial failure, which is different from the solder fatigue failure of the high-lead materials.

6.5 Summary

Solder materials, silver-sintering materials, and TLPB materials are three types of potential candidates as lead-free, high-temperature, die-attachment materials. Table 6.1 summarizes the physical and mechanical properties, the material forms, and the cost of these materials relative to Pb-5Sn-2.5Ag. With the emergence of the need for lead-free bonding materials, different packages may require

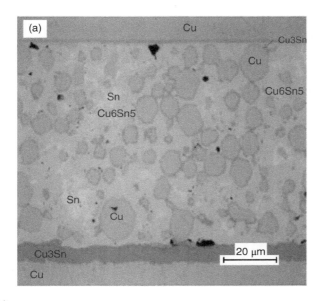

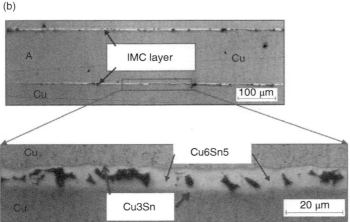

Figure 6.20 The morphology of the TLPB joint from (a) copper-tin paste between two copper blocks; (b) Tin/Copper/Tin composite ribbon between two copper blocks. [11].

different materials based on the individual properties of the target device and its reliability requirements. The appropriate die-attach material may be selected from a combination of the material properties, the material forms, the material cost, and the processing capability. Each material discussed may have merits and weaknesses. BiAg/BiAgX and BiCu/BiCu-X, with a final joint solidus around

Table 6.1 Physical and mechanical properties of some potential high temperature lead-free die-attach materials [11].

	Physical and mechanical properties					Material forms			Material cost
	Melting temperature (°C)	Shear strength (MPa) @RT	Shear strength (MPa) @200 °C	Electrical resistivity (μΩ*cm)	Thermal conductivity (W mK⁻¹)	Paste	Wire	Preform	Relative to Pb5Sn2.5Ag
Pb5Sn2.5Ag	298	28	7.5	19	26	Yes	Yes	Yes	1
ZnAl	>360	120	70	7.50	100	No	Yes	Yes	0.2~1
BiAgX	>260	45	22	86.00	14	Yes	Yes	Yes	2–5
SnSbCuAg	>320 °C	35 MPa	>30 MPa	N/A	34	Yes	N/A	N/A	3–10
AuSn	280	130	100	16.40	59	Yes	Yes	Yes	>2000
TLP bonding	>300	20–40	20–40	Material dependent	Material dependent	Yes	No	Yes	Material dependent
Ag sintering	>900	20–80	20–40	5	>80	Yes	No	No	>40

262–270 °C, as well as their good thermal-mechanical performance, allow them to be considered as potential high-lead replacements in smaller die semiconductor power (discrete and small module) devices. SnSbCuAg high-temperature solders, showing a melting temperature above 300 °C, have been investigated targeting at higher junction temperature power modules with positive results, although they are still in the early stage of development.

Silver-sinter bonding, through solid interdiffusion, has been extended to die-attachment with the desire of good stability, good thermal conductivity, and good mechanical property, for high power module applications. The intrinsic pores and the slow process may necessitate pressure-assisted bonding technology, but pressure-less sintering is feasible for some specific packages.

TLPB materials, converting the low-melting phase into high-melting IMCs through solid-liquid interdiffusion, were being tested and employed for certain interconnection joint formation. However, the pores, the remaining low temperature phase, and the processing feasibility are still in investigation.

One material to satisfy all the requirements in die-attachment is not realistic in practice. However, each solution has the potential to satisfy a niche within this broader categorization. Recently, INEMI initiated a project to evaluate the current high temperature, lead-free bonding materials for die-attachment. This project is expected to provide the industry with a clearer picture of the applicability of each material candidate and the end-use area [10].

Acknowledgment

The authors would like to thank Dr. Liang Ying from GE Research Center and Dr. Harry Schoeller from Universal Instruments Corporation for their work and discussions on BiAgX. The authors would also like to acknowledge the colleagues from Indium Corporation, Jonathan Minter, Christine LaBarbera, Dr. Sihai Chen and, Dr. Andy Mackie for their continuous support, help, and discussion on the development of the high-temperature lead-free bonding materials. This chapter was made possible by the editing work of Angela Hulbert-Wojcik and Anita Brown, and it is very much appreciated. The constructive comments and suggestions from Jasbir Bath are much appreciated as well.

References

1 Coyle, R., Sweatman, K., and Arfaei, B. (2015). *JOM* 67: 2394–2415.
2 Laurila, T., Vuorinen, V., and Kivilahti, J. (2005). *Materials Science and Engineering R* 49: 1–60.

3 Shangguan, D. (2005). *Lead-Free Solder Interconnect Reliability*. Materials Park, OH: ASM International.

4 Bradley, E., Handwerker, C.A., Bath, J., Parker, R.D., and Gedney, R. (2007). *Lead-Free Electronics*. Hoboken, NJ: Wiley.

5 Suganuma, K. (2004). *Lead-Free Soldering in Electronics*. Basel, NY, USA: Marcel Dekker.

6 Hwang, J. (2001). *Environment-Friendly Electronics: Lead-Free Technology*. Bristol, England: Arrowsmith: Electrochemical Publications.

7 Tegehall, E. (2006/2007) Review of the impact of intermetallic layers on the brittleness of tin-lead and lead-free solder joints. IVF Project Report.

8 Lee, T.K., Bieler, T.R., Kim, C.U., and Ma, H. (2015). *Fundamentals of Lead-Free Solder Interconnect Technology from Microstructure to Reliability*. New York, NY: Springer.

9 Siewert, T., Liu, S., Smith, D.R., and Madeni, J.C. (2002). *Properties of Lead-free Solders, R.4*. National Institute of Standards and Technology and Colorado School of Mines.

10 Lim, S.P., Pan, B., Zhang, H., et al. (2017). *Proceeding of 2017 International Conference on Electronics Packaging*, Yamagata, Japan, TA4-4.

11 Zhang, H., Minter, J., and Lee, N.-C. (2019). *Journal of Electronic Materials* 48: 201–210.

12 Siow, K.S. (2019). *High Temperature Die-Attach Materials for Microelectronics Packaging*. Springer.

13 Manikam, V.R. and Cheong, K.Y. (2011). *IEEE Transactions on Components, Packaging and Manufacturing Technology* 1: 457.

14 Eilken, B. (2019). Die Attach 5 Project, http://www.infineon.com/da5customerpresentation.

15 Zhang, H. and Lee, N.-C (2018). *Proceeding of 51st International Symposium on Microelectronics*, TP3-096, Pasadena, CA.

16 Zhang, H., Minter, J., and Lee, N.-C. (2017). *Proceeding of the International Conference on Soldering & Reliability*. Markham, Ontario, Canada: SMTA.

17 Zeng, G., McDonald, S., and Nogita, K. (2012). *Microelectronics and Reliability* 52: 1306.

18 Baker, H., Okamoto, H., Henry, S.D. et al. (1990). *ASM Handbook, Volume 3: Alloy Phase Diagram* (eds. H. Okamoto, M.E. Schlesinger and E.M. Mueller), 2.26, 2.56, 2.69, 2.76 and 2.99. ASM International.

19 Holaday, J.R. and Handwerker, C.A. (2019). Transient liquid phase bonding. In: *High Temperature Die-Attach Materials for Microelectronics Packaging* (ed. K.S. Siow), 197–249. Springer.

20 Bosco, N.S. and Zok, F.W. (2004). *Acta Materialia* 52: 2965.

21 Chu, K., Sohn, Y., and Moon, C. (2015). *Scripta Materialia* 109: 113.

22 Cook, G.O. III, and Soresen, C.D. (2011). *Journal of Materials Science* 46: 5305.

23 Yoon, S.W., Glover, M.D., Mantooth, H.A., and Shiozaki, K. (2012) *Proceeding of 2012 IMAPS High Temperature Electronics Conference*, Albuquerque, NM.

24 Sharif, A., Gan, C.L., and Chen, Z. (2014). *Journal of Alloys and Compounds* 587: 365.

25 Dudek, R. (2013) *Proceeding of 14th International Conference on Thermal, Mechanical and Multi-physics Simulation and Experiments in Microelectronics and Microsystems (TLP)*.

26 Shearer, C., Shearer, B., Matijasevic, G., and Gandhi, P. (1999). *Journal of Electronic Materials* 28: 1319.

27 Mackenzie, J.K. and Shuttleworth, R. (1949). *Proceedings of the Physical Society. Section B* 62: 833.

28 Schwarzbauer, H., (1989) U.S. Patent 4810672.

29 Schwarzbauer, H., (1990) US Patent 4903885.

30 Schwarzbauer, H., (1991) U.S. Patent 5058796.

31 Thummler, F. and Oberacker, R. (1993). *Introduction to Powder Metallurgy*. London: Institute of Materials.

32 Tan, Y., Lin, X., Chen, G. et al. (2015). *Journal of Electronic Materials* 44: 761.

33 Khazaka, R., Mendizabai, L., and Henry, D. (2014). *Journal of Electronic Materials* 43: 2459.

34 Li, J.F., Agyakwa, P.A., and Johnson, C.M. (2010). *Acta Materialia* 58: 3429.

35 Bai, J., Zhang, Z.Z., Calata, J.N., and Lu, G.-Q. (2006). *IEEE Transactions on Components, Packaging and Manufacturing Technology* 29: 589.

36 Siow, K.S. (2014). *Journal of Electronic Materials* 43: 947.

37 Lei, T.G., Calata, J.N., Lu, G.-Q. et al. (2010). *IEEE Transactions on Components and Packaging Technologies* 33: 98.

38 Chen, S. and Lee, N.C., (2018) US Patent 9875983 B2.

39 Chen, S., Shambach, W., LaBarbera, C., and Lee, N.C. (2018) *Proceeding of International Conference on High Temperature Electronics*, Albuquerque, NM.

40 Chen, S., Fan, G., Yan, X. et al. (2014) *Proceeding of International Symposium on Microelectronics*, San Diego, CA, 92.

41 Nikolic, N.A., (2005) U.S. Patent US2005/0137340A1.

42 Henkel. Dia attach film adhesives. http://www.henkel-adhesives.com/die-attach-film-adhesives-27373.htm (accessed 27 October 2017).

43 Tanimoto, S., Matsui, K., Murakami, Y. et al. (2010) *Proceedings of 2010 International Conference and Exhibition of High Temperature Electronics*, Albuquerque, NM, TA1.

44 Tanimoto, S., Matsui, K., Zushi, Y. et al. (2012). *Materials Science Forum* 717–720: 853.

45 Tanimoto, S., Hirama, N., Watanabe, K. et al. (2013) *Annual Spring Meeting of Applied Physics*, Japan.

46 Zhang, H., Mao, R.S., Lee, N.C., and Yin, L. (2013) *TMS Annual Meeting*, San Antonio, TX.

47 Geng, J., Zhang, H., Mutuku, F., and Lee, N.C. (2018). *IEEE Power Electronics Magazine* 5 (3): 56.

48 Bultitude, J., Jones, L., McConnell, J. and Gurav, A. (2014) *Proceeding of International Conference on High Temperature Electronics*, Albuquerque, NM, 112.

49 Zhang, H., Lim, S.P., Geng, J. et al. (2018) *China Semiconductor Packaging Test Seminar*, HeFei, China.

50 Shen, Z., Johnson, R.W., and Hamilton, M.C. (2015). *IEEE Transactions on Electron Devices* 62: 346.

51 Shen, Z., Johnson, R.W., and Hamilton, M.C. (2014) *Proceeding of 2014 IMAPS High Temperature Electronics*, Albuquerque, NM.

52 Shen, Z., Fang, K., Hamilton, M.C. et al. (2013) *Proceeding of 2013 IMAPS High Temperature Electronics Networks*, Oxford, UK.

53 Scholes, H. (2014) *AREA Consortium Spring Meeting*, Binghamton, NY, March.

54 Yin, L. (2012) *AREA Consortium Meeting*, Binghamton, NY (March).

55 Kim, B., Lee, C.-W., Lee, D., and Kang, N. (2014). *Journal of Alloys and Compounds* 592: 207.

56 Zhang, H., Minter, J., and Lee, N.C. (2016) *Proceeding of 2016 IMAPS High Temperature Electronics*, Albuquerque, NM, USA.

57 Cui, J.Z., Johnson, R.W., and Hamilton, M. (2017). *IEEE Transactions on Components, Packaging and Manufacturing Technology* 7: 1598.

58 Lee, Y. and Link, D. (2016) *Proceeding of 2016 IMAPS 49th International Symposium on Microelectronics*, Pasadena, CA. USA.

59 Jiang, Q., Mukherjee, S., Dasgupta, A. et al. (2016) Mechanical constitutive properties of a Bi-rich high temperature solder alloy. *15th IEEE Intersociety Conference on Thermal and Thermomechanical Phenomena in Electronic Systems (ITherm)*, Las Vegas, NV, 1236–1239. doi: https://doi.org/10.1109/ITHERM.2016.7517688.

60 Liu, W.P. and Lee, N.C., (2013) U.S. Patent 8348139 B2.

61 Teo, J.W.R., Ng, F.L., Goi, L.S.K. et al. (2008). *Microelectronic Engineering* 85: 512.

62 Haque, A., Lim, B.H., Haseeb, A.S.N.A., and Masjuki, H.H. (2012). *Journal of Materials Science: Materials in Electronics* 23: 115.

63 Lalena, J.N., Dean, N.F., and Weiser, M.W. (2002). *Journal of Electronic Materials* 31: 1244.

64 Shi, Y., Fang, W., Xia, Z. et al. (2010). *Journal of Materials Science: Materials in Electronics* 21: 875.

65 Yamada, Y., Takaku, Y., Yagi, Y. et al. (2006) R&D Review of Toyota CRDL 43: 43.

66 Nakako, K. (2004) *Proceedings of 10th Symposium on Microjoining and Assembly Technology in Electronics*, Japan.

67 Zhang, H. and Lee, N.C., U.S. Patent 9017446 B2 (2015).

68 Zhang, H. and Lee, N.C., U.S. Patent 9636784 B2 (2017).

69 Zhang, H. and Lee, N.C., U.S. Patent 9802274 B2 (2017).

70 Zhang, H. and Lee, N.C., J.P. Patent 5938032 (2016).

71 Zhang, H. and Lee, N.C., C.N. 2011/8/0022297.6 (2011).

72 Zhang, H. and Lee, N.C. (2013). *Journal of Surface Mount Technology* 26: 28.

73 Zhang, H. and Lee, N.C. (2018) Korean Patent No. 10-1913994.

74 Zhang, H. and Lee, N.C. (2017) US Patent Apn 15/725265.

75 Zhang, H., Minter, J., Wu, J. and Lee, N.C. (2018) US Patent Apn 15/992102.

76 Zhang, H., Minter, J. and Lee, N.C. (2016) BiAgX® A drop-in HTLF solder and the reliability for 200°C applications. *Electronics Packaging Symposium, GE Research Center*, Albany, NY, USA.

77 Zhang, H. and Lee, N.C. (2012) *Proceeding of IMAPS 45th International Symposium on Microelectronics*, San Diego, CA.

78 Zhang, H. and Lee, N.C. (2014) *Proceeding of IMAPS High Temperature Electronics*, Albuquerque, NM.

79 Zhang, H. and Lee, N.C. (2012) *Proceeding of IMAPS High Temperature Electronics*, Albuquerque, NM.

80 Schoeller, H., Mallampati, S. and Cho, J. (2015) *Proceedings of SMTAI*, Chicago, IL.

81 Motoyasu, G., Kadowaki, H., Soda, H., and McLean, A. (1999). *Journal of Materials Science* 34: 3893.

82 Yin, L., Wentlent, L., Yang, L. et al. (2011). Recrystallization and precipitate coarsening in Lead-free solder joints during thermomechanical fatigue. *Journal of Electronic Materials* 41 (2): 241–252.

83 Rettenmayr, M., Lambracht, P., Kempf, B., and Tschudin, C. (2002). *Journal of Electronic Materials* 31: 278.

84 Yamaguchi, T., Ikeda, O., Oda, Y. et al. (2015). *Journal of Electronic Materials* 44: 751.

85 Takaku, Y., Felicia, L., Ohnuma, L. et al. (2008). *Journal of Electronic Materials* 37: 314.

86 Kang, S.J.L. (2005). *Sintering: Densification, Grain Growth and Microstructure*. New York: Elsevier.

7

Lead (Pb)-Free Solders for High Reliability and High-Performance Applications

Richard J. Coyle

Nokia Bell Laboratories, Murray Hill, NJ, USA

7.1 Evolution of Commercial Lead (Pb)-Free Solder Alloys

7.1.1 First Generation Commercial Pb-Free Solders

The European Union RoHS Directive (Restriction on the use of hazardous substances in electrical and electronic equipment) that drove the industry conversion to lead (Pb)-free manufacturing was implemented in 2006 [1]. However, before implementation became required by law or by market pressure, multiple national and international research projects were formed between 1991 and 2006 in the United States, the European Union, and Japan to examine Pb-free alternatives to eutectic tin-lead (63Sn37Pb) solder [2]. Among the most-referenced studies are the NCMS (US), IDEALS (Europe), DTI summary report (UK), JEITA and NEDO (Japan), and iNEMI (international) [3–8]. Handwerker et al. [2] provide a summary of the results from these studies, albeit some of the findings are dated based on the current understanding of solder alloying effects and advances in microstructural analysis.

The NCMS project [8] was one of the earliest of these investigations but considered one of the most influential because it set the industry direction for development of Pb-free solder alloys. The NCMS alloy screening methodology considered toxicology, economics, manufacturing, and reliability. Because the team consisted primarily of high reliability end users, thermomechanical or thermal fatigue reliability was emphasized. Thermal fatigue requirements always have been a priority for the products of many high reliability end users [9]. Solder joints age and degrade during service and eventually fail by the common wear-out mechanism

Lead-free Soldering Process Development and Reliability, First Edition. Edited by Jasbir Bath.

Table 7.1 Composition and melting range of high-Ag, near eutectic first generation solder alloys.

Alloy	Nominal composition (wt%)			Melting range (°C)
	Sn	Ag	Cu	
SAC387	95.5	3.8	0.7	217
SAC405	95.5	4.0	0.5	217–225
SAC305	96.5	3.0	0.5	217–221
SAC396	95.5	3.9	0.6	217–221
SnAg	96.5	3.5	0.0	221

of thermally activated solder fatigue (creep fatigue) [10]. Solder fatigue is the major wear-out failure mode and major source of failure for surface mount technology (SMT) components in electronic assemblies [11].

In work following the NCMS study, the high-Ag alloys given consideration were the ternary eutectic (Sn3.8Ag0.7Cu), hypereutectic (Sn4.0Ag0.5Cu), hypoeutectic (Sn3.0Ag0.5Cu), hypereutectic NEMI (Sn3.9Ag0.6Cu), and binary eutectic (Sn3.5Ag) as shown in Table 7.1. This high-Ag family of SnAgCu alloys are known as SAC alloys and now are referred to as first generation commercial Pb-free alloys.

The implementation of overmolded ball grid array (BGA) technologies coincided approximately with the start of the NCMS study, but these early BGA packages could not withstand elevated temperature Pb-free soldering, thus were not included in that study. Later studies, however, confirmed the beneficial effect of high Ag content on thermal fatigue reliability. Results from thermal cycling tests of flip chip and BGA packages as a function of Ag content are shown in Figures 7.1–7.3.

A multi-year program conducted by the IPC Solder Product Value Council (SPVC) lead eventually to the universal adoption of SAC305 (Sn3.0Ag0.5Cu) in manufacturing. The comprehensive SPVC program included differential scanning calorimetry (DSC) melt analysis, solderability testing, visual and X-ray inspection of solder joint voids, thermal cycling and thermal shock tests for reliability analysis, and metallurgical analysis of time zero and failed solder joints [15].

7.1.2 Second Generation Commercial Pb-Free Solders

Since the implementation of the European Union RoHS Directive and introduction of those first generation commercial Pb-free alloys, there have been significant innovations in Pb-free solder alloy formulations. The relatively short transition period to Pb-free solders has not allowed sufficient time for systematic

Figure 7.1 Thermal cycling test data from Terashima et al. showing the direct relationship between Ag content and thermal fatigue lifetime. The test vehicle was flip chip (first level) interconnects [12]. Copyright 2003 by Springer Nature. Used with permission.

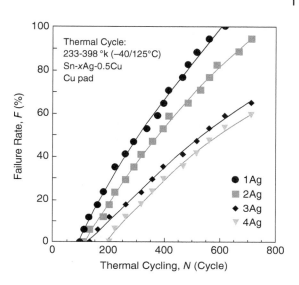

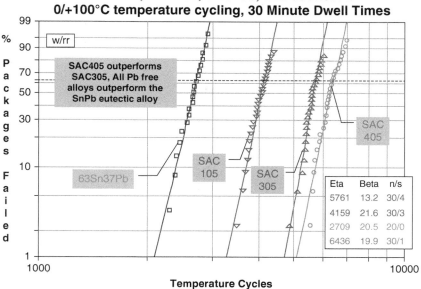

Figure 7.2 Thermal cycling test data from Coyle et al. showing the direct relationship between Ag content and thermal fatigue lifetime. The test vehicle was a 35 mm body ball grid array [13].

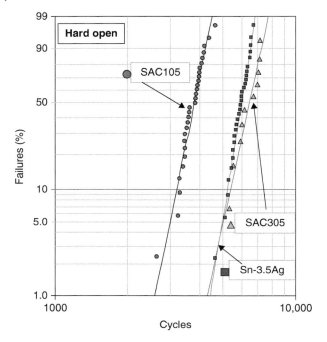

Figure 7.3 Thermal cycling test data from Henshall et al. showing the direct relationship between Ag content and thermal fatigue lifetime. The test vehicle was a ball grid array [14]. Figure used with permission of IPC International, Inc.

alloy design and comprehensive testing [16]. Consequently, alloy development has been driven primarily by experience gathered through volume manufacturing and increased deployment of a variety of Pb-free products of increasing complexity. This experience has resulted in an increased number of Pb-free solder alloy choices beyond the first generation near-eutectic SAC alloys that were established initially as replacements for eutectic SnPb [17]. Second generation, lower Ag alloys have been developed and introduced to address the shortcomings of the first generation near-eutectic SAC alloys, such as poor mechanical shock performance, higher cost, and copper erosion and plated-through-hole (PTH) damage during wave soldering and repair and rework operations [9, 17].

Failures due to mechanical loading in high-volume handheld consumer products provided a significant motivation for development of second generation solder alloys. Second generation commercial alloys typically are characterized by lower Ag content and microalloy additions to improve mechanical properties of the bulk solder or the intermetallic interfaces [17]. Figure 7.4 illustrates the improved mechanical drop test performance of lower Ag alloys compared to the

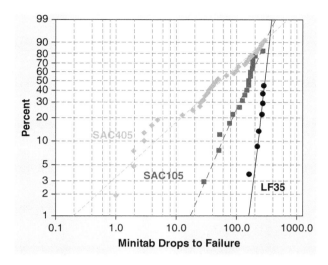

Figure 7.4 Data of Kim et al. showing cumulative failures versus number of drops to failure for high silver Sn4.0Ag0.5Cu (SAC405), low silver Sn1.0Ag0.5Cu (SAC105) and microalloyed low silver Sn1.2Ag0.5Cu+Ni(LF35) [18].

Figure 7.5 Data of Syed et al. showing improved 1st failure drop test performance of a 0.8 mm pitch ball grid array (BGA) with lower Ag content and microalloy additions [19].

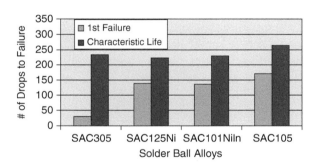

higher Ag SAC405 [18], and Figure 7.5 shows that the first failure in drop testing is delayed significantly in alloys with lower Ag content [19].

The advent of second generation commercial Pb-free alloys stimulated research and development efforts to fill the gap in knowledge associated with thermal fatigue resistance of first- and second generation alloys. Much of this work has been sponsored by industrial consortia such as the International Electronics Manufacturing Initiative (iNEMI), High Density Package User Group (HDPUG), the Center for Advanced Life Cycle Engineering (CALCE) at the University of Maryland, Universal Advanced Research in Electronics Assembly and the Center for Advanced Vehicle and Extreme Environment Electronics (CAVE[3]) at Auburn University [20–24]. The iNEMI Alloy Alternatives Characterization Program has been particularly active in publishing results from thermal fatigue studies of these two initial generations of Pb-free solder alloys [9, 17, 25–38].

7.1.3 Third Generation Commercial Pb-Free Solders

Lead-free solder alloy development is continuing to evolve with the emergence of third generation commercial alloys designed to provide additional alternatives as Pb-free manufacturing becomes pervasive, designs continue to evolve in complexity, and operating environments become increasingly more aggressive. The impetus for development of third generation Pb-free solders has been the dramatic increase in electronic content in automobiles. Automotive electronic assemblies must perform in environments characterized by increasing temperatures, thermal and power cycling, vibration, and thermal and mechanical shock. Automotive electronics no longer can be characterized simply as "under the hood."

Many automotive control modules, sensors, and components are mounted in areas that experience high operating temperatures, and rapid thermal and power cycling, in combination with vibration and shock [39]. Figure 7.6 illustrates locations of sensors and control modules and the anticipated range of aggressive thermal exposures [40, 41]. Software and electronics are now considered core competencies of automotive manufacturing, and this is driving innovation and an increase in electronic content in automotive. Figure 7.7 offers two perspectives on the cost of automotive electronics relative to total car cost with Figure 7.7a showing various areas of innovation driving the increase in electronic content. The projected growth of electric vehicles shown in Figure 7.8 is expected to accelerate the growth of electronic content. Furthermore, Figure 7.9 shows the compound annual growth rate (CAGR) for automotive electronics is twice that of the consumer electronics market [42]. The prevailing opinion is that common SAC alloys cannot satisfy the reliability requirements for these applications and use environments.

7.2 Third Generation Alloy Research and Development

7.2.1 Limitations of Sn-Ag-Cu Solder Alloys

Solder alloys based on the SnAgCu system (SAC) are more resistant to thermal fatigue than the eutectic SnPb alloy, but they have reliability limitations at higher operating temperatures [43] such as those illustrated in Figure 7.6. During solidification of SAC solders, the Ag and Sn react to form networks of Ag_3Sn precipitates at the primary Sn dendrite boundaries [44, 45]. These intermetallic precipitates are recognized as the primary strengthening mechanism in SAC

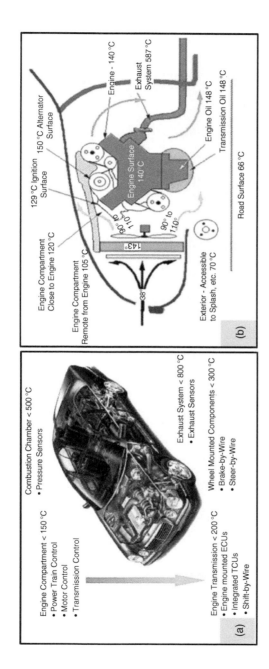

Figure 7.6 Illustrations of (a) sensor and electronic control module locations and anticipated thermal exposures [40], and (b) an engine compartment thermal profile [41].

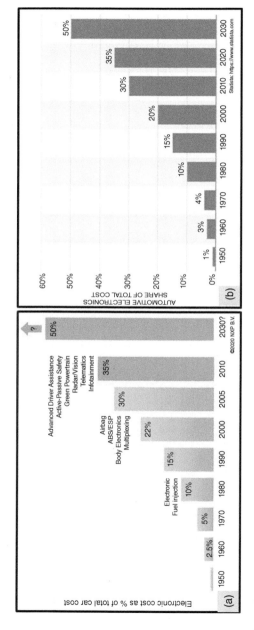

Figure 7.7 (a) Areas of innovation that are driving the increase in electronic content Source: Courtesy/Used with permission of EDN Network and NXP and, (b) the projected growth of automotive electronics and impact on cost per car. Source: Courtesy/Used with permission of Statista Inc.

Figure 7.8 Data projecting the growth of electric vehicle sales. Source: Courtesy Roskill Information Services.

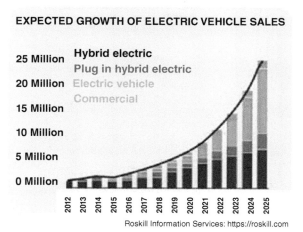

EXPECTED GROWTH OF ELECTRIC VEHICLE SALES

Hybrid electric
Plug in hybrid electric
Electric vehicle
Commercial

25 Million
20 Million
15 Million
10 Million
5 Million
0 Million

2012 2013 2014 2015 2016 2017 2018 2019 2020 2021 2022 2023 2024 2025

Roskill Information Services: https://roskill.com

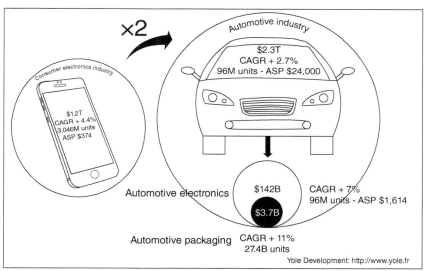

Figure 7.9 The need for advanced technologies is driving increased automotive electronic content with a compound annual growth rate (CAGR) of 7%, which is twice the CAGR for the consumer electronics market [42].

solders [12, 33, 44–46]. During thermal or power cycling and extended high temperature exposure, the Ag_3Sn precipitates coarsen and become less effective in inhibiting dislocation movement and slowing damage accumulation. Eventually, this leads to recrystallization of Sn grains in areas of stress concentration during thermal aging or thermal cycling [47–50]. This pattern of microstructural evolution is characteristic of the thermal fatigue failure process in these Sn-based

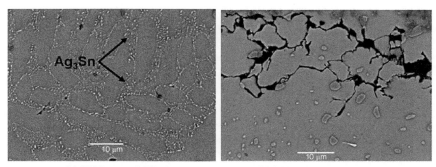

SAC305 as solidified   **SAC305 after thermal cycling**

Figure 7.10 Scanning electron micrographs illustrating Ag$_3$Sn intermetallic precipitate coarsening that precedes recrystallization and fatigue crack propagation during thermal cycling of SAC305 [39].

Pb-free alloys and was described originally in detail by Dunford et al. [51]. Figure 7.10 shows scanning electron micrographs illustrating coarsening of the Ag$_3$Sn precipitates in SAC305 solder caused by thermal cycling. Figure 7.11 illustrates the combined effects of strain and temperature on precipitate coarsening and fatigue damage during thermal cycling.

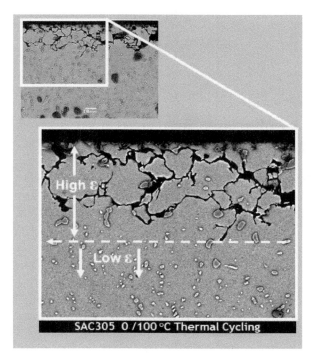

Figure 7.11 A scanning electron micrograph showing the accelerated intermetallic precipitate coarsening in the strain-localized region of a SAC305 BGA sample after 0/100 °C thermal cycling. The combination of strain (ε) and temperature in this region promotes recrystallization and fatigue crack propagation. In the absence of higher strain, coarsening is much slower as evidenced by the smaller particles and higher particle density in the region adjacent to (below) the crack [35].

Figure 7.12 Data from the iNEMI 3rd Generation Alloy test program demonstrating the dramatic reduction in characteristic lifetime of a SAC305 BGA as the thermal cycling profiles become more aggressive [52].

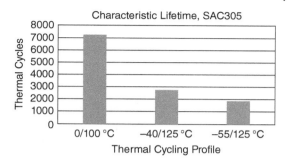

Figure 7.12 is a bar chart comparing the thermal cycling performance of SAC305 BGA tested with 0/100 °C (TC1), −40/125 °C (TC3), and −55/125 °C (TC4) thermal profiles as defined by the IPC-9701 attachment reliability guidelines [53]. The chart illustrates the dramatic reduction of 60–70% in characteristic lifetime of the SAC305 solder when tested with the more aggressive thermal cycling profiles. The TC3 and TC4 test profiles are required to qualify products that will be deployed into automotive and military/defense, and avionic use environments.

While the economic motivation for third generation Pb-free alloy development clearly is coming from the automotive sector, high reliability electronic challenges exist in medical, military/defense, and avionics. Thermal and mechanical requirements in aerospace/defense applications are similar or more demanding than automotive, and product lifetimes typically are greater. The aerospace/defense industry has increasing pressure from its supply chain as they continue in their effort to maintain traditional SnPb components and manufacturing. Figure 7.13 shows the small market share for electronics in this sector, which provides minimal leverage in cost or availability of components and hence, the latest

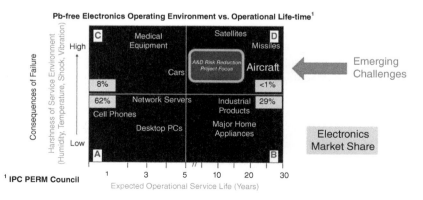

Figure 7.13 Aerospace/defense applications are characterized by harsh use environments and long operational service lifetimes. The aerospace/defense sector has minimal influence on the global supply chain due to its small market share.

technologies [54]. Because of concerns regarding the impact of technology limitations on designs, the aerospace/defense sector, in conjunction with the IPC PERM Council has begun to explore steps needed for Pb-free conversion in that sector [55].

7.2.2 Emergence of Commercial Third Generation Alloys

7.2.2.1 The Genesis of 3rd Generation Alloy Development

Early in the twenty-first century, as it appeared certain that the European Union RoHS Directive [1] would proceed to implementation, a task group formed of forward-looking solder suppliers, end users, and academic researchers to develop a commercial Pb-free alloy that could meet the performance challenges of higher temperature automotive applications [56–61] The investigation that ensued was known initially as the hotEL (hot ELectronics) Project and more commonly as the Innolot Project. After several years of metallurgical experimentation and mechanical and thermal testing, an alloy was developed based on the SAC387 (Sn3.8Ag0.7Cu) nominal ternary eutectic composition, with alloying additions of 3.0 wt% bismuth (Bi) and 1.5 wt% antimony (Sb) substituted for Sn, along with a microalloy addition of 0.15 wt% nickel (Ni). This complex alloy was identified by the project code name 90iSC and idiomatically as the *six-part alloy* since its composition consisted of tin (Sn) plus five alloying elements. The metallurgical concept of substituting elements such as bismuth (Bi), antimony (Sb), and indium (In) for Sn had been explored extensively in the original NCMS study [3], but this alloy development is regarded as the first commercial, third generation Pb-free alloy [60], with a European patent issued [62].

7.2.2.2 An Expanding Class of 3rd Generation Alloys

Third generation high reliability solder alloys may contain significant major alloy and microalloy additions to promote better high temperature performance. Most of these alloys are based on the Sn-Ag-Cu (SAC) system, although a few alloys are based on the Sn-Cu system and do not contain Ag. Harsh environment applications have various requirements for resistance to damage from high strain rate shock and vibration loading, in addition to the requirement for superior resistance to thermal fatigue damage [63]. The need for higher overall reliability performance has driven the development of an expanding class or group of commercial third generation, high reliability solder alloys [39, 52].

This new group of solder alloys is so different than conventional solder alloys that it can be considered an emergent class of specialty or special-purpose materials. These alloys are proprietary formulations and typically trademarked, patent-protected, or held as a trade secret. A list of the names and compositions of many of these commercial alloys is shown in Table 7.2. A notable metallurgical

Table 7.2 A list of the trade names, alloy developers, and chemical compositions of third generation, commercial high reliability solder alloys.

Trade names and nominal composition (wt%) of high reliability solder alloys								
Alloy	Developer	Sn	Ag	Cu	Bi	Sb	In	Other
405Y	Inventec	95.5	4.0	0.5				0.05 Ni; Zn
Cyclomax (SAC-Q)	Accurus	92.8	3.4	0.5	3.3			
Ecalloy	Accurus	97.3		0.7	2.0			0.05 Ni
HT1	Heraeus	95.0	2.5	0.5			2.0	Nd
Indalloy 272	Indium	90.0	3.8	1.2	1.5	3.5		
Indalloy 277	Indium	S9.0	3.8	0.7	0.5	3.5	2.5	
Indalloy 279	Indium	89.3	3.8	0.9		5.5	0.5	
Innolot	Heraeus	91.3	3.8	0.7	3.0	1.5		0.12 Ni
LF-C2	Nihon	92.5	3.5	1.0	3.0			
M794	Senju	89.7	3.4	0.7	3.2	3.0		Ni
M758	Senju	93.2	3.0	0.8	3.0			Ni
MaxRel plus	Alpha	91.9	4.0	0.6	3.5			
PS48BR[a]	Harima	Bal.	3.2	0.5	4.0	3.5		Ni, Co
REL22[a]	AIM	Bal.	3.0	0.7	3.0	0.6		0.05Ni; other
REL61[a]	AIM	Bal.	0.6	0.7	2.0			
SB6NX	Koki	89.2	3.5	0.8	0.5		6.0	
SN100CV	Nihon	97.8		0.7	1.5			0.05Ni
SN100CW1	Nihon	95.8		0.7	1.5	2.0		
Violet	Indium	91.25	2.25	0.5	6.0			
Viromet 347	Asahi	88.4	4.1	0.5			7.0	
Viromet 349	Asahi	91.4	4.1	0.5			4.0	

a) Nominal values; actual composition proprietary.

parallel in alloy development is found with the class of special-purpose materials that includes specialty stainless steels and heat-resistant alloys (superalloys) [64].

7.2.3 Metallurgical Considerations

The addition of Ag strengthens Sn and improves the creep resistance of the SAC solder by precipitation hardening as illustrated in Figure 7.14. The addition of other alloying elements can improve the creep resistance of the solder by means of two other well-known metallurgical strengthening mechanisms, solid solution hardening and dispersion hardening. The introduction of solute atoms into solid

Small Solute atom   Large Solute atom

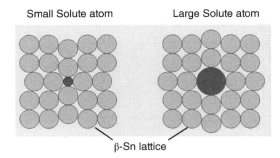

β-Sn lattice

Figure 7.14 A simple schematic illustrating lattice distortion due to substitutional solute atoms.

Solid Solution   Dispersion
Strengthening   Strengthening

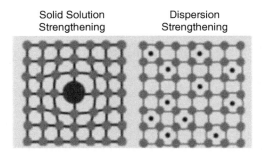

Figure 7.15 A simple schematic comparing solid solution (left) and dispersion strengthening (right).

solution of a solvent-atom lattice invariably produces an alloy that is stronger than the pure metal [65]. Figures 7.14 and 7.15 show a simplified schematic illustration of substitutional solid solution strengthening. Substitutional or interstitial solute atoms strain the lattice and dislocation movement or deformation is inhibited by interaction between dislocations and solute atoms incorporated into the β-Sn lattice.

Even when solute atoms precipitate from solution during thermal excursions in service, the solder alloy may strengthen due to subsequent dispersion hardening. Dispersion strengthening occurs when insoluble particles are finely dispersed in a metal matrix. Typical dispersion-strengthened alloys employ an insoluble, incoherent second phase that is thermally stable over a large temperature range (Figure 7.15) [66]. For Sn-based SAC solder alloys, the strength could be derived from a combination of increased solid solution strengthening at higher temperatures due to increased solubility, and dispersion strengthening that would supplement the solid solution effect at lower temperatures where solubility has decreased.

The commercial development of the Sn3.8Ag0.7Cu3.0Bi1.5Sb0.15Ni (Innolot) alloy provides evidence that substitutional solid solution strengthening can improve resistance to creep and fatigue at higher temperatures in Sn-based, Pb-free solders. Figure 7.16 shows that this alloy outperforms a SnAgCu (SAC) alloy in a combined thermal cycling and shear test [61]. The current working

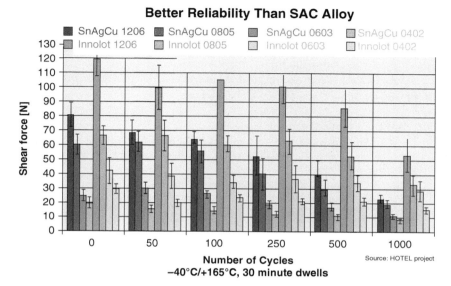

Figure 7.16 Data of Miric showing Sn3.8Ag0.7Cu3.0Bi1.5Sb0.15Ni (Innolot) outperforms SnAgCu (SAC) in a combination thermal cycling/shear test [61]. The test vehicles were four different size chip resistors.

hypothesis is that solid solution and dispersion strengthening not only can supplement the Ag_3Sn precipitate hardening found in SAC solders, but continue to be effective once precipitate coarsening reduces the effectiveness of the intermetallic Ag_3Sn precipitates [67].

The elements proposed most commonly for improving high temperature properties in third generation high reliability solders are bismuth (Bi) and antimony (Sb). The element indium (In) is used to a lesser extent, at least partly due to its high cost. The elements Bi and In, when used as major alloying elements (not as microalloying), also reduce the melting point of most solder alloy formulations, while the addition of Sb tends to increase the melting point [61, 68]. These modified SAC alloys have off-eutectic compositions, are characterized by non-equilibrium solidification, and often have significant melting or so-called pasty ranges [2, 3, 69, 70].

Although many third generation Pb-free solders have been commercialized, the concept of using major element alloying to improve mechanical properties or to alter melting behavior is not novel. Formulations incorporating Bi, Sb, and In into basic Sn-Ag or Sn-Ag-Cu eutectics were studied by the NCMS consortium of industrial partners in 1997 [3] and the properties were documented by NIST and the Colorado School of Mines beginning in 2002 [69]. The NCMS study was considered a comprehensive study at the time, but the thermal fatigue aspect of the work ultimately was limited because the study predated the widespread introduction of area array technology. A more recent, general discussion of the effects of alloying on solidification, melting behavior, and properties can be found in reference [2]. A review of the effects of alloying Sb, In, and Bi with Sn and Sn-based solders, mostly based on observations from the respective binary phase diagrams, is provided in the following sections.

7.2.3.1 Antimony (Sb) Additions to Tin (Sn)

The binary Sn-Sb phase diagram in Figure 7.17 shows solubility of Sb in Sn of approximately 0.5 wt% at room temperature to 1.5 wt% at 125 °C [71, 72]. Thus,

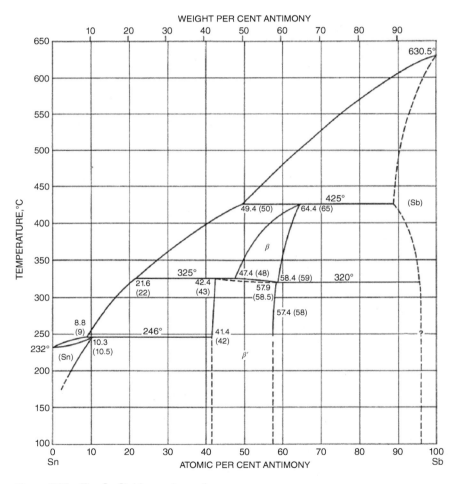

Figure 7.17 The Sn-Sb binary phase diagram.

some contribution might be expected from solid solution strengthening due to Sb dissolved in Sn-based Pb-free solders.

Alloying with Sb may improve performance through other strengthening mechanisms. Studies by Li et al. [73, 74] and Belyakov et al. [75] show that Sb slows the growth rate of Cu_6Sn_5 intermetallic compound (IMC) layers at attachment interfaces. Fast interfacial IMC growth on Cu surfaces tends to produce irregular and non-uniform IMC layers. This can lead to reduced mechanical reliability by inducing fractures at IMC interfaces or through the IMC in drop/shock loading [76].

Figure 7.17 also shows that Sb has the potential to form multiple different intermediate phases or IMCs with Sn (Sb_2Sn_3, SbSn, Sb_4Sn_3, Sb_5Sn_4, and $SbSn_2$) in the bulk solder [71]. Lu et al. [77] and El-Daly et al. [78] identified SbSn intermediate phase precipitates <5 μm in size and distributed throughout the Sn dendrites. Beyer et al. showed that Sn5Sb and Sn8Sb alloys have increased shear strength and ductility compared to conventional SAC solders and maintain their shear strength with good ductility after isothermal aging [79]. El-Daly et al. suggested that alloying Sn with Sb can improve creep performance and tensile strength [80]. In this case, the SbSn precipitates form within the Sn dendrites, unlike the well-known SAC Ag_3Sn mechanism, where the precipitates form at the Sn dendrite boundaries. Presumably, the SbSn precipitates work to resist recrystallization by strengthening the Sn dendrites [65].

7.2.3.2   Indium (In) Additions to Tin (Sn)

The binary In-Sn phase diagram is shown in Figure 7.18. While there is some disagreement over the solid solubility of In in Sn, a reasonable estimate is ∼7 wt% at room temperature and as much as 12 wt% at 125 °C [71]. Because of its range of solubility in Sn, In has been explored as a solid solution strengthening agent in Sn-based, Pb-free solders [68, 81]. The equilibrium diagram shows that indium forms two intermediate phases (β and γ) of variable composition with Sn [71], but does not appear to form any true stoichiometric compounds with Sn.

Results from multiple solder alloy studies indicate that In additions can improve drop and shock resistance by slowing the growth of interfacial IMC layers. Yu et al. report improved drop [82] and thermal shock [83] performance by adding as little as 0.4% indium, and Amagai et al. report improved drop performance at or below 0.5% indium [84]. Hodúlová et al. report that indium slows growth of Cu_3Sn and that the hybrid IMC phase $Cu_6(Sn, In)_5$ forms [85]. Sharif et al. also observed the formation of $Cu_6(Sn, In)_5$ as well as formation of $(Cu, Ni)_3(Sn, In)_4$ on Ni substrates with Sn-Ag-Cu solder alloyed with In [86]. Those IMCs were found in the bulk as well as the soldered interfaces. In these hybrid IMCs, In substitutes for Sn which fundamentally is different than the common modified IMCs $(Cu, Ni)_6Sn_5$ or the $(Ni, Cu)_3Sn_4$ where Cu and Ni exchange.

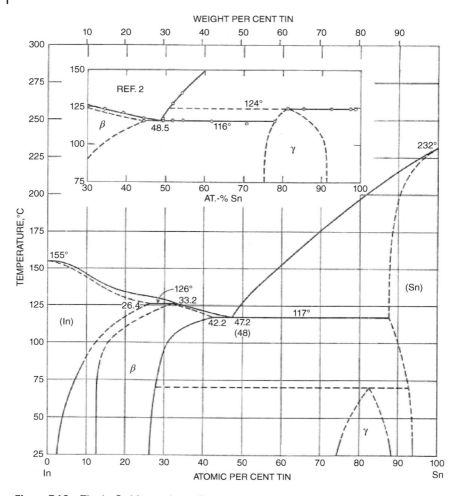

Figure 7.18 The In-Sn binary phase diagram.

Other reactions can occur when indium is added to SAC-based solders, and this complicates the ability to understand the effect of indium content on solder joint reliability. In a study by Chantaramanee et al. additions of 0.5% In and Sb in combination with indium was found to promote formation of $Ag_3(Sn, In)$ and SbSn [87]. They reported that small precipitates reduced the Sn dendrite size by 28%, but they were unable to determine the relative influence of indium versus Sb on this reaction. With alloys containing indium of the order of 10%, Sopoušek et al. found that some of the Ag_3Sn transforms to $Ag_2(Sn,In)$ and Ag_2Sn [88]. These observations are consistent with the Ag-In binary phase diagram that shows Ag_3In, Ag_2In, and

AgIn$_2$ [71]. Wang et al. reported that an addition of 1% indium to Sn-Ag-Cu solder resulted in larger (coarser) Ag$_3$Sn precipitates [89]. This is a very interesting observation, since larger or coarser Ag$_3$Sn precipitate at time zero could shorten the solder joint lifetime in thermal cycling. In principle there is a large solid solubility of indium in Sn, but the effective indium content in a SAC-based solder may be diminished by interactions with other elements to form multiple phases.

It is noteworthy that many of the studies were conducted using laboratory bulk solder samples with microstructures that may be atypical of microelectronic solder joints. Some studies include more than one significant alloy addition [87], which makes it difficult to isolate effects due to individual alloying elements. The work by Wada et al. [68, 81], while it includes tensile testing with relatively large, bulk samples, also includes thermal cycling and drop testing with surface mount components. Their microstructural analysis included X-ray diffraction and they found InSn$_4$, In$_4$Ag$_9$, Ag$_3$(Sn,In), and possibly αSn in addition to βSn. Wada et al. concluded that the optimum ductility and reliability was achieved with an indium content of 6 wt%.

7.2.3.3 Bismuth (Bi) Additions to Tin (Sn)

The binary Sn-Bi phase diagram is shown in Figure 7.19. The solubility of Bi in Sn is approximately 1.5 wt% at room temperature and increases to almost 7 wt% at 100 °C, and as much as 15 wt% at 125 °C [71]. There is virtually no solubility of Sn in Bi, and no intermediate phases or IMCs found in the Sn-Bi system.

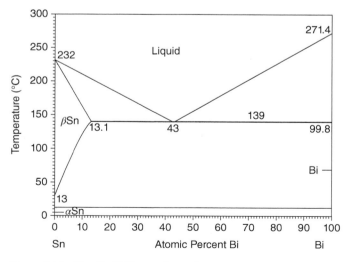

Figure 7.19 The Sn-Bi binary phase diagram.

Multiple studies have shown that Bi improves the mechanical properties of Sn and SAC solders [43, 61, 67, 90–100]. Vianco [90, 91] and Witkin [94, 97, 98] have done extensive mechanical testing and microstructural analysis and discuss the dual strengthening mechanisms of Bi in solid solution and Bi precipitated within Sn dendrites and at Sn boundaries. Recently, Delhaise et al. [99] reported results from their study of the effects of thermal preconditioning (aging) on microstructure and property improvement in an alloy containing 6 wt% Bi (see Table 7.2, Violet). They suggest that strain from Bi precipitation induces recrystallization and an increase in the amount of Sn grain boundaries which in turn, are pinned by the Bi precipitates at those boundaries. These microstructural features work in conjunction with Bi in solid solution to resist creep deformation.

The results from the fundamental studies by Vianco [90, 91] and Witkin [94, 97, 98] leave no doubt that Bi additions can have a positive effect on the physical properties of Sn and Sn-based solder alloys. However, those studies used cast, bulk alloy samples and it is debatable if those results can be scaled effectively to smaller, microelectronic solder joints. Nishimura et al. for example, recommend a maximum Bi content of only 1.5 wt% because of the uncertainty that the alloying effect will be sustained as the microstructure evolves in response to the thermal cycling in normal service [96]. Delhaise has shown that the Bi distribution and microstructure depend on solidification conditions and subsequent thermal exposure, which ultimately determine the relative contributions of Bi to solid solution and dispersion strengthening (Figure 7.20). Furthermore, it is possible that adding enough Bi to take advantage of the Bi solubility limit at higher temperatures may have a negative effect because Bi does not always precipitate homogeneously. Clustering of Bi is known to occur [99] and in the extreme case, stratification or segregation may induce brittle behavior [101, 102].

7.3 Reliability Testing Third Generation Commercial Pb-Free Solders

7.3.1 Thermal Fatigue Evaluations

The reliability challenges for third generation Pb-free solder alloys are similar to, but even more demanding than, those encountered with the earlier proliferation of second generation alloys [17]. Second generation alloys fundamentally are Sn-Ag-Cu (SAC) alloys, whereas the third generation alloys while SAC-based, are modified heavily with additional alloying elements. The potential for multiple temperature dependent strengthening mechanisms in the newer alloys increases the complexity of microstructural response to thermal and mechanical service conditions (see Section 7.2.3).

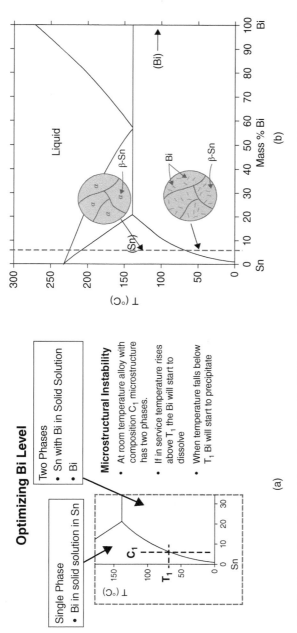

Figure 7.20 Emphasis on the Sn-rich regions of the Sn-Bi binary phase diagram showing: (a) Factors to consider when optimizing the Bi level [101], and (b) Schematic microstructures shown for solid solution (upper) and dispersion strengthening (lower) with a 6 wt% Bi alloy [99].

Second generation alloy development was driven by experience gathered through volume manufacturing, increased deployment of a variety of Pb-free products of increasing complexity, and analysis of failures in service [17]. In that case, there was a clear need to address certain performance deficiencies of first generation solder alloys. Third generation alloys are being developed to address emerging challenges in automotive applications or in the case of aerospace/defense applications, there is a *perception* that that Sn-based Pb-free alloys will struggle to perform as well as the traditional SnPb eutectic solder alloy. For safety-related and mission-critical requirements, there is no luxury of gathering information from a learning curve during product deployment. Consequently, there is an urgency among high reliability end users to begin characterizing and understanding the long-term attachment performance of these new solder alloys prior to widespread deployment.

Most of the test and evaluation data for third generation alloys has been collected through various combinations of thermal and mechanical tests on solder attachments or bulk property measurements. Furthermore, much of the performance data in the literature specifically is for the Sn3.8Ag0.7Cu3.0Bi1.5Sb0.15Ni (Innolot) alloy. This alloy was developed for high reliability applications and it was the first alloy of its type introduced commercially. Testing has been limited mostly to simple components such as ceramic capacitors or chip resistors. Trodler et al. [103] compared SAC305 to Sn3.8Ag0.7Cu3.0Bi1.5Sb0.15Ni (Innolot) and Sn2.5Ag0.5Cu2.0In+Nd (HT1), another highly alloyed solder. Trodler showed that both high reliability alloys outperformed SAC305 by testing solder joints of ceramic capacitors with thermal cycling preconditioning followed by shear to failure [103]. Miric also showed that Innolot outperformed SAC305 (Figure 7.16) using thermal cycling and shear [61], and Chada et al. [104] demonstrated that the alloy has higher creep resistance and improved reliability in −40 to +150 °C thermal cycling compared to Sn3.8Ag0.7Cu (SAC387). The thermal cycling data reported were for chip resistors, but the test did not include in situ monitoring to enable development of Weibull statistics [104]. Dudek et al. [105] also provided additional confirmation of superior performance compared to SAC305 using electrically functional test modules from production, but they did not have in situ resistance monitoring of the solder joints. Solder joint cracking was determined through X-ray computed tomography and metallographic analysis [105]. Barry reported superior tensile properties of Innolot compared to SAC as part of a high cycle fatigue study, but was unable to conclude definitively that the highly alloyed material outperformed eutectic SnPb in high cycle fatigue [106]. Results from the Barry study suggest that an increase in solder alloy strength does not guarantee an increase in high cycle fatigue performance.

Su et al. [107] used ball shear as a function of strain rate and mechanical shear fatigue and Chowdhury et al. [108] used uniaxial tensile tests over the range of room temperature to 200 °C to compare performance of three proprietary solder alloys: Sn3.8Ag0.7Cu3.0Bi1.5Sb0.15Ni (Innolot), Sn3.4Ag0.5Cu3.3Bi (Cyclomax or SAC-Q), and Sn0.9Cu2.5Bi (Ecolloy or SAC-R). Their results showed better fatigue reliability and superior high temperature tensile strength with higher Ag and Bi contents but results from this type of testing cannot be used to predict the thermally-driven microstructural evolution that occurs during temperature or power cycling. Snugovsky et al. [109] and Kosiba et al. [110] used thermal cycling to screen various alloys with high Bi content. Their testing demonstrated that adding Bi can improve thermal cycling performance of a basic SAC alloy. However, they were unable to quantify the effect of Bi content because the test did not run long enough to develop statistical reliability data.

The thermal cycling study by Hillman and Wilcoxon [111] was likely the first to use in situ monitoring to evaluate high reliability alloys and report statistical data for reliability comparisons. Although the study focused on comparisons of Pb-free and SnPb solders as well as mixed metallurgy joints (Pb-free components soldered with SnPb solder), the authors had the foresight to include a comparison of Sn3.9Ag0.6Cu (SAC396) to Sn3.0Ag0.5Cu3.4Bi (SAC+3.4 wt% Bi). In their experiments using the −55/125 °C (TC4) thermal cycle defined in IPC-9701 [53], the SAC+3.4%Bi alloy outperformed SAC396 with a ceramic leadless chip carrier (CLCC), thin small outline package (TSOP), and thin quad flat pack (TQFP) components [111].

Until very recently, the work by Sanders et al. was one of a very few investigations that used a BGA test vehicle fabricated with third generation high reliability solder spheres [112]. While the Sanders study included multiple components and test variables, the relevant findings are from the −40/125 °C (TC3) thermal cycling comparison of SAC305 and Innolot using a chip array BGA (CABGA208) test vehicle. In this test, the high reliability alloy outperformed SAC305 by roughly a factor of two based on Weibull two-parameter characteristic lifetimes.

7.3.2   iNEMI/HDPUG Third Generation Alloy Pb-Free Thermal Fatigue Project

Because there are very few published reports of thermal cycling data for third generation alloys, a thorough treatment and an understanding of thermal fatigue performance is a major knowledge gap for these new alloys. To address this knowledge gap, the high reliability end users and solder suppliers from the iNEMI Alloy Alternatives Characterization Project [9, 17] agreed to extend the project to encompass thermal fatigue evaluations of third generation high reliability solder alloys [39]. The Third Generation Alloy Evaluation project was launched jointly by iNEMI and the HDPUG, and includes active participation from both the CALCE (the Center for Advanced Life Cycle Engineering) [22],

Table 7.3 Nominal solder compositions and estimated melting ranges for the high reliability solder alloys included in the iNEMI/HDPUG Third Generation Alloy study [39].

Alloy	Nominal composition (wt%)							Melting range (°C)
	Sn	Ag	Cu	Bi	Sb	In	Other	
SAC305	96.5	3.0	0.5					217–221
Innolot	91.3	3.5	0.7	3.0	1.5		0.12 Ni	206–218
HT	95.0	2.5	0.5			2.0	Nd	206–218
MaxRel plus	91.9	4.0	0.6	3.5				212–220
M794	89.7	3.4	0.7	3.2	3.0		Ni	210–221
M758	93.2	3.0	0.8	3.0			Ni	205–215
SB6NX	89.2	3.5	0.8	0.5		6.0		202–206
Violet	91.25	2.25	0.5	6.0				205–215
Indalloy 272	90.0	3.8	1.2	1.5	3.5			216–226
Indalloy 277	89.0	3.8	0.7	0.5	3.5	2.5		214–223
Indalloy 279	89.3	3.8	0.9		5.5	0.5		221–228
LF-C2	92.5	3.5	1.0	3.0				208–213
SN100CV	97.8		0.7	1.5			0.05Ni	221–225
405Y	95.5	4.0	0.5				0.05 Ni; Zn	217–221

and the AREA (Universal Advanced Research in Electronic Assembly) consortia [23]. These consortia collectively are supported by members from high reliability markets including telecom, automotive, avionics, and aerospace/defense end users, as well as multiple solder suppliers, and contract electronic manufacturers.

The alloys included currently in this investigation are shown in Table 7.3. The test matrix contains SAC305 as the performance baseline alloy. The table shows that Bi is the alloying element used most frequently, which is consistent with the attention given to Bi in the Pb-free alloy literature [43, 61, 67, 90–96].

This project uses the basic experimental approach developed in the original iNEMI Alloy Alternatives study of second generation solders to enable development of thermal fatigue data for third generation Pb-free solders. The investigation utilizes the components and printed circuit board (PCB) developed as the test vehicle for the earlier study [17]. Figure 7.21 shows a populated test board and the two daisy-chained BGAs, a 192 I/O chip array BGA (192CABGA), and an 84 I/O thin core chip array (84CTBGA) [113–115]. The test boards for each alloy were built using the same alloy composition for the BGA spheres and the solder paste [39]. The thermal fatigue test plan incorporates several thermal cycling profiles used by high reliability end users in telecom, aerospace/defense and avionics, and

Figure 7.21 A fully populated, daisy-chained test vehicle and BGA components used in the Third Generation Alloy project for developing thermal fatigue data.

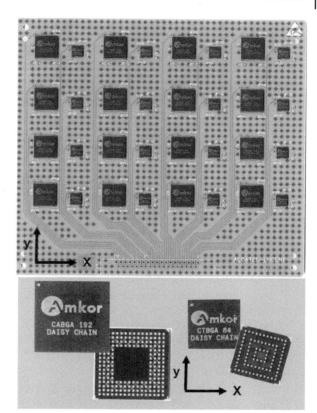

Table 7.4 Thermal cycling profiles used in the iNEMI Third Generation Alloy project: telecommunications (TC1), automotive and consumer (TC3), and avionics and aerospace/defense (TC4).

Thermal cycle	Minimum temp. (°C)	Maximum temp. (°C)	Temp. range ΔT (°C)	Dwell time (min)
TC1	0	100	100	10
TC3	−40	125	165	10
TC4	−55	125	180	10

automotive markets as shown in Table 7.4. The experimental matrix includes 165 populated test boards and nearly 5000 components for thermal cycling tests and time zero microstructural characterization.

The initial results from this thermal cycling study have been reported [113–115]. One of the comparisons made in these papers demonstrates the effect of adding high amounts of three different individual alloying elements, bismuth (Bi), antimony (Sb), and indium (In), on thermal cycling performance. The alloys include Violet (6 wt% Bi), SB6NX (6 wt% In), and Indalloy 279 (5.5 wt% Sb) from Table 7.3. Summaries of the thermal cycling data are shown in Table 7.5 and Figure 7.22. For the 192CABGA component, the high reliability alloys outperformed SAC305 in all thermal cycles with two notable exceptions. The performance of Violet was no better than SAC305 in 0/100 and −40/125 °C and the performance of Violet was substantially worse than SAC305 in −55/125 °C. The thermal cycling results for the 84CTBGA component were significantly different as illustrated in Figure 7.22. While all high reliability alloys outperformed SAC305 in all thermal cycles, the performance of Violet nominally was best in the group, which is in sharp contrast to the performance of that alloy when tested with the 192CABGA package.

The alloy containing 5.5 wt% Sb, performed consistently well with both components and all thermal cycling profiles. The alloy containing 6 wt% In, performed better than the alloy containing 6 wt% Bi with the larger 192CABGA component but that trend was reversed with the smaller 84CTBGA. However, those differences are not always substantial and may not be statistically significant because comparisons using these Weibull statistics have limitations due to variations in Weibull slopes (β) across the data sets.

Metallographic failure analysis of thermally cycled 192CABGA packages showed the failure mode to be predominantly fatigue in the bulk solder as shown in Figure 7.23. However, interfacial or mixed mode fracture was detected in the Violet (6 wt% Bi) and Indalloy 279 (5.5 wt% Sb) alloys as shown in Figure 7.24. No interfacial cracking was reported for the SB6NX alloy (6 wt% In). Because there was no evidence of interfacial cracking at time zero in these alloys, it is assumed that the cracking initiated during thermal cycling. Interfacial cracking was detected more often in samples from the aggressive −55/125 °C thermal cycling profile and in the Violet alloy, and some interfacial cracking was detected in Violet samples tested with the 0/100 °C profile. Considering the range of elemental additions in these alloys, no clear correlation has emerged between composition and interfacial fracture.

While it is not obvious if the interfacial cracking impacts the Weibull statistics for either alloy, the propensity for interfacial fracture and possible early failures could represent a reliability risk. More detailed failure analysis and Weibull statistics are provided in the publications [113–115].

Metallographic failure analysis of thermally cycled 84CTBGA packages confirmed fatigue cracking in the bulk solder. Figure 7.25 shows backscattered scanning electron images of examples of crack propagation in the four alloys. None of the 84CTBGA samples that were analyzed from the three thermal cycles contained the type of clear interfacial separation found in the 192CABGA images shown in Figures 7.23 and 7.24. The absence of interfacial cracking could account

Table 7.5 A summary of accelerated temperature cycling failure statistics for SAC305 and three high reliability solder alloys [113].

192CABGA Thermal Cycling Data

Alloy	Temperature Cycle	Characteristic Lifetime η cycles	Slope β	Correlation Coefficient ρ
SAC305 (Sn3Ag0.5Cu)	0/100 °C	5542	6.6	0.98
	-40/125 °C	1692	3.6	0.99
	-55/125 °C	1123	6.1	0.98
Violet (Sn2.25Ag0.5Cu6Bi)	0/100 °C	5607	5.0	0.95
	-40/125 °C	1557	5.4	0.99
	-55/125 °C	834	2.8	0.99
SB6NX (Sn3.5Ag0.8Cu0.5Bi6In)	0/100 °C	6183	6.0	0.98
	-40/125 °C	2042	3.6	0.96
	-55/125 °C	1290	2.9	0.96
Indalloy 279 (Sn3.8Ag0.9Cu5.5Sb0.5In)	0/100 °C	6917	6.8	0.98
	-40/125 °C	2751	7.1	0.96
	-55/125 °C	1765	7.2	0.96

84CTBGA Thermal Cycling Data

Alloy	Temperature Cycle	Characteristic Lifetime η cycles	Slope β	Correlation Coefficient ρ
SAC305 (Sn3Ag0.5Cu)	0/100 °C	7286	8.4	0.99
	-40/125 °C	2758	8.9	0.97
	-55/125 °C	1946	8.3	0.99
Indium Violet (Sn2.25Ag0.5Cu6Bi)	0/100 °C	20447	3.6	0.98
	-40/125 °C	5580	7.6	0.98
	-55/125 °C	3584	6.1	0.98
Koki SB6NX (Sn3.5Ag0.8Cu0.5Bi6In)	0/100 °C	9852	5.1	0.98
	-40/125 °C	3444	2.8	0.98
	-55/125 °C	2269	4.4	0.99
Indalloy 279 (Sn3.8Ag0.9Cu5.5Sb0.5In)	0/100 °C	14997	5.9	0.97
	-40/125 °C	5034	9.8	0.98
	-55/125 °C	3151	6.3	0.99

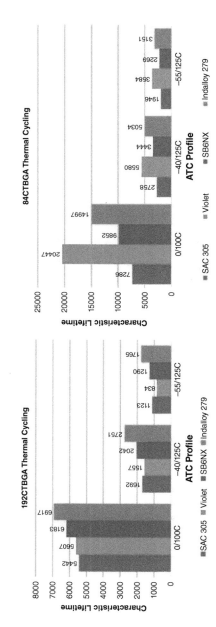

Figure 7.22 Bar charts comparing the characteristic lifetimes (N63) of the 192CABGA and 84CTBGA packages with SAC305, Violet, SB6NX, and Indalloy 279. Data are presented for three different thermal cycling profiles: 0/100, −40/125, and −55/125 °C thermal cycling profiles [113, 115].

Figure 7.23 Solder fatigue failures in the 192CABGA component with the SB6NX (6 wt% In), Violet (6 wt% Bi), and Indalloy 279 (5.5 wt% Sb) alloys [113].

for the distinctly better performance of the Violet alloy with the smaller 84CTBGA package (Figure 7.22).

7.3.3 Microstructure and Reliability of Third Generation Alloys

After more than a decade of experience, there is a relatively good understanding of the microstructure of SAC305 and its evolution during thermal cycling. The as-solidified SAC305 microstructure shown in Figure 7.10 consists of primary Sn dendrites with secondary Ag_3Sn precipitates (shown as the lighter phase) at the cell or dendrite boundaries. The large amount of undercooling required typically to nucleate the β–Sn phase tends to suppress the equilibrium ternary eutectic structure [116] even in near-eutectic alloys such as SAC305 [117]. The mechanism of thermal fatigue in SAC solders is well-documented and is illustrated in Figures 7.10 and 7.11.

High reliability solders typically contain significant additions of Ag (Table 7.3), so coarsening of Ag_3Sn precipitates and recrystallization is expected to occur. However, the potential for solid solution strengthening and formation of additional phases that can influence performance complicates the understanding and analysis of these alloys. Figure 7.26 shows the baseline (before thermal cycling) microstructures of three high reliability alloys to illustrate the effect of In, Bi, and Sb on microstructure.

The alloy containing 6 wt% In also contains 3.5 wt% Ag and its basic microstructure is similar to SAC305. There are primary Sn cells with secondary Ag_3Sn precipitates at the Sn dendrite boundaries. It is reasonable to assume that most of the In has dissolved into the β–Sn, but the morphology of some precipitates in the higher magnification image could indicate the presence of $Ag_3(Sn, In)$ [87] or In_4Ag_9 [81]. A more sophisticated analytical method such as electron backscatter diffraction (EBSD) is required to identify these phases.

The microstructure of the alloy containing 6 wt% Bi has less visible Ag_3Sn precipitates due to its lower Ag content (2.25 wt%) and because the image is dominated by numerous bright white Bi precipitates. Some Bi undoubtedly remains in

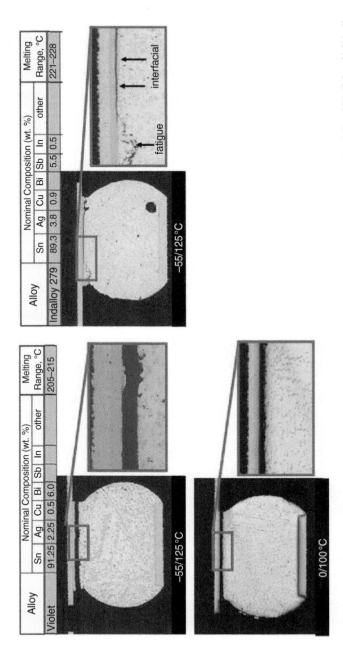

Figure 7.24 Interfacial or mixed mode fracture in the 192CABGA component with the Violet (6 wt% Bi) and Indalloy 279 (5.5 wt% Sb) alloys [113].

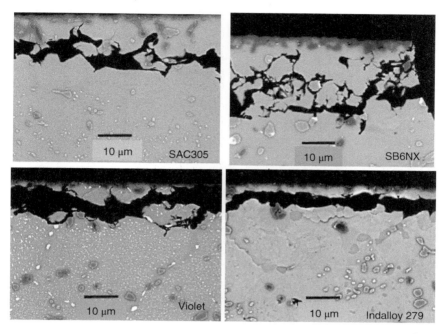

Figure 7.25 Backscattered electron micrographs showing examples of representative thermal fatigue damage in the 84CTBGA component with the SAC305, SB6NX (6 wt% In), Violet (6 wt% Bi), and Indalloy 279 (5.5 wt% Sb) alloys [115].

solid solution. Researchers have studied alloys with similar Bi content and have explored the stability of Bi in solution with Sn after thermal preconditioning [99], precipitation and morphology of Bi at room temperature [118], and effect on creep rate [119]. More work is needed to understand the reaction of Bi to various thermal cycling profiles and to understand the interaction between strain and temperature on reliability.

The alloy containing 5.5 wt% Sb has 3.8 wt% Ag, which is a higher Ag content than the alloys containing high Bi or In content and this may explain its higher density of Ag_3Sn precipitates. There is limited solubility of Sb in Sn, but no obvious evidence of precipitation of Sn-Sb intermetallic phases. The morphologies and contrast (density) differences of the precipitates in the alloy prior to thermal cycling suggest some precipitates are not simply Ag_3Sn. Based on reports from the literature, these precipitates most likely are SbSn [77, 78, 80], but the precipitates are extremely small and additional analysis is needed to confirm the composition.

Fundamental microstructural analysis of as-solidified microelectronic-sized solder joints is needed to assess solubility of alloying elements and phase identification before any in depth analysis can be attempted on failed solder joints from

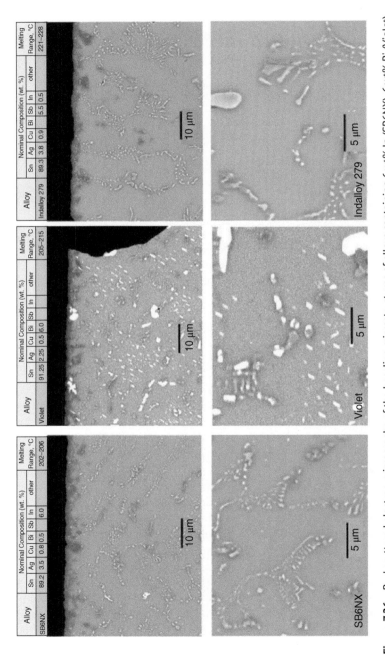

Figure 7.26 Backscattered electron micrographs of the baseline microstructures of alloys containing 6 wt% In (SB6NX), 6 wt% Bi (Violet), and 5.5 wt% Sb (Indalloy 279) [113].

thermal cycling tests. However, for the microstructural analysis of these highly alloyed, third generation solders, there are clear limitations with conventional analytical methods such as optical microscopy and scanning electron microscopy with energy dispersive X-ray analysis (SEM-EDX).

7.4 Reliability Gaps and Suggestions for Additional Work

New high reliability solders continue to be developed and introduced, but a fundamental understanding of alloy behavior and characterization of solder attachment reliability is lagging. The iNEMI/HDPUG Third Generation Alloy project was launched to begin the process of generating thermal fatigue data for high reliability solders and establishing an industry state of knowledge for performance and its relationship to alloy metallurgy and microstructure. After almost two years of thermal cycle testing some interesting findings are emerging, but the results currently are limited to only one of the BGA components and a fraction of the solder alloys in the test matrix [113, 114]. This is the first large-scale project in the industry to address thermal fatigue of these alloys, and projects of this scope and detail are needed to close knowledge gaps. Even though the early results are incomplete, preliminary data analysis has uncovered additional gaps and suggestions are being made for further investigations. The following sections discuss reliability gaps for third generation Pb-free solders and suggest additional investigations to close those gaps.

7.4.1 Root Cause of Interfacial Fractures

The observation of non-solder fatigue failures shown in Figure 7.24 requires further destructive analysis to identify failure modes. Beyond complicating interpretation of the thermal cycling test data, non-fatigue failure modes could introduce a reliability risk in practice. Interfacial cracking was not anticipated, and it adds another significant parameter to the evaluation of attachment reliability. It is important to develop a better understanding of the interfacial cracking phenomenon in terms of root cause and the impact on reliability. Interfacial accumulation of Bi can occur and in the extreme case, may induce brittle behavior by the process known as bismuth stratification [101, 102]. However, Bi cannot be the sole cause of the interfacial cracking because some interfacial cracking was detected in an alloy containing no Bi (Figure 7.24). Therefore, a careful microscopic analysis should be performed at soldered interfaces to identify segregation of Bi or other species that could initiate cracking.

7.4.2   Effect of Component Attributes on Thermal Fatigue

The iNEMI/HDPUG project includes numerous alloys, several thermal cycling profiles, and large sample sizes, but the investigation is restricted to two relatively small BGA components. Solder fatigue is driven by strain as well as temperature and is dependent strongly on the coefficient of thermal expansion (CTE) mismatch (difference) between the component and the PCB as well as the distance from neutral point (DNP) [120]. The BGA packages in the iNEMI/HDPUG study tend to minimize the DNP effect, and their relatively large die to package ratios (DPRs) result in substantial CTE mismatch [121]. The interfacial cracking observed with the 192CABGA package may be more a consequence of its very large DPR than a specific alloying effect.

The small volume of the solder balls used with these packages is expected to result in more Sn undercooling, which is known to influence solidification, microstructure, and potentially reliability [116, 122]. It is not safe to assume that the thermal fatigue, interfacial cracking behavior, and overall performance of alloys from this BGA study will translate or scale to other packages or components. Thus, future studies should include BGA packages with larger DNP, larger solder balls, and various DPR, and wafer-level chip scale packages (WLCSPs) with much smaller solder balls. Future studies should also include other surface mount components in common use such as resistors, quad flat no-leads (QFN), and connectors, which would also provide opportunities to evaluate performance of high reliability alloy solder pastes.

7.4.3   Effect of Surface Finish on Thermal Fatigue

Interfacial cracking has been reported at the PCB side of Pb-free and SnPb solder joints. Typically, this has been attributed to a surface finish anomaly such as black pad or micro-voiding [123–126]. It is not known if similar phenomena occur during the testing of components assembled with highly alloyed SAC-based solders. The PCB surface finish in the iNEMI/HDPUG study is predominantly organic solderability preservative (OSP), which results in soldering onto a copper (Cu) substrate. A single thermal cycling test cell in the experiment used electroless nickel/immersion gold (ENIG), which results in soldering onto a nickel (Ni) substrate. This provides a limited view of possible finish effects within this study, and there are multiple other PCB surface finishes utilized in practice [127–129]. To characterize any possible interaction between final finishes and highly alloyed solders, future investigations should consider incorporating PCB finish as a major test variable.

Interfacial cracking was induced at the package/solder interface during thermal cycling of some iNEMI/HDPUG samples. The BGA packages in the study used a traditional electrolytic nickel/gold finish. Some current package designs use other finishes such as OSP to avoid joint embrittlement at the interface [130, 131]. Similarly, an OSP BGA pad finish might mitigate the interfacial fracture reported for several high reliability alloys shown in Figure 7.24 [113]. The addition of gold (Au) dissolved from the surface finish was enough to alter Sn grain morphology and improve reliability in very small volume solder joints [29, 132]. Package surface finish should be a variable in future thermal cycling experiments but acquiring daisy-chained packages with other finishes may be costly, so this is likely to have lower priority for thermal cycling studies.

7.4.4 Thermomechanical Test Parameters and Test Outcomes

7.4.4.1 Thermal Cycling Dwell Time

The thermal cycling profiles used in the iNEMI/HDPUG study are typical of those used across the industry, and have relatively short hot and cold dwell times of 10–15 minutes. The effect of longer dwell times in thermal cycling of SAC solders was explored in considerable detail by many researchers [13, 30, 41, 42, 47, 133–141]. Although these studies demonstrated that longer dwell times reduce reliability of SAC solders, shorter dwells eventually were accepted for SAC-type solders where Ag_3Sn precipitate coarsening is the only significant solid-state reaction during thermal cycling. The findings from dwell time studies of SAC solders will not be directly applicable to high reliability solders due to the overlaid effects of solid solution and dispersion strengthening mechanisms, in some cases from multiple alloying elements (described in Section 7.2.3). The microstructure and performance of these high reliability alloys, particularly alloys containing Bi, can be very sensitive to thermal exposure [75, 99, 117, 119]. Future evaluations should consider incorporating test profiles with extended dwell times to develop data to enhance the understanding of alloy behavior in service environments. A suggested starting point, based on the results for SAC alloys from the literature, would be 60 minutes at the upper and lower temperature extremes.

7.4.4.2 Preconditioning (Isothermal Aging)

The effect of preconditioning, often referred to as aging, is another phenomenon that has been studied exhaustively for SAC alloys [10, 26, 35, 42, 112, 142–160]. In most of these studies, preconditioning exposure lead to a modest decrease in characteristic thermal fatigue life, which is attributed to changes in the Ag_3Sn precipitate distribution. Coarsening of this microstructural hardening phase reduces the resistance to creep deformation during the thermal cycle. While the magnitude of this effect typically is only moderate in SAC alloys, Delhaise has shown a significant response to preconditioning in SAC alloyed with Bi [99, 161]. This could present an additional reliability risk for products subject to isothermal exposure in addition to thermal cycling for applications with long-term storage exposure [142]. Similar to Ag_3Sn precipitate coarsening, intermediate phase precipitates of In and Sb would also be expected to coarsen and lose their strengthening effect with aging. Investigations of aging behavior will be needed to develop a complete understanding of the reliability of third generation solder alloys, but interpretation of the results is expected to be more difficult due to the complex metallurgy and multiple strengthening mechanisms.

7.4.4.3 Thermal Cycling of Mixed Metallurgy BGA Assemblies

The iNEMI/HDPUG test matrix uses matching solder paste and ball compositions to evaluate the situation of a so-called pure high reliability alloy. Today, the vast majority of BGA packages are fabricated with high-Ag solder balls, generally SAC305. If a high reliability alloy solder paste was proposed for implementation, the alloy compositions of the solder paste and all the BGA components would not be matched. This is known as mixed metallurgy or mixed alloy assembly and is illustrated schematically in Figure 7.27 [43]. Evaluations of assembly quality and thermal fatigue attachment reliability of the mixed assemblies are required to demonstrate alloy compatibility.

7.4.4.4 Thermal Shock or Aggressive Thermal Cycling

Historically, the thermal cycling parameters in IPC-9701 have defined the accepted practice for evaluating solder attachments [53]. For applications in harsh environments, a more aggressive cycling profile such as −40/150 °C may be specified [103]. Thermal shock, which typically is not specified for long-term board-level attachment evaluations, may be required for automotive or defense applications. Thermal shock is characterized by hot and cold ramp rates that are greater than $20\,°C\,min^{-1}$. This contrasts with the ramp rates in thermal cycling that are less

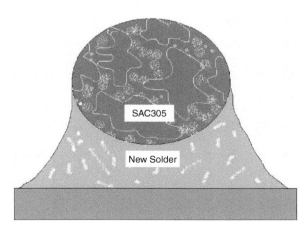

Figure 7.27 Illustration of mixed metallurgy BGA assembly with SAC305 BGA ball and a different solder paste [43].

than $20\,°C\,min^{-1}$ and are more typically $10\,°C\,min^{-1}$. It also is necessary to explore the effect of power cycling on attachment reliability [162], but power cycling test vehicle design is challenging and the experiments are costly and difficult to execute. An industry workshop on next generation solder alloys highlighted the need for aggressive testing to evaluate new alloys [163].

A learning curve is expected with this new class of alloys because there is limited experience using aggressive thermal cycling and thermal shock profiles for long-term testing. One of the challenges of using aggressive testing profiles is the potential for introducing multiple failure modes. Figure 7.28 shows multiple damage mechanisms in a BGA sample from a $-40/150\,°C$ thermal cycling test. Solder attachment reliability cannot be characterized and comparisons of alloy performance cannot be made when multiple damage mechanism are acting. Test vehicle design and laminate material selection must be explored to eliminate such a significant noise factor from the experiments.

7.4.5 Reliability Under Mechanical Loading: Drop/Shock, and Vibration

The primary performance focus for the evolving class of high reliability solder alloys has been to promote better high temperature performance. To achieve this, solder developers have modified basic Sn-Ag-Cu or Sn-Cu alloys with significant major and micro alloying additions as illustrated in Table 7.3. In addition to the requirement for superior resistance to thermal fatigue damage, targeted applications for high reliability alloys in automotive, avionics, and defense also have requirements for resistance to damage from high strain rate or repetitive mechanical loading. The electronics industry, including its major consortia, does not have an active, systematic program in place to address the performance of these high reliability solders under conditions of mechanical shock and vibration. This is a major gap in the overall reliability risk assessment for third generation, high reliability solder alloys [163].

The development of shock and vibration data to complement acceptable thermal fatigue performance is necessary for implementation of high reliability solders in automotive, avionics, and defense applications. However, the development of shock and vibration data is a challenging task for a variety of reasons. IPC-9701 provides detailed guidance for thermal cycling test methods and requirements for surface mount solder attachments [53]. IPC-9701 has been in use since 2002 and is recognized and used universally by device and solder suppliers, OEMs (original equipment manufacturers), OSATs (outsourced assembly and tests) and contract electronic manufacturers. Documents with a similar legacy do not exist for drop/shock or vibration test methods and requirements for Pb-free board-level solder attachments.

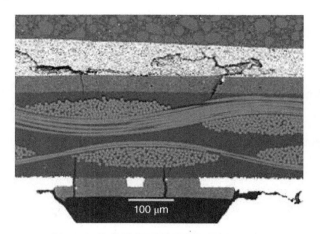

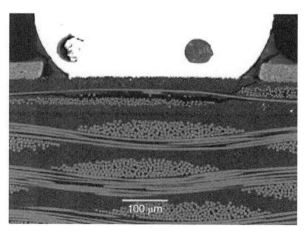

Figure 7.28 Multiple fracture modes in a single BGA sample after approximately 2000 thermal cycles using a −40/150 °C profile. Cracking is detected in the die-attach and package substrate (above) and in the solder (center), and pad cratering is detected in PCB (below) Source: Data courtesy of the iNEMI/HDPUG Third Generation Alloy project.

The test vehicle design from the latest JEDEC standard for board-level drop test [164, 165] is being used in a consortium study evaluating low temperature solders [166], but the JEDEC drop method and test vehicle have not yet achieved the widespread usage or acceptance that IPC-9701 has for thermal cycling.

There is even less clarity with vibration testing standards. A task force is evaluating revisions to the current JEDEC test specification (JESD22-B103-B), which defines test parameters but does not define a test vehicle [167]. Because vibration depends strongly on the characteristics of the PCB or test vehicle design, some companies including avionics and defense contractors have designed test vehicles representative of their component mix and product [168, 169]. Automotive OEMs perform vibration testing for qualification, product verification, and research and development, and often require testing by their component manufacturers. Automotive testing tends to be very application-specific and generally utilizes actual product boards not general test vehicles. To summarize, vibration testing protocols often are performed at a system level and target application performance requirements and specific use conditions, whereas IPC-9701 testing protocols are performed specifically to evaluate solder attachments under various conditions of thermal cycling.

In the case of aggressive thermal cycling, the greatest challenge for testing and ranking relative performance of solder alloys is to generate failures within the bulk solder (Figure 7.28). This may likewise be the challenge for drop and vibration testing. Multiple fracture modes in BGA assemblies were identified early in the transition to Pb-free manufacturing. This provided the incentive to develop and introduce second generation, commercial Pb-free solders as described in Section 7.1.2. Figure 7.29 shows the different failure or fracture modes that can arise from mechanical stresses [170]. Figure 7.29 is taken from a paper published in 2005, which demonstrates the early awareness of these phenomena with solders containing at least 3 wt% Ag. Clearly this should be considered a threat for mechanical testing of third generation alloys, since Figure 7.28 shows that several of these fracture modes have been detected in aggressive thermal cycling tests of these alloys.

The review of the literature in previous sections leaves no doubt that certain alloy additions in third generation solders result in higher tensile, shear, and mechanical fatigue strength than SAC alloys. This increased solder strength resists deformation and transfers loads to other areas of the interconnect structure, thereby causing failure or fracture outside the bulk solder. This compromises test results, restricts the ability to rank alloy performance, and could have serious implications in practice. Because of their higher strength, third generation solders may not meet certain application requirements for mechanical loading despite showing superior thermal cycling performance. The industry faced a similar situation with high Ag content first generation SAC alloys that could not meet drop requirements.

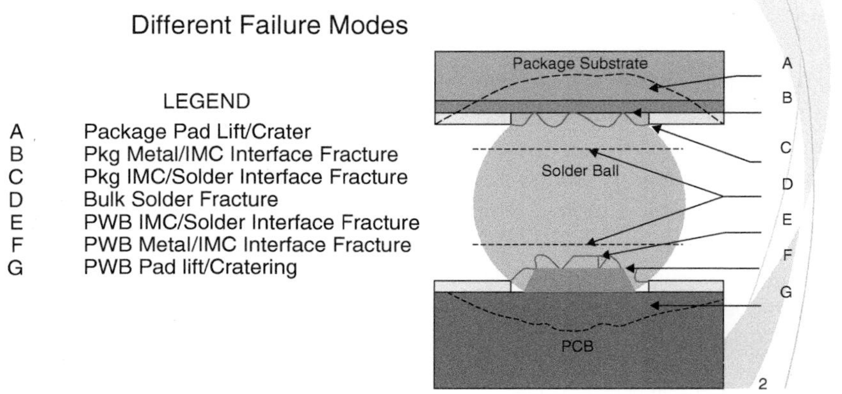

Different Failure Modes

LEGEND

A   Package Pad Lift/Crater
B   Pkg Metal/IMC Interface Fracture
C   Pkg IMC/Solder Interface Fracture
D   Bulk Solder Fracture
E   PWB IMC/Solder Interface Fracture
F   PWB Metal/IMC Interface Fracture
G   PWB Pad lift/Cratering

Figure 7.29 An illustration of multiple fracture modes in a BGA sample caused typically by mechanical loading [170].

Eventually they were replaced by second generation alloys that had lower overall strength and thermal fatigue resistance, but resisted catastrophic interconnect damage. Third generation alloys, unlike their second generation counterparts, will be required to provide satisfactory reliability in thermal cycling as well as mechanical loading. Ultimately, testing will be necessary in combined environments of thermal and mechanical exposure that will bring a substantial level of added complexity to the process of testing and evaluation. Therefore, it is critical to begin evaluations of the response of third generation solder alloys to drop/shock and vibration as soon as possible. It also is critical to partner with high-performance end users to develop effective, representative, and realistic test protocols and requirements.

7.4.6 Solder Alloy Microstructure and Reliability

A review of the literature including recent work by Belyakov et al. [171] indicates that multiple phases can appear in Sn-based alloys when significant amount of Bi, Sb, and In are present. The initial results from the iNEMI/HDPUG project indicate phases other than the common Ag_3Sn and Cu_6Sn_5 are present based on the density differences of precipitates in the SEM backscattered images of these alloys. Some of these precipitates are too small to be identified by the SEM Energy Dispersive X-ray Spectroscopy (EDS or EDX) used in that investigation but those precipitates may be critical to performance. Electron backscatter diffraction (EBSD), electron probe microanalysis (EPMA), and Wavelength Dispersive X-ray Spectroscopy (WDS or WDX) can be used to identify these phases. Belyakov et al. investigated the effects of alloying additions to the Sn-Cu system, and showed that Bi is an effective solid solution strengthening agent, moderate additions of Sb improved ductility and modified the IMC layer but did not alter the βSn microstructure [75]. The root cause of interfacial cracking in some alloys during thermal cycling is a major concern. It is not obvious that interfacial cracking impacts thermal cycling Weibull statistics, but the propensity for interfacial fracture and possible early failures could represent a reliability risk [113–115]. Detailed microstructural analysis of the bulk solder and IMC layers will be required to gain a better understanding of thermal fatigue and mechanical reliability and failure modes of these new solders.

7.4.7 Summary of Suggestions for Additional Investigation

Based on the gap analysis for Pb-free, third generation solder alloys described in the previous sections, the topics for additional detailed exploration are as follows:

- *Investigate the root cause of interfacial fracture.* If third generation Pb-free alloys are susceptible in practice to interfacial solder joint fractures, the consequences could be catastrophic. Investigate the root cause of interfacial fractures in thermal cycling, with an emphasis on the influence of alloy composition and thermal cycling profile. Quantify the effect of interfacial fractures on attachment reliability.
- *Expand the available data for thermal cycling.* Thermal cycling studies of third generation Pb-free alloys are very limited in scope. Develop thermal cycling data for additional BGA components of different sizes, different package constructions, and different attachment pad surface finishes. Develop data for other surface mount components in common use such as resistors, QFN, and connectors.
- *Assess the effect of PCB and component surface finish on reliability.* Determine if the surface finish on the PCB side or on the component side of the solder joint affects reliability by altering microstructure through microalloying or by altering the strength of the intermetallic layer and the strength of the soldered interface.
- *Expand thermal fatigue test protocols.* The current thermal cycling data are limited to short dwell times that are not representative of the targeted use environments. Longer dwell times should be incorporated into thermal cycling test plans to enable a better determination of alloy stability and reliability. Thermal shock and power cycling tests may be required in addition to thermal cycling. The ability of component and board test vehicles to survive harsh thermomechanical testing must be resolved to differentiate alloy performance.
- *Assess the effectiveness of thermal preconditioning (aging) on thermal fatigue life.* Harsh use conditions include operating temperatures that could reach or exceed 100 °C. Isothermal preconditioning should be explored to provide additional

acceleration to thermal cycling. Experiments are needed to establish the relationship between alloy composition and the effects of preconditioning time and temperature on microstructural evolution and subsequently on thermal cycling performance.

- *Mixed metallurgy or mixed alloy assembly.* Evaluations of assembly quality and thermal fatigue attachment reliability of the mixed assemblies are required to demonstrate compatibility of highly alloyed, 3rd generation solders with current Pb-free materials and assembly processes.
- *Drop/shock and vibration testing.* Applications for high reliability alloys in automotive, avionics, and defense also have requirements for resistance to damage from high strain rate or repetitive mechanical loading. Data must be developed for mechanical reliability in parallel with the thermomechanical fatigue reliability data for the class of high reliability Pb-free solder alloys.
- *Relationship between microstructure and the thermal fatigue performance.* The strengthening effects associated with various alloying additions to Sn-Ag-Cu and Sn-Cu systems are recognized. However, the identification of specific phases, the dissolution of the alloying elements in the β-tin (Sn) matrix, and the relationship between microstructure, microstructural evolution, and, ultimately, the thermal fatigue performance have not been determined. Advanced analytical techniques including EBSD, EPMA, and WDS (wavelength dispersive X-ray spectroscopy) are needed to characterize phase formation in the bulk composition, and structure of intermetallic layers, and dissolution of alloying elements in the β-tin (Sn).

7.5 Conclusions

The drivers, benefits and concerns associated with the development and eventual deployment of third generation, high reliability Pb-free solders has been discussed. These alloys have the potential to solve reliability issues arising from the need to transition from SnPb to Pb-free manufacturing or to meet performance requirements in aggressive use environments defined by automotive, avionics, and defense applications. However, if not managed properly, implementation of these alloys could introduce new reliability risks. Based on a review of the literature and input from many active researchers in the field, the following conclusions can be made at this time.

- There is a long and expanding list of commercialized high reliability alloys based on the Sn-Ag-Cu and Sn-Cu systems. These alloys are characterized by singular, major alloying additions of bismuth (Bi), antimony (Sb), or indium (In) or combinations of these major alloying elements. Some alloys also contain microalloy additions of a variety of elements to improve performance. The wide range of solder alloys and combinations of elemental additions adds to the complexity of test programs and to the subsequent risk analyses. Although these are classified as commercial alloys, volume implementations at the time of this writing are scarce and solder suppliers continue to modify compositions in response to emerging test data. This adds uncertainty and risk within the supply chain.

- All these new alloys appear to outperform the near-eutectic SAC305 alloy in thermal cycling. However, early results from the iNEMI/HDPUG project reveal much remains to be learned about their thermal cycling behavior. Early thermal cycling test results along with other test data from the literature indicate Bi and In are effective solid solution strengthening agents. Bi may also strengthen by dispersion hardening. Additions of Sb can improve performance in thermal cycling, probably through precipitation hardening. However, alloys containing major additions of Bi or Sb exhibited some interfacial cracking during thermal cycling. In contrast, an alloy with high In content showed slightly lower thermal cycling performance but no evidence of interfacial cracking. The observation of interfacial cracking raises concerns about alloy performance under high strain rate mechanical loading.

- Other areas related to thermal cycling reliability that need further attention include effects due to thermal cycling profile and dwell time, isothermal preconditioning, power cycling, surface finish, and mixed metallurgy assembly. Test programs should begin to address these issues to enable a complete analysis and understanding of the performance behavior of this class of alloys.

- A better understanding of the relationships between composition, microstructure and alloy performance and failure modes during thermal cycling are needed to assess the roles and effectiveness of the various alloying elements. This includes developing a better understanding of microstructural stability and evolution in the bulk and at soldered interfaces.

- The performance requirements in targeted high reliability applications address a variety of demanding thermal and mechanical use environments. A comprehensive evaluation of the performance of third generation, high reliability Pb-free solders has not been done for thermal shock, mechanical drop/shock, or vibration. In fact, test protocols for mechanical testing of these solders are yet to be developed, which is a significant gap.

- Investigations of existing and new third generation, high reliability alloys are expected to accelerate as more high reliability end users become engaged. This work is taking place at solder suppliers, OEMs, contract manufactures, and universities. Industrial consortia and working groups are expected to play a major role because of the vast scope, significant knowledge gaps, and breadth and depth of expertise needed to address the issues and execute these programs.

Acknowledgments

Many people contributed to the development and assessment of the information in this chapter. The author wants to thank his current and former colleagues at Nokia Bell Labs who helped to develop and analyze test data over several years: Charmaine Johnson, Richard Popowich, Pete Read, and Debbie Fleming. The author also is indebted to the iNEMI/HDPUG Third Generation, Pb-Free Alloy Team. A large core group of dedicated collaborators has supported investigations of alternative alloys vigorously starting with second generation alloys in 2009 to the present. Special thanks go to Rich Parker, current Project Co-Chair and to the two past Project Chairs, Dr. Greg Henshall formerly of Hewlett-Packard and Elizabeth Benedetto of HP Inc. The author thanks the staffs at iNEMI and HDPUG, especially Grace O'Malley, Marc Benowitz, and Marshall Andrews for their help in establishing and nurturing the collaborative agreement and Larry Marcanti, and Robert Smith for their continued support in coordinating the work between the cooperating consortia. The author also thanks his colleagues Dave Hillman, Keith Howell, Keith Sweatman, Tetsuro Nishimura, Chris Gourlay, Sergei Belyakov, Babak Arfaei, Joe Smetana, Jean-Paul Clech, and Andre Delhaise for numerous insights and technical guidance. Special thanks go to Melissa Young, Nokia Bell Labs Reference Librarian, for her expert support of literature reviews for the project.

References

1 Annex to Directive 2002/95/EC. (2006) Restriction on the use of hazardous substances (RoHS) in electrical and electronic equipment. Official Journal of the European Union, 14.10.2006, L283/48-49, October 12, 2006.
2 Handwerker, C.A., de Kluizenaar, E.E., Suganuma, K., and Gayle, F.W. (2004). Major international lead (Pb)-free solder studies. In: *Handbook of Lead-Free Solder Technology for Microelectronic Assemblies* (eds. K.J. Puttlitz and K.A. Stalter), 665–728. New York: Marcel Dekker.
3 Lead-Free Solder Project Final Report. (1997) NCMS Report 0401RE96, Section 2.4: Properties Assessment and Alloy Down Selection. Ann Arbor, MI: National Center for Manufacturing Sciences.
4 Marconi Materials Technology (1999) Improved Design Life and Environmentally Aware Manufacturing of Electronics Assemblies by Lead-Free Soldering: 'IDEALS'. Contract number BRPR-CT96-0140, 30 June.
5 Richards, B., Levogner, C.L., Hunt, C.P. et al. (1999) Lead-Free Soldering –An Analysis of the Current Status of Lead-Free Soldering. London: Department of Trade and Industry.

6 Lead-Free Soldering Roadmap Committee, Technical Standardization Committee on Electronics Assembly Technology (2002) JEITA Lead-free Roadmap 2002 for Commercialization of Lead-free Solder. JEITA.

7 JEIDA (2002) NEDO Research and Development on Lead-Free Soldering. Report No. 00-ki-17. JEIDA.

8 Bradley, E., Handwerker, C.A., Bath, J. et al. (eds.) (2007). *Lead-Free Electronics: iNEMI Projects Lead to Successful Manufacturing*. Hoboken, NJ: Wiley.

9 Henshall, G., Sweatman, K., Howell, K. et al. (2009). iNEMI lead-free alloy alternatives project report: thermal fatigue experiments and alloy test requirements. In: *Proceedings of SMTAI*, 317–324. San Diego CA.

10 Smetana, J., Coyle, R., Read, P. et al. (2011). Variations in thermal cycling response of Pb-free solder due to isothermal preconditioning. In: *Proceedings of SMTAI*, Fort Worth, TX, 641–654.

11 Engelmaier, W. (1989). Surface mount solder joint long-term reliability: design, testing, prediction. *Soldering and Surface Mount Technology* 1 (1): 14–22.

12 Terashima, S., Kariya, Y., Hosoi, M., and Tanaka, M. (2003). Effect of silver content on thermal fatigue life of Sn-xAg-0.5Cu flip-chip interconnects. *Journal of Electronic Materials* 32 (12): 1527–1533.

13 Coyle, R., McCormick, H., Osenbach, J. et al. (2011). Pb-free alloy silver content and thermal fatigue reliability of a large plastic ball grid array (PBGA) package. *Journal of Surface Mount Technology* 24 (1): 27–33.

14 Henshall, G., Bath, J., Sethuraman, S. et al. Comparison of thermal fatigue performance of SAC105 (Sn-1.0Ag-0.5Cu), Sn-3.5Ag, and SAC305 (Sn-3.0Ag-0.5Cu) BGA components with SAC305 solder paste. Proceedings IPC APEX, S05-03, 2009.

15 IPC Solder Products Value Council (2005) Round Robin Testing and Analysis of Lead-Free Solder Pastes with Alloys of Tin, Silver and Copper, Final Report. IPC International.

16 Henshall, G. (2011). Lead-free alloys for BGA/CSP components. In: *Lead-Free Solder Process Development* (eds. G. Henshall, J. Bath and C.A. Handwerker), 95–124. Hoboken, NJ: Wiley.

17 Henshall, G., Miremadi, J., Parker, R. et al. (2012). iNEMI Pb-free alloy characterization project report: Part I – program goals, experimental structure, alloy characterization, and test protocols for accelerated temperature cycling. In: *Proceedings of SMTAI*, 335–347. Orlando, FL.

18 Kim, D., Suh, D., Millard, T. et al. (2007). Evaluation of high compliant low Ag solder alloys on OSP as a drop solution for the 2nd level Pb-free interconnection. In: *Proceedings 57th Electronic Component and Technology Conference* Reno, NV, 1614–1619.

19 Syed, A., Kim, T.-S., Cha, S.-W. et al. (2007). Effect of Pb free alloy composition on drop/impact reliability of 0.4, 0.5, and 0.8 mm pitch Chip scale packages with NiAu pad finish. In: *Proceedings 57th Electronic Component and Technology Conference*, Reno, NV (May 28 – June 1), 951–956.

20 International Electronics Manufacturing Initiative (iNEMI). www.inemi.org (accessed 4 February 2020).

21 HDP User Group, International, Inc. (HDPUG). https://hdpug.org (accessed 4 February 2020).

22 The Center for Advanced Life Cycle Engineering (CALCE), University of Maryland. https://calce.umd.edu (accessed 4 February 2020).

23 Universal Advanced Research in Electronics Assembly. http://www.uic.com/solutions/apl/area-consortium (accessed 4 February 2020).

24 The Center for Advanced Vehicle and Extreme Environment Electronics (CAVE[3]), Auburn University. http://cave.auburn.edu (accessed 4 February 2020).

25 Coyle, R., Sweatman, K., and Arfaei, B. (2015). Thermal fatigue evaluation of Pb-free solder joints: results, lessons learned, and future trends. *Journal of the Minerals, Metals & Materials Society* 67 (10): 2394–2415.

26 Coyle, R., Parker, R., Smetana, J. et al. (2015). iNEMI Pb-free alloy characterization project report: PART IX – summary of the effect of isothermal preconditioning on thermal fatigue life. In: *Proceedings of SMTAI*, Chicago, IL, 743–755.

27 Coyle, R., Parker, R., Benedetto, E. et al. (2014). iNEMI Pb-free alloy characterization project report: PART VIII – thermal fatigue results for high-Ag alloys at extended dwell times. In: *Proceedings of SMTAI*, Chicago, IL, 547–560.

28 Sweatman, K., Coyle, R., Parker, R. et al. (2014). iNEMI Pb-free alloy characterization project report: PART VII – thermal fatigue results for low-Ag alloys at extended dwell times. In: *Proceedings of SMTAI*, Chicago, IL, 561–574.

29 Coyle, R., Parker, R., Arfaei, B. et al. (2014). The effect of nickel microalloying on thermal fatigue reliability and microstructure of SAC105 and SAC205 solders. In: *Proceedings of Electronic Components Technology Conference*, 425–440. Orlando, FL: IEEE.

30 Coyle, R., Parker, R., Osterman, M. et al. (2013). iNEMI Pb-free alloy characterization project report: part V – the effect of dwell time on thermal fatigue reliability. In: *Proceedings of SMTAI*, Ft. Worth, TX, 470–489.

31 Coyle, R., Parker, R., Arfaei, B. et al. (2013). iNEMI Pb-free alloy characterization project report: Part VI – the effect of component surface finish and solder paste composition on thermal fatigue of SN100C solder balls. In: *Proceedings of SMTAI*, Ft. Worth, TX, 490–414.

32 George, E., Osterman, M., Pecht, M. et al. (2013) Thermal cycling reliability of alternative low-silver tin-based solders. *Proceedings of IMAPS 2013, 46th International Symposium on Microelectronics*, Orlando, FL (October 2013).

33 Parker, R., Coyle, R., Henshall, G. et al. (2012). iNEMI Pb-free alloy characterization project report: Part II – thermal fatigue results for two common temperature cycles. In: *Proceedings of SMTAI*, Orlando, FL, 348–358.

34 Sweatman, K., Howell, K., Coyle, R. et al. (2012). iNEMI Pb-free alloy characterization project report: Part III – thermal fatigue results for low-Ag alloys. In: *Proceedings of SMTAI*, Orlando, FL, 359–375.

35 Coyle, R., Parker, R., Henshall, G. et al. (2012). iNEMI Pb-free alloy characterization project report: Part IV – effect of isothermal preconditioning on thermal fatigue life. In: *Proceedings of SMTAI*, Orlando, FL, 376–389.

36 Henshall, G., Healey, R., Pandher, R.S. et al. (2009) Addressing the opportunities and risks of Pb-free solder alloy alternatives. *Proceedings of Microelectronics and Packaging Conference*, 1–11 June, Rimini, Italy.

37 Henshall, G., Healy, R., Pander, R.S. et al. (2008). iNEMI Pb-free alloy alternatives project report: state of the industry. *Journal of Surface Mount Technology* 21 (4): 11–23.

38 Henshall, G., Healy, R., Pander, R.S. et al. (2008). iNEMI Pb-free alloy alternatives project report: state of the industry. In: *Proceedings of SMTAI*, Orlando, FL, 109–122.

39 Coyle, R., Parker, R., Howell, K. et al. (2016). A collaborative industrial consortia program for characterizing thermal fatigue reliability of third generation Pb-free alloys. In: *Proceedings of SMTAI*, Rosemont, IL, 188–196.

40 Thompson, R. (2003) *Proceedings of the SMTA/CAVE Workshop Harsh Environment Electronics*, Dearborn, MI (24–25 June, 2003).

41 Fairchild, M.R., Snyder, R.B., Berlin, C.W., and Sarma, D.H.R. (2002) Emerging substrate technologies for harsh-environment automotive electronics applications. SAE Technical Paper Series 2002-01-1052.

42 Jolivet, E. and Dhond, P. (2019) Challenges in automotive packaging technologies. Chip Scale Review 15–16 May – June 2019.

43 Snugovsky, P., Bagheri, S., Romansky, M. et al. (2012). New generation of Pb-free solder alloys: possible solution to solve current issues with main stream Pb-free soldering. *Journal of Surface Mount Technology* 25 (3): 42–52.

44 Coyle, R., Read, P., McCormick, H. et al. (2011). The influence of alloy composition and temperature cycling dwell time on the reliability of a quad flat no lead (QFN) package. *Journal of Surface Mount Technology* 25 (1): 28–34.

45 Coyle, R., Osenbach, J., Collins, M. et al. (2011). Phenomenological study of the effect of microstructural evolution on the thermal fatigue resistance of Pb-free solder joints. *IEEE Transactions on Components, Packaging, and Manufacturing Technology* 1 (10): 1583–1593.

46 Henshall, G., Bath, J., Sethuraman, S. et al. (2009) Comparison of thermal fatigue performance of SAC105 (Sn-1.0Ag-0.5Cu), Sn-3.5Ag, and SAC305 (Sn-3.0Ag-0.5Cu) BGA components with SAC305 solder paste. *Proceedings IPC APEX*, S05-03.

47 Zhang, Y., Cai, Z., Suhling, J.C. et al. (2008). The effects of aging temperature on SAC solder joint material behavior and reliability. In: *Proceedings of Electronic Components and Technology Conference*, Lake Buena Vista, FL, 99–112.

48 Bieler, T., Jiang, H., Lehman, L. et al. (2008). Influence of Sn grain size and orientation on the thermomechanical response and reliability of Pb-free solder joints. *IEEE Transactions on Components, Packaging, and Manufacturing Technology* 31 (2): 370–381.

49 Darveaux, R., Reichman, C., Berry, C. et al. (2008). Effect of joint size and pad metallization on solder mechanical properties. In: *Proceedings of Electronic Components and Technology Conference*, Lake Buena Vista, FL, 113–122.

50 Kang, S.K., Lauro, P., Shih, D.-Y. et al. (2004). Evaluation of thermal fatigue life and failure mechanisms of Sn-Ag-Cu solder joints with reduced Ag contents. In: *Proceedings of Electronic Components and Technology Conference*, Las Vegas, NV, 661–667.

51 Dunford, S., Canumalla, S., and Viswanadham, P. (2004). Intermetallic morphology and damage evolution under thermomechanical fatigue of lead (Pb)-free solder interconnections. In: *Proceedings of Electronic Components Technology Conference*, Las Vegas, NV, 726–736.

52 Coyle, R., Hillman, D., Johnson, C. et al. (2018) Alloy composition and thermal fatigue reliability of high reliability Pb-free solder alloys. *Proceedings of SMTAI*, Rosemont, IL (October 2018).

53 IPC-9701A (2006) *Performance test methods and qualification requirements for surface mount solder attachments*. IPC International.

54 IPC Pb-Free Reliability Risk Management (PERM) Council, www.ipc.org/ContentPage.aspx?pageid=PERM-Council (accessed 4 February 2020).

55 Rafanelli, A.J. (2017) Addressing the challenges of Pb-free technology in high performance (aerospace/defense) products. *SMTA New England Expo and Technical Forum*, Worcester, MA (16 November).

56 Nowottnick, M., Scheel, W., and Wittke, K. (eds.) (2005). *Innovative Production Processes for High-temperature Electronics in Automotive Electronics Systems: Construction and Connection Technology*, vol. 2. Germany: M. Detert Publishing.

57 Albrecht, J. (2008) Final presentation BMBF project LIVE; Acceptance criteria of thermally highly stressed miniaturized solder joints. Siemens AG, Corporate Technology, MM6, Berlin, Germany, 17 September.

58 Trodler, J. (2009) Summary Innolot – Project March. 2000 to February 2004. W.C. Heraeus GmbH; CMD-AM-AT, Hanau, Germany, January 2009 (complete final report confidential).

59 Steen, H. and Toleno, B. (2009) Development of a lead-free alloy for high-reliability, high temperature applications. SMT, January.

60 Albrecht, H.-J., Frühauf, P., and Wilke, K. (2009). Pb-free alloy alternatives: reliability investigation. In: *Proceedings SMTAI*, San Diego, CA, 308–316.

61 Miric, A.-Z. (2010). New developments in high-temperature, high-performance lead-free solder alloys. *Journal of Surface Mount Technology* 23 (4): 24–29.

62 Albrecht, H.-J., Bartl, K.H.G., Kruppa, W. et al. (2017) Soldering material based on Sn Ag and Cu. EP1617968B1, European Patent Office, 1 March.

63 Steller, A., Zimmermann, A., Eisenberg, S. et al. (2009). Reliability testing and damage analysis of lead-free solder joints: new assessment criteria for laboratory methods. *SAE International Journal of Materials and Manufacturing* 2 (1): 502–510.

64 ASM International Handbook Committee (ed.) (1990). Specialty steels and heat – resistance alloys. In: *Metals Handbook 10* (ed. ASM International Handbook Committee), 822–1006. Materials Park, OH: ASM International.

65 Dieter, G.E. (1961). Plastic deformation of polycrystalline aggregates, solid solution hardening. In: *Mechanical Metallurgy*, 128. McGraw-Hill.

66 Haasen, P. (1996). *Physical Metallurgy*, 3e, 375–378. Cambridge University Press.

67 Delhaise, A., Snugovsky, L., Perovic, D. et al. (2014). Microstructure and hardness of Bi-containing solder alloys after solidification and ageing. *Journal of Surface Mount Technology* 27 (3): 22–27.

68 Wada, T., Tsuchiya, S., Joshi, S. et al. (2017) Improving thermal cycle reliability and mechanical drop impact resistance of a lead-free tin-silver-bismuth-indium solder alloy with minor doping of copper additive. *Proceedings of IPC APEX*, San Diego, CA (14–16 February).

69 Siewert, T., Liu, S., Smith, D.R., and Madeni, J.C. (2002) Database for solder properties with emphasis on new lead-free solders: properties of lead-free solders release 4.0. NIST and Colorado School of Mines, February, 2002.

70 Handwerker, C.A., Kattner, U., Moon, K. et al. (2007). Alloy selection. In: *Lead-Free Electronics* (eds. E. Bradley, C.A. Handwerker, J. Bath, et al.), 9–46. Piscataway, NJ: IEEE Press.

71 Hansen, M. (1958). *Constitution of Binary Alloys*, 2e, 1175–1177. McGraw-Hill.

72 Elliot, R.P. (1965). *Constitution of Binary Alloys*, First Supplement, 802. McGraw-Hill.

73 Li, G.Y., Chen, B.L., and Tey, J.N. (2004). Reaction of Sn-3.5Ag-0.7Cu-xSb solder with Cu metallization during reflow soldering. *IEEE Transactions on Electronics Packaging Manufacturing* 27 (1): 77–85.

74 Li, G.Y., Bi, X.D., Chen, Q., and Shi, X.Q. (2011). Influence of dopant on growth of intermetallic layers in Sn-Ag-Cu solder joints. *Journal of Electronic Materials* 40 (2): 165–175.

75 Belyakov, S.A., Nishimura, T., Akaiwa, T. et al. (2019) Role of Bi, Sb, and In in microstructural formation and properties of Sn-0.7Cu-0.05Ni-X BGA interconnections. *IEEE International Conference on Electronic Packaging (ICEP)*, Niigata, Japan (17–20 April).

76 Tegehall, P.-E. (2006) Review of the impact of intermetallic layers on the brittleness of tin-lead and lead-free solder joints, Section 3, Impact of intermetallic compounds on the risk for brittle fractures. IVF Project Report 06/07. IVF Industrial Research and Development Corporation.

77 Lu, S., Zheng, Z., Chen, J., and Luo, F. (2010). Microstructure and solderability of Sn-3.5Ag-0.5Cu-xBi-ySb solders. In: *Proceedings 11th International Conference on Electronic Packaging Technology and High Density Packaging, ICEPT-HDP*, 410–412.

78 El-Daly, A.A., Swilem, Y., and Hammad, A.E. (2008). Influences of Ag and au additions on structure and tensile strength of Sn-5Sb lead free solder alloy. *Journal of Materials Science and Technology* 24 (6): 921–925.

79 Beyer, H., Sivasubramaniam, V., Hajas, D. et al. (2014). Reliability improvement of large area soldering connections by antimony containing lead-free solder. In: *PCIM Europe Conference Proceedings*, 1069–1076.

80 El-Daly, A.A., Swilem, Y., and Hammad, A.E. (2009). Creep properties of Sn-Sb based lead-free solder alloys. *Journal of Alloys and Compounds* 471: 98–104.

81 Wada, T., Mori, K., Joshi, S., and Garcia, R. (2016). Superior thermal cycling reliability of Pb-free solder alloy by addition of indium and bismuth for harsh environments. In: *Proceedings of SMTAI*, Rosemont, IL, 210–215.

82 Yu, A.-M., Jang, J.-W., Lee, J.-H. et al. (2014). Microstructure and drop/shock reliability of Sn-Ag-Cu-In solder joints. *International Journal of Materials and Structural Integrity* 8 (1–3): 42–52.

83 Yu, A.-M., Jang, J.-W., Lee, J.-H. et al. (2012). Tensile properties and thermal shock reliability of Sn-Ag-Cu solder joint with indium addition. *Journal of Nanoscience and Nanotechnology* 12 (4): 3655–3657.

84 Amagai, M., Toyoda, Y., Ohnishi, T., and Akita, S. (2004). High drop test reliability: lead-free solders. In: *Proceedings 54th Electronic Components and Technology Conference*, 1304–1309.

85 Hodúlová, E., Palcut, M., Lechovič, E. et al. (2011). Kinetics of intermetallic phase formation at the interface of Sn-Ag-Cu-X (X = Bi, In) solders with Cu substrate. *Journal of Alloys and Compounds* 509 (25): 7052–7059.

86 Sharif, A. and Chan, Y.C. (2006). Liquid and solid state interfacial reactions of Sn-Ag-Cu and Sn-In-Ag-Cu solders with Ni-P under bump metallization. *Thin Solid Films* 504 (1–2): 431–435.

87 Chantaramanee, S., Sungkhaphaitoon, P., and Plookphol, T. (2017). Influence of indium and antimony additions on mechanical properties and microstructure of Sn-3.0Ag-0.5Cu lead free solder alloys. *Solid State Phenomena* 266: 196–200.

88 Sopoušek, J., Palcut, M., Hodúlová, E., and Janovec, J. (2010). Thermal analysis of the Sn-Ag-Cu-In solder alloy. *Journal of Electronic Materials* 39 (3): 312–317.

89 Wang, J., Yin, M., Lai, Z., and Li, X. (2011). Wettability and microstructure of Sn-Ag-Cu-In solder. *Hanjie Xuebao/Transactions of the China Welding Institution* 32 (11): 69–72.

90 Vianco, P.T. and Rejent, J.A. (1999). Properties of ternary Sn-Ag-Bi solder alloys: Part I – thermal properties and microstructural analysis. *Journal of Electronic Materials* 28 (10): 1127–1137.

91 Vianco, P.T. and Rejent, J.A. (1999). Properties of ternary Sn-Ag-Bi solder alloys: Part I – wettability and mechanical properties analyses. *Journal of Electronic Materials* 28 (10): 1138–1143.

92 Zhao, J., Qi, L., and Wang, X.-M. (2004). Influence of Bi on microstructures evolution and mechanical properties in Sn-Ag-Cu lead-free solder. *Journal of Alloys and Compounds* 375 (1–2): 196–201.

93 Hillman, D., Pearson, T., and Wilcoxon, R. (2010). NASA DOD −55 °C to +125 °C thermal cycle test results. In: *Proceedings of SMTAI*, Orlando, FL, 512–518.

94 Witkin, D. (2013) Mechanical properties of Bi-containing Pb-free solders. *Proceedings IPC APEX*, S11-01, San Diego, CA (February 2013).

95 Juarez, J.M. Jr., Snugovsky, P., Kosiba, E. et al. (2015). Manufacturability and reliability screening of lower melting point Pb-free alloys containing bismuth. *Journal of Microelectronics and Electronic Packaging* 12 (1): 1–28.

96 Nishimura, T., Sweatman, K., Kita, A., and Sawada, S. (2015). A new method of increasing the reliability of lead-free solder. In: *Proceedings of SMTAI*, Rosemont, IL, 736–742.

97 Witkin, D. (2012). Creep behavior of Bi-containing lead-free solder alloys. *Journal of Electronic Materials* 41 (2): 190–203.

98 Witkin, D.B. (2012). Influence of microstructure on quasi-static and dynamic mechanical properties of bismuth-containing lead-free solder alloys. *Materials Science and Engineering A* 532: 212–220.

99 Delhaise, A.M., Snugovsky, P., Matijevic, I. et al. (2018). Thermal preconditioning, microstructure restoration and property improvement in Bi-containing solder alloys. *Journal of Surface Mount Technology* 31 (1): 33–42.

100 Sweatman, K. (2017) Nihon Superior, private communication, November 2017.

101 Raeder, C.H., Felton, L.E., Knott, D.B. et al. (1993). Microstructural evolution and mechanical properties of Sn-Bi based solders. In: *Proceedings of International Electronics Manufacturing Technology Symposium*, Santa Clara, CA, 119–127.

102 Coyle, R., Aspandiar, R., Osterman, M. et al. (2016). Thermal cycle reliability of a low silver ball grid array assembled with tin bismuth solder paste. In: *Proceedings of SMTAI*, Rosemont, IL, 72–83.

103 Trodler, J., Dudek, R., and Röllig, M. (2016). Risk for ceramic component cracking dependent on solder alloy and thermo-mechanical stress. In: *Proceedings of SMTAI*, Rosemont, IL, 197–203.

104 Chada, S., Currie, M., Toleno, B. et al. (2011). Developing a Pb-free solder through micro-alloying. In: *Proceedings of SMTAI*, Fort Worth, TX, 724–730.

105 Dudek, R., Hildebrandt, M., Doering, R. et al. (2014) Solder fatigue acceleration prediction and testing results for different thermal test- and field cycling environments. *Proceedings 5th Electronics System-integration Conference (ESTC)*, Helsinki, Finland (16–18 September).

106 Barry, N. (2008) Lead-free solders for high-reliability applications: high-cycle fatigue studies. Ph.D. thesis. Department of Metallurgy and Materials, School of Engineering, The University of Birmingham.

107 Su, S., Jian, M., Akkara, F.J. et al. (2018) Fatigue and shear properties of high reliable solder joints for harsh applications. *Proceedings of SMTAI*, Rosemont, IL (October 2018).

108 Chowdhury, M.R., Ahmed, S., Fahim, A. et al. (2016). Mechanical characterization of doped SAC solder materials at high temperature. In: *15th IEEE Intersociety Conference on Thermal and Thermomechanical Phenomena in Electronic Systems (ITherm)*, Las Vegas, NV, 1202–1208.

109 Snugovsky, P., Kosiba, E., Kennedy, J. et al. (2013). Manufacturability and reliability screening of lower melting point Pb-free alloys containing Bi. In: *Proceedings, IPC APEX*, San Diego, CA, 171–208.

110 Kosiba, E., Bagheri, S., Snugovsky, P., and Perovic, D. (2013) Microstructure and reliability of low Ag Bi-containing solder alloys. *International Conference on Soldering and Reliability (ICSR)*, Toronto, ON.

111 Hillman, D. and Wilcoxon, R. (2006) JCAA/JG-PP No-lead solder project: −55 °C to +125 °C thermal cycle testing final report. Rockwell Collins Advanced Manufacturing Technology Group, Contract: GST 0504BM3419, May 28, 2006.

112 Sanders, T., Thirugnanasambandam, S., Evans, J. et al. (2015). Component level reliability for high temperature power computing with SAC305 and alternative high reliability solders. In: *Proceedings of SMTA International*, Rosemont, IL, 144–150.

113 Coyle, R., Hillman, D., Parker, R. et al. (2018) The effect of bismuth, antimony, or indium on the thermal fatigue of high reliability Pb-free solder alloys. *Proceedings of SMTAI*, Rosemont, IL (October).

114 Coyle, R., Hillman, D., Johnson, C. et al. (2018) Alloy composition and thermal fatigue of high reliability Pb-free solder alloys. *Proceedings of SMTAI*, Rosemont, IL, (October).

115 Coyle, R., Johnson, C., Hillman, D. et al. (2019). Thermal cycling reliability and failure mode of two ball grid array packages with high reliability Pb-free solder alloys. In: *Proceedings of SMTAI*, Rosemont, IL, 439–456.

116 Kinyanjui, R., Lehman, L.P., Zavalij, L., and Cotts, E. (2005). Effect of sample size on the solidification temperature and microstructure of SnAgCu near eutectic alloys. *Journal of Materials Research* 20 (11): 2914–2918.

117 Yin, L., Wentlent, L., Yang, L.L. et al. (2011). Recrystallization and precipitate coarsening in Pb-free solder joints during thermomechanical fatigue. *Journal of Electronic Materials* 41 (2): 241–252.

118 Belyakov, S.A., Xian, J., Zeng, G. et al. (2019). Precipitation and coarsening of bismuth plates in Sn-Ag-Cu–Bi and Sn-Cu-Ni–Bi solder joints. *Journal of Materials Science: Materials in Electronics* 30: 378–390.

119 Sweatman, K., Akaiwa, T., and Nishimura, T. (2018) Effect of creep rate of alloying additions to Ni-stabilised Sn-Cu eutectic solder. *Proceedings of SMTAI*, Rosemont, IL, (October).

120 Engelmaier, W. (1990). The use environments of electronic assemblies and their impact on surface mount solder attachment reliability. *IEEE Transactions on Components, Hybrids, and Manufacturing Technology* 13 (4): 903–908.

121 Syed, A., Panczak, T., Darveaux, R. et al. (1999). Solder joint reliability of chip array BGA. *Journal of Surface Mount Technology* 12 (2): 1–7.

122 Arfaei, B., Wentlent, L., Joshi, S. et al. (2012). Improving the thermomechanical behavior of lead free solder joints by controlling the microstructure. In: *Proceedings of ITHERM*, 392–398.

123 Zeng, K., Stierman, R., Abbott, D., and Murtuza, M. (2006). Root cause of black pad failure of solder joints with electroless nickel/immersion gold plating. In: *Proceedings of Thermal and Thermomechanical Phenomena in Electronics Systems, IEEE ITHERM '06*, 1111–1119.

124 Ejim, T.I., Hollesen, D.B., Holliday, A. et al. (1997). *Proceedings of the 21st IEMT*, Austin, TX, 25–31.

125 Mei, Z., Callery, P., Fisher, D. et al. (1997). *Proceedings Pacific Rim/ASME Inter. Intersociety Electronic, and Photonic Packaging Conference, Advances in Electronic Packaging*, 1543–1550.

126 Cullen, D.P. (2006) Characterization, reproduction, and resolution of solder joint microvoiding. *Proceedings IPC APEX*, S26-2-1.

127 Thein, G., Geiger, D., and Kurwa, M. (2014) Study of various PCBA surface finishes. *Proceedings IPC APEX*, S13-03 (March 25–27 2014).

128 The PCB Magazine (2015) Featured content surface finishes. February. http://iconnect007.uberflip.com/i/457216-pcb-feb2015

129 Shen, C., Hai, Z., Zhao, C. et al. (2017). Packaging reliability effect of ENIG and ENEPIG surface finishes in board level thermal test under long-term aging and cycling. *Materials* https://www.mdpi.com/1996-1944/10/5/451.

130 Chang, D., Bai, F., Wang, Y.P., and Hsiao, C.S. (2004). The study of OSP as reliable surface finish of BGA substrate. In: *Proceedings 6th Electronics Packaging Technology Conference (EPTC)*, Singapore, 149–153.

131 Arfaei, B., Anselm, M., Mutuku, F., and Cotts, E. (2014). Effect of PCB surface finish on Sn grain morphology and thermal fatigue performance of lead-free solder joints. In: *Proceedings of SMTAI*, Rosemont, IL, 406–414.

132 Arfaei, B., Mutuku, F., Sweatman, K. et al. (2014). Dependence of solder joint reliability on solder volume, composition and printed circuit board surface finish. In: *64th Electronic Component and Technology Conference*, Orlando, FL, 655–665.

133 George, E., Osterman, M., Pecht, M., and Coyle, R. (2012) Effects of extended dwell time on thermal fatigue life of ceramic chip resistors. *Proceedings of IMAPS 2012 45th International Symposium on Microelectronics*, San Diego, CA (September).

134 Coyle, R., Reid, M., Ryan, C. et al. (2009). The influence of the Pb-free solder alloy composition and processing parameters on thermal fatigue performance of a ceramic chip resistor. In: *Proceedings of Electronic Components Technology Conference*, Piscataway, NJ, 423–430. IEEE.

135 Manock, J., Coyle, R., Vaccaro, B. et al. (2008). Effect of temperature cycling parameters on the solder joint reliability of a Pb-free PBGA package. *Journal of Surface Mount Technology* 21 (3): 36.

136 Osterman, M., Dasgupta, A., and Han, B. (2006). A strain range based model for life assessment of Pb-free SAC solder interconnects. In: *Proceedings of Electronic Components and Technology Conference*, Piscataway, NJ, 884. IEEE.

137 Bath, J., Sethuraman, S., Zhou, X. et al. (2005) Reliability evaluations of lead-free SnAgCu PBGA676 components using tin-lead and lead-free SnAgCu solder paste. *Proceedings of SMTAI*, Edina, MN, 891.

138 Clech, J.-P. (2005). Acceleration factors and thermal cycling test efficiency for lead-free Sn-Ag-Cu assemblies. In: *Proceedings of SMTA International*, Chicago, IL, 902–917.

139 Fan, X., Raiser, G., and Vasudevan, V. (2005). Effects of dwell time and ramp rate on lead-free solder joints in FCBGA packages. In: *Electronic Components and Technology Conference*, 901–906.

140 Lee, J. and Subramanian, K. (2003). Effect of dwell times on thermomechanical fatigue behavior of Sn-Ag–based solder joints. *Journal of Electronic Materials* 32 (6): 523–530.

141 Bartelo, J., Cain, S., Caletka, D. et al. (2001) Thermomechanical fatigue behavior of selected Pb-free solders. *Proceedings of IPC APEX 2001*, LF2-2, Bannockburn, IL.

142 Raj, A., Sridhar, S., Gordon, S. et al. (2018) Long term isothermal aging of BGA packages using doped lead free solder alloys. *Proceedings of SMTAI*, Rosemont, IL (October).

143 Thirugnanasambandam, S., Sanders, T., Evans, J. et al. (2014). Component level reliability for high temperature power computing with SAC305 and alternative high reliability solders. In: *Proceedings of SMTAI*, 262–270.

144 Zhang, Y., Cai, Z., Suhling, J.C. et al. (2009). The effects of SAC alloy composition on aging resistance and reliability. In: *Proceedings of Electronic Components and Technology Conference*, San Diego, CA, 370–389.

145 Ma, H., Suhling, J.C., Zhang, Y. et al. (2007). The influence of elevated temperature aging on reliability of lead free solder joints. In: *Proceedings of Electronic Components Technology Conference 2007*, Reno, NV, 653–668.

146 Ma, H., Suhling, J.C., Zhang, Y. et al. (2006). Reliability of the aging lead free solder joint. In: *Proceedings of Electronic Components Technology Conference*, San Diego, CA, 849–864.

147 Snugovsky, L., Perovic, D., and Rutter, J. (2005). Experiments on the aging of Sn-Ag-Cu solder alloys. *Powder Metallurgy* 48: 193–198.

148 Hasnine, M., Mustafa, M., and Suhling, J.C. (2013). Characterization of aging effects in lead free solder joints using nanoindentation. In: *Proceedings of Electronic Components Technology Conference*, 166–178.

149 Chavali, S., Singh, Y., Kumar, P. et al. (2011). Aging aware constitutive models for SnAgCu solder alloys. In: *Proceedings of Electronic Components Technology Conference*, 701–705.

150 Mysore, K., Chan, D., Bhate, D. et al. (2008). Aging-informed behavior of Sn3.8Ag0.7Cu solder alloys. In: *Proceedings of ITHERM*, 870–875.

151 Venkatadri, V., Yin, L., Xing, Y. et al. (2009). Accelerating the effects of aging on the reliability of lead free solder joints in a quantitative fashion. In: *Proceedings of Electronic Components Technology Conference*, 398–405.

152 Lee, T.-K., Ma, H., Liu, K.-C., and Xue, A.J. (2010). Impact of isothermal aging on long-term reliability of fine-pitch ball grid array packages with Sn-Ag-Cu solder interconnects: surface finish effects. *Journal of Electronic Materials* 39 (12): 2564–2573.

153 Wilcox, J., Coyle, R., Lehman, L., and Smetana, J. (2014). Effect of isothermal preconditioning on thermal fatigue life and microstructure of a SAC305 BGA. In: *Proceedings of SMTAI*, Chicago, IL, 122–133.

154 Dompierre, B., Aubin, V., Charkaluk, E. et al. (2001). Cyclic mechanical behaviour of Sn3.0Ag0.5Cu alloy under high temperature isothermal ageing. *Materials Science and Engineering A* 528: 4812–4818.

155 Zhang, Y., Cai, Z., Suhling, J., et al. (2009) Aging effects in SAC solder joints. *Proceedings of the SEM International Congress and Exposition on Experimental and Applied Mechanics*, Albuquerque, NM (June).

156 Bhate, D., Chan, D., Subbarayan, G. et al. (2007). Constitutive behavior of Sn3.8Ag0.7Cu and Sn1.0Ag0.5Cu alloys at creep and low strain rate regimes. *IEEE Transactions on Components, Packaging, and Manufacturing Technology* 31 (3): 621–633.

157 Xiao, Q., Nguyen, L., and Armstrong, W. (2004). Aging and creep behavior of Sn3.9Ag0.6Cu solder alloy. In: *Proceedings of Electronic Components Technology Conference*, Las Vegas, NV, 1325–1332.

158 Xiao, Q., Bailey, H., and Armstrong, W. (2004). Aging effects on microstructure and tensile property of Sn3.9Ag0.6Cu solder alloy. *Journal of Electronic Packaging* 126: 208–212.

159 Mysore, K., Chan, D., Bhate, D. et al. (2008). Aging-informed behavior of Sn3.8Ag0.7Cu solder alloys. In: *Proceedings Thermal and Thermomechanical Phenomena in Electronic Systems (ITHERM)*, Florida, FL, 870–875.

160 Kim, D.H., Lee, T.-K., Kim, S.H. et al. (2008). Study on dynamic shock performance of SAC305 solder joint after different aging conditions. In: *Proceedings of SMTAI*, Orlando, FL, 182–186.

161 Delhaise, A.M., Brillantes, M., Tan, I. et al. (2018) Restoration of microstructure and mechanical properties of lead-free bismuth containing solder joints after accelerated reliability testing using a thermal treatment. *Proceedings SMTAI*, Rosemont, IL (October).

162 Lutz, J., Hermann, T., Feller, M. et al. (2011) Power cycling induced failure mechanism in the viewpoint of rough temperature environment. *5th International Conference on Integrated Power Electronics Systems*, Nuremberg, Germany (11–13 March).

163 iNEMI Next generation Solder Materials Workshop. Norman Armendariz (2019) Defense industry circuit card assembly lead (Pb)-free transition challenges, Jie Geng, Hongwen Zhang, and Ning-Cheng Lee, "Die attach soldering," Richard Coyle, "Recommendations for improvements in reliability testing for next generation lead free solder alloys," and Dock Brown, "Next generation solder technology". *IPC APEX EXPO 2019* (31 January). www .inemi.org/solder-workshop-apex-2019-get-presentations (accessed 4 February 2020).

164 JESD22-B111A (2016) *Board level drop test method of components for hand-held electronic products.* Arlington, VA: JEDEC.

165 Thukral, V., Zaal, J.J.M., Roucou, R. et al. (2018). Understanding the impact of PCB changes in the latest published JEDEC board level drop test method. In: *Proceedings of the 2018 IEEE 68th Electronic Components and Technology Conference*, 756–763.

166 Fu, H., Radhakrishnan, J., Ribas, M. et al. (2018) iNEMI project on process development of Bi-Sn-based low temperature solder pastes – Part IV: comprehensive mechanical shock tests on POP components having mixed BGA BiSn-SAC solder joints. *Proceedings of SMTAI*, Rosemont, IL (October 2018).

167 JESD22-B103-B (2010) *Vibration, variable frequency test method.* Arlington, VA: JEDEC.

168 Wong, S.F., Malatkar, P., Rick, C. et al. (2007). Vibration testing and analysis of ball grid array package solder joints. In: *Proceedings of 57th Electronic Components Technology Conference*, Reno, NV, 373–380.

169 Woodrow, T. (2010) NASA-DoD lead-free electronics project: vibration test. Boeing Electronics Materials and Process Report – 603 Rev. A, 18 November.

170 Mukadam, M., Long, G., Butler, P., and Vasudevan, V. (2005). Impact of cracking beneath solder pads In printed circuit boards on reliability of ball grid array packages. In: *Proceedings SMTAI*, Rosemont, IL, 324–329.

171 Belyakov, S.A., Arfaei, B., Johnson, C. et al. (2019). Phase formation and solid solubility in high reliability Pb-free solders containing Bi, Sb, or In. In: *Proceedings of SMTAI*, Rosemont, IL, 492–508.

8

Lead-Free Printed Wiring Board Surface Finishes
Rick Nichols

Atotech, Berlin, Germany

8.1   Introduction: Why a Surface Finish Is Needed

The layman's idea of soldering is with a soldering iron, flux, and soldering wire. In the field of Printed Wiring Boards (PWBs), independently of scale, the flux and solder wire is replaced by solder paste and the soldering iron is replaced by a reflow oven or a remote heat source. Typically the former is low volume whereas the second procedure is to some extent mass production. The principle is, however, the same, and is used to form a solid bond between metals. The principle still holds true but the applications and methodology have changed significantly. The transition to lead-free soldering is only part of the story.

The first step to successful soldering, which is valid in the electronics industry, is a clean copper surface, as the interfacial tension values are sensitive to contaminants, even those present at seemingly insignificant concentrations [1]. Interfacial tensions have a direct impact on wettability or the lack of wettability. The metal that is usually employed to form circuitry is copper which, while being abundant and conductive, has a tendency to oxidize and corrode. To overcome this issue the industry has introduced final finishes that protect the copper and maintain the solderability.

Ultimately the choice of a final finish (FF) is controlled by the original equipment manufacturer (OEM) or the electronic manufacturing service (EMS) both of which are effectively end users. The choice is made by what is available on the market, how it fits the requirements and whether the finish can fulfill the specifications generated. For newer final finishes overcoming specifications that have been established based on historical data can be a challenge and may have to provide a very specific performance that other finishes cannot provide. Whether it is because of cost reasons, or performance limitations, no one final finish can fulfill all the expectations in the field.

Lead-free Soldering Process Development and Reliability, First Edition. Edited by Jasbir Bath.
© 2020 John Wiley & Sons, Inc. Published 2020 by John Wiley & Sons, Inc.

While the aim of a soldering operation is relatively simple: to solder on to copper, the printed circuit board (PCB) fabrication operation has many variables to work in unison. These include:

- Poorly executed previous production steps resulting in residues or even galvanic cells
- Solder mask types
- Discipline during start-up and running-down
- Analysis and control
- On site infrastructure such as rinse water pressure consistency, pneumatic pressure consistency and exhaust capabilities
- Extreme dimensions
- Ever-evolving base materials
- Historical testing procedures that are no longer valid

Solderable finishes were originally associated with lower-end product and were championed by Hot Air Solder Leveling (HASL), but with the emergence of Surface Mount Technology (SMT) soldering evolved to accommodate for the ever-decreasing form factors and more specifically, to overcome the non-planar surfaces created in the HASL process. In addition to this, lead-free soldering was introduced in the new millennium, for better or worse, mainly due to the European Union Restriction of Hazardous Substances (EU RoHS). This move was championed in Europe after 1 July 2006.

8.2 Surface Finishes in the Market

Based from information extrapolated from a report [2], the global distribution final finishes is represented in Figure 8.1.

Electroless nickel/immersion gold (ENIG) is still the dominant finish in the PWB market. This is a "work horse" that is established and well understood. ENIG, however, is facing testing times with the introduction of the latest version of IPC 4552 [3]; specifically with respect to corrosion. A standoff has emerged between the end users who want to stipulate zero corrosion tolerance and the fabricators and suppliers who want to pursue high-volume production. As yet there has been no concrete link established between the degree of corrosion incidents and poor solderability [3]. ENIG is a significant all-round performance surface finish for mass production as it is solderable and capable for aluminum wire bonding. Issues such as black pad are addressed by the IPC 4552 specification.

Immersion tin (I-Sn) and organic solderability preservative (OSP) have both benefitted from the recent surge in electronic components/products in the automotive industry. Immersion tin is an established finish in the automotive industry due

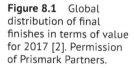

Figure 8.1 Global distribution of final finishes in terms of value for 2017 [2]. Permission of Prismark Partners.

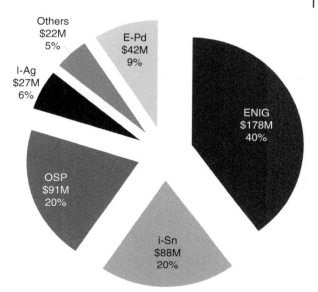

to the finishes unchallenged suitability for press fit; until now. The US automotive market has historically endorsed immersion silver (I-Ag). The finishes in this paragraph occupy roughly the same arena of competence in terms of soldering. They are not wire bondable and their respective niches will be discussed later.

HASL is a favorite for the conservative military and aerospace electronics market that is dominated by reliability and where change is slow. In many cases the solder applied is still eutectic or lead-containing. This finish is also experiencing a change in fortune as HASL equipment manufacturers are also evolving to meet the market trends in a positive fashion.

Electroless palladium containing final finishes were deemed significant enough to be included as an independent entity in the 2017 report [2].

This report [2] has been included because it was the source data for the extrapolated 2018 distribution in Figure 8.2. Based on the predicted growth and percentage distribution per region the data for 2018 was generated. The extrapolated data is in Table 8.1.

Figure 8.2 demonstrates that continued growth was seen for all the finishes apart from I-Ag in 2018. Immersion silver, while being an excellent finish is not as resistant to environmental corrosion, which may eliminate it as a suitable finish for humid, sulfur-rich environments such as Asia. In this respect I-Ag has no significant advantage over OSP and is more expensive.

The US automotive industry has been a historical bastion of I-Ag, yet influence from the competing Japanese and European manufacturers who favor OSP and I-Sn respectively may have made the US uncertain of their preferred automotive

Table 8.1 Extrapolated market situation based on growth data provided by 2017 report in M$ [2]. Permission of Prismark Partners.

	Global	Americas		Europe		Japan		Korea		Taiwan and China		ROA	
	WW 2017	2017	Market %	2017	Market %	2017	Market %	2017	Market %	2017	Market %	2017	Market %
OSP	90,00	1,00	1,1%	1,00	1,1%	11,00	12,2%	4,00	4,4%	62,00	68,9%	11,00	12,2%
I-Ag	27,00	2,00	7,4%	1,00	3,7%	1,00	3,7%	0,00	0,0%	22,00	81,5%	1,00	3,7%
i-Sn	87,00	1,00	1,1%	4,00	4,6%	2,00	2,3%	8,00	9,2%	64,00	73,6%	8,00	9,2%
ENIG	178,00	10,00	5,6%	6,00	3,4%	41,00	23,0%	6,00	3,4%	106,00	59,6%	9,00	5,1%
E-Pd	42,00	4,00	9,5%	1,00	2,4%	11,00	26,2%	2,00	4,8%	23,00	54,8%	1,00	2,4%
Others	22,00	1,00	4,5%	2,00	9,1%	4,00	18,2%	0,00	0,0%	13,00	59,1%	2,00	9,1%

	Delta 17–18	WW 2018	Americas 2018	Market %	Europe 2018	Market %	Japan 2018	Market %	Korea 2018	Market %	Taiwan and China 2018	Market %	ROA 2018	Market %
OSP	+11%	99,90	1,11		1,11		12,21		4,44		68,82		12,21	
I-Ag	−7%	25,11	1,86		0,93		0,93		0,00		20,46		0,93	
i-Sn	+11%	96,57	1,11		4,44		2,22		8,88		71,04		8,88	
ENIG	+8%	192,24	10,80		6,48		44,28		6,48		114,48		9,72	
E-Pd	+10%	46,20	4,40		1,10		12,10		2,20		25,30		1,10	
Others	+16%	25,52	1,16		2,32		4,64		0,00		15,08		2,32	
Total		448,00	21,00	4,7%	16,00	3,6%	69,00	15,4%	20,00	4,5%	289,00	64,5%	33,00	7,4%
Total		485,54	22,76		17,34		74,78		21,68		313,22		35,77	

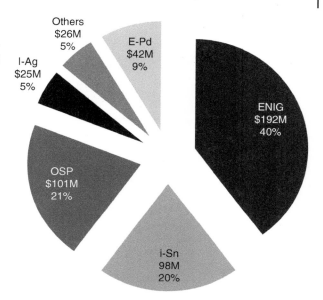

Figure 8.2 Global distribution of final finishes in terms of value for 2018. Source: Derived from data supplied by 2017 report [2]. Permission of Prismark Partners.

final finish. Therefore it is reasonable to conclude that the automotive industry is most likely the catalyst for the downturn in I-Ag in favor of OSP and I-Sn. From Figure 8.2 it can be seen that the 1% drop in the adjusted 2018 figures for I-Ag has been absorbed by the increase in OSP and I-Sn. In Europe the automotive market has focused on I-Sn but the influence of the Japanese market has also led to the traditional strong hold of I-Sn starting to investigate OSP as an alternative.

Geographical locations have tendencies that are quite individual depending on their primary manufacturing goals. Figure 8.3 represents the breakdown of final finishes according to countries or regions. This notion will be elaborated upon in the following paragraphs with a brief overview of graphs (a) to (f).

The nomenclature of Americas includes both North and South America. These regions have disparities in terms of production goals. The OSP is driven largely by the Central and South American countries with a small influence from the experimentation of the North American automotive producers and the domination of the precious metal finishes is due to the North American military requirements.

Europe is dominated by military/aerospace and the high end automotive industry. These industries prefer ENIG and I-Sn respectively. Additionally, these fields are very conservative because an integral part of their expectations is safety.

The Korean market is a dichotomy of mobile phone production and consumable electronics all primarily dominated by a single end user. The automotive industry is a contributor but not a significant player. The mobile phone industry accounts for the precious metal finishes, while the remaining market accounts for the OSP and I-Sn.

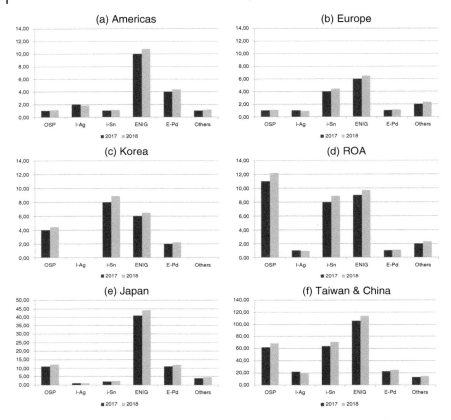

Figure 8.3 The evolution of the final finish market from 2017 to 2018 in M$. Source: Based on 2017 report data [2]. Permission of Prismark Partners.

The rest of Asia (ROA) is mainly represented by Thailand for automotive and low end consumables such as printers, India for automotive and military, and Malaysia for packaging. The relocation to Thailand and Vietnam of the larger Japanese and Korean concerns has fueled the development in all areas. Automotive production is a dominant force in the regions with the strongest contributor being Thailand. Malaysia is planning to step into this market.

Japan is very strong in the automotive market but unlike Europe favors OSP as the appropriate automotive final finish. The infotainment systems and car safety oriented electronics are represented by the ENIG values.

China is an all-round supplier of products and global leader in PWB production. Naturally, the consumption of final finishes is also greatest in China.

This section of the chapter is only seeking to give a rough outline from a regional perspective and, as such, may not have captured all relevant alternative interpretations.

8.3 Application Perspective

Previously it was alluded to that no one final finish has the attributes to fulfill the full range of requirements in the market, including the fiscal ones.

The following processes will be the final finishes included in this chapter:

- Lead-free Hot Air Solder Leveling (HASL)
- High temperature Organic Solderability Preservative (OSP)
- Immersion tin (I-Sn)
- Immersion silver (I-Ag)
- Electroless Nickel/Immersion Gold (ENIG)
- Electroless Nickel/Electroless Palladium/immersion Gold (ENEPIG)
- Electroless Nickel/Autocatalytic Gold (ENAG)
- Electroless Palladium Autocatalytic Gold (EPAG)
- Electrolytic Nickel Electrolytic Gold

Most of the categories discussed in the previous section are clearly defined; however, the Semi-Autocatalytic Gold definition has newly emerged as a significant consideration and can apply to all the universal final finishes in Table 8.2. There are even smaller sub-sets within some of the more clearly defined categories. These will be elaborated upon later.

It must also be stressed that the capabilities are based on the outer layer only. This is especially important when referring to line and space capabilities.

The categories "Horizontal" and "Reel to Reel" are process related and feature due to the emergence of an enhanced/increased flexible and flex-rigid market. As the primary purpose of a final finish is to protect the integrity of the underlying copper circuit dimensionally and functionally, the headings chosen are based on physical characteristics. Potentially there are more but these are the most pertinent and are represented in Table 8.2. This table is helpful to focus in on what kind of equipment is applicable to which final finish process.

Form factor and resolution are key concerns when selecting a final finish. For example; a highly corrosion-resistant finish employing a 4–5 μm nickel diffusion barrier will have limitations for fine line applications.

Figure 8.4 is a scaled visual representation of the comparative thicknesses of the final finishes. It must be stressed that in Figure 8.4 the dimensional differences are

Table 8.2 An overview of the performance characteristics for some of the recognized final finishes.

Final finish	Solderable				Solderable and wire bondable	Universal finish			
	HASL Pb-free	HT OSP	Immersion tin	Immersion silver	ENIG	ENEPIG	ENAG	EPAG	E'lytic Ni/Au
Intermetallic compound (IMC)	Cu/Sn	Cu/Sn	Cu/Sn	Cu/Sn	Ni/Sn	Ni/Sn	Ni/Sn	Cu/Sn	Ni/Sn
Typical layer thickness (μm)	>3	0.2–0.4	1–1.5	0.08–0.12	3–6	3–7	3–7	0.1–0.3	5.5–7.5
Corrosion protection	••	○	••	○	•	•	•	•	••
Planarity	○○	••	••	••	••	••	••	••	••
Thickness distribution	○	•	••	••	•	••	••	••	•
Compatibility lead-free	••	•	••	••	••	••	•	••	••
Compatibility solder mask	•/○	••	•/○	••	•	•	•	••	••
Multi-soldering	••	•	••	••	••	••	••	••	••
Press fit	•	•/○	••	•	○	○	○	••	••
Shelflife	••	○	••	•	••	••	••	••	••
Horizontal processing	•	••	••	••	○	○	○	•	○○
Vertical processing	••	•	•	○	••	••	••	••	••
Reel to reel processing	○○	••	••	••	•	•	•	••	•

•• = very good, • = good, •/○ = neutral, ○ = poor, ○○ = very poor.

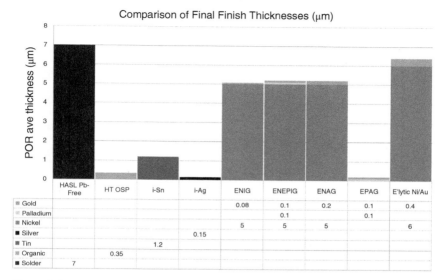

	HASL Pb-Free	HT OSP	i-Sn	i-Ag	ENIG	ENEPIG	ENAG	EPAG	E'lytic Ni/Au
Gold					0.08	0.1	0.2	0.1	0.4
Palladium						0.1		0.1	
Nickel					5	5	5		6
Silver				0.15					
Tin			1.2						
Organic		0.35							
Solder	7								

Figure 8.4 A comparison of the final finishes with respect to thickness after process of record (POR) layer deposition (μm).

only in the Z-axis while in reality the disparities in dimensions will be in the x axis as well, which will inevitably impact line and space capability. In Figure 8.5, less gap loss is preferred.

In Table 8.3 the fine line capabilities are outlined according to final finish. There are process modifications that exist to manipulate a finish to be more capable for fine line applications, usually at the expense of other attributes.

The most desirable situation, excluding cost, is a finish that does not require front end engineering compensations while being able to fulfill the task at hand. The two finishes that stand out in Table 8.3 are the OSP and EPAG finishes. While both finishes are suitable for advanced roadmap resolution predictions, OSP can only be soldered and is not wire bondable, has a relatively short shelf life and is not regarded as corrosion-resistant whereas EPAG can be soldered, is multi-medium wire bondable, and is moderately corrosion-resistant. EPAG is, however, relatively expensive compared to OSP.

When to apply a final finish and for what technology is ultimately the most relevant topic. In Table 8.4, an overview has been composed to show the general areas of application. Again the categories are broad and only serve as a general guide.

Communications is a topic that encompasses mobile networks through to server and satellite technology. In addition, 5G technology has pushed communication to the forefront of the electronics industry and by association final finish compatibility is subsequently in focus. The key differentiations in the field of communication

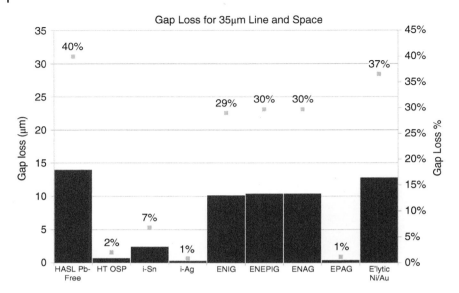

Figure 8.5 Gap losses of the finishes at typical layer thicknesses.

Table 8.3 Final finish with reference to fine line capabilities.

	Fine line capabilities								
	HASL Pb-free	HT OSP	Immersion tin	Immersion silver	ENIG	ENEPIG	ENAG	EPAG	E'lytlc Nl/Au
Panel level packaging (PLP) (<8 μm)		O						O	
Semi-additive process (SAP) (>8 μm)		O		O				O	
Advanced modified SAP (>18 μm)		O	O	O				O	
Modified SAP (>25 μm)		O	O	O				O	
Anylayer (>35 μm)		O	O	O	O	O	O	O	O
Tent & etch (>40 μm)	O	O	O	O	O	O	O	O	O

Table 8.4 Final finish use with reference to technology.

Technology	Solderable HASL Pb-free	Solderable HT OSP	Solderable Immersion tin	Solderable Immersion silver	Solderable and wire bondable ENIG	Universal finish ENEPIG	Universal finish ENAG	Universal finish EPAG	Universal finish E'lytic Ni/Au
Communication (network/ infrastructure)		O	O	O	O	O	O	O	O
Consumables (cellphones, laptops etc)	O	O	O						
Automotive		O	O	O	O	O			O
Industrial	O	O	O		O				O
Test & measurement						O	O	O	
Medical					O	O	O	O	O
Military/ aerospace	O		O		O	O	O	O	O

electronics are highest reliability, high frequency and fiscal concerns. Highest reliability is catered for by the precious metal finishes containing nickel such as ENIG, Electrolytic Nickel Electrolytic Gold and ENEPIG. ENIG is commonly found as a functional finish for backplane panels or server boards for heat dispersion and in mobile phones in the case of aluminum wire bonding and soldering. Both ENIG and ENEPIG are common finishes for satellite components for maximum reliability built on historically generated perceptions. Immersion silver (I-Ag) and EPAG are ideally suited to high frequency applications, due to low signal losses, and are able to accommodate high density circuitry, which are often co-existing requirements. The OSP is cost driven and is confined typically to mobile phone production as Second Image Technology (SIT) with ENIG. It must be noted that OSP also has a role in very high density circuitry that requires wide bandwidth chip to chip connection. Immersion tin (I-Sn) is being championed as a final finish for aerials for military and to provide for the 5G market.

Although the information in Figure 8.6 is a comparison of the relative processing times for the respective final finishes, it is not intended to be used as a design template for plating equipment layouts as individual system and suppliers have specialized specifications.

Typically, a horizontal process covers less area or has a smaller production "footprint" than a vertical process depending on throughput. This being said, vertical equipment is more versatile with manipulation to cater for variations in a

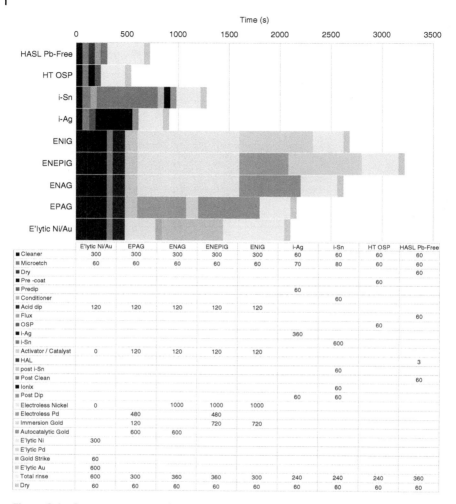

	E'lytic Ni/Au	EPAG	ENAG	ENEPIG	ENIG	i-Ag	i-Sn	HT OSP	HASL Pb-Free
■ Cleaner	300	300	300	300	300	60	60	60	60
■ Microetch	60	60	60	60	60	70	80	60	60
■ Dry									60
■ Pre -coat								60	
■ Predip						60			
▦ Conditioner							60		
■ Acid dip	120	120	120	120	120				
▦ Flux									60
■ OSP								60	
■ i-Ag						360			
▦ i-Sn							600		
▦ Activator / Catalyst	0	120	120	120	120				
■ HAL									3
▦ post i-Sn							60		
■ Post Clean									60
■ Ionix							60		
■ Post Dip						60	60		
▦ Electroless Nickel	0		1000	1000	1000				
▦ Electroless Pd		480		480					
▦ Immersion Gold		120		720	720				
▦ Autocatalytic Gold		600	600						
▦ E'lytic Ni	300								
▦ E'lytic Pd									
▦ Gold Strike	60								
▦ E'lytic Au	600								
▦ Total rinse	600	300	360	360	300	240	240	240	360
▦ Dry	60	60	60	60	60	60	60	60	60

Figure 8.6 Process sequence with relative generic processing times in seconds for different final finishes.

production portfolio. In a vertical system each tank is effectively a separate process where dwell times can be set according to need.

The dwell times for a horizontal process are significant shorter than those for a vertical process because in a horizontal process the solution is brought to the panel by a fluid delivery system whereas in vertical processing equipment the panel is brought to the solution where soak time is fundamental for reaction and rinsing to take place. Both horizontal and vertical equipment can be augmented by ultrasonic devices in key steps. Ultrasonic transducers have a greater impact in horizontal processing equipment. Ultrasonic transducers are not unilaterally suitable.

8.4   A Description of Final Finishes

This section highlights the core board plating process details of each finish individually. A numerical score has been allocated according to the complexity of the process. According to the topic, a score is allocated according to complexity level which is given as in the example in Table 8.5. A low score is superior to a high score and no score shows that a variable has been excluded from impacts on the process. The complexity scores are added up to give an overall value for the process.

A Complexity A is scored as one point, a Complexity B is scored as two points and a Complexity C is scored as three points. The point allocation is based on an experience based judgment. For example a process that requires titration to be controlled will score lower than a process that requires High Pressure Liquid Chromatography (HPLC).

By using the scoring system outlined in Table 8.5, it is possible to judge the demands that will be placed on a board plating facility when choosing a final finish as shown in Figures 8.7 and 8.8.

Functional chemical finishes inherently have an element of complexity, but the complexity is not unilaterally equivalent. The implication of complexity is that some processes are better suited to specific applications or fabricators.

Table 8.5   How each final finish was scored in terms of complexity.

		Complexity A	Complexity B	Complexity C
Maintenance and handling	Analysis	1		
	Dummying		2	
	Maintenance			3
Most suitable production environment	Prototype			3
	Volume		2	
Amenities required	Compressed air	1		
	Water	1		
	Heating		2	
	Waste water			3
	Extraction		2	
	Filtration	1		
Process complexity	Score	4	8	9
	Final score		**21**	

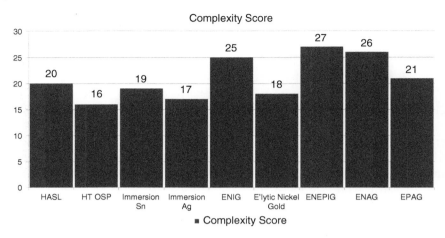

Figure 8.7 The total complexity score by finish.

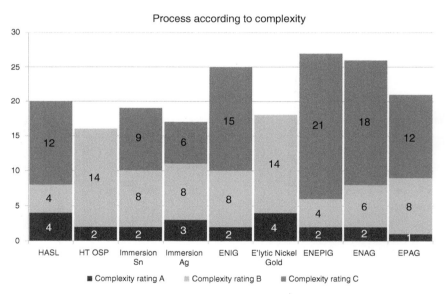

Figure 8.8 Final finishes according to process complexity.

In terms of "ease of handling," a process with only complexity scores of 1 would be the preferred. Complexity, however, is not the only measure of a process because as demands on the process are amplified the tendency is for the process to become more complex. A process must suite the application no matter how complex the application is.

8.4.1 Hot Air Solder Leveling (HASL)

Lead-free solder leveling is showing a resurgence due to the improvement in the physical manipulation of the physical dynamics. The relatively poorer wetting performance of the lead-free solder, compared to the eutectic tin-lead solder, is an advantage when removing the excess solder and therefore results in improved coplanarity, especially when combined with state of the art equipment. The equipment has also evolved significantly by introducing automation to reduce the impact of human error. Timing and copper content is automatically controlled and monitored.

Despite the negative publicity, which focused on the increase in temperature and handling procedures required to apply lead-free solder compared to eutectic tin-lead solder, the simplicity and reliability of the finish coupled with the realization that fine pitch surface mount technology is not universally required is proving to be technologically and fiscally alluring. Currently fine pitch is approximately 300 μm but is not restricted to this value.

Currently the default lead-free solder is Sn3Ag0.5Cu (SAC305) which is capable of depositing between 1.75 and 15 μm thickness of coating on the board [4]. The major benefit of the finish is the metallurgic similarities between the finish and the solder medium used for assembly.

8.4.1.1 Process Complexity

The process complexity score for HASL is 20 according to the classification outlined previously (Table 8.6). This score has been given because the physical requirements, especially maintenance, are significant for the process. The score is in itself only an overall ranking comparison. More relevant is the allocation of the scores for an understanding of ownership benefits. These benefits or complexities are better represented in Figure 8.9.

8.4.1.1.1 Maintenance and Handling

HASL is primarily a physical or mechanical process, and therefore does not require the degree of chemical control that is associated with the other finishes that are addressed in this chapter. HASL is a process very well suited to stop and go environments as the system can handle periods of sporadic activity in combination with prolonged idle time. Even the typical chemically based pretreatments are

Table 8.6 Complexity score for HASL.

		Complexity A	Complexity B	Complexity C
Maintenance and handling	Analysis	1		
	Dummying			
	Maintenance			3
Most suitable production environment	Prototype	1		
	Volume	1		
Amenities required	Compressed air			3
	Water		2	
	Heating			3
	Waste water		2	
	Extraction			3
	Filtration	1		
Process complexity	Score	4	4	12
	Final score		**20**	

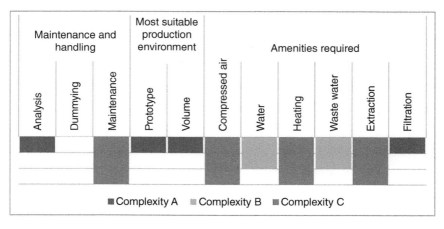

Figure 8.9 A visual breakdown of the complexity scores for HASL.

stable chemical baths and need minimum analysis. Analysis aside the process is also dummy-free which means no activation panels are required to "wake up" the process as is the case with alternative electroless processes. The key to successful production when employing HASL is regular maintenance. The air knives in the HASL equipment are the top priority for regular maintenance.

8.4.1.1.2 *Most Suitable Production Environment*

Other than for materials that have a low threshold for high temperature and asymmetrical build-ups, as these panels have a tendency to warp, HASL is suitable for all production environments and volumes. This capability has further been enhanced by the evolutionary sophistication of the associated equipment. HASL processes employ predominantly vertical (dip) soldering tanks. Horizontal processing is also mature and provides a significant improvement in coplanarity versus vertical systems.

8.4.1.1.3 *Amenities Required*

The process, does, however, depend on amenities such as compressed air, extraction, and most significantly heating power. The operating temperature of a HASL line is significantly higher than all other final finish application systems. Single-sided and simple double-sided boards can be hot air solder leveled with a solder bath temperature of 260 °C, but to get the surface of a heavier multilayer board to wetting temperature in a reasonable time the solder bath temperature might have to be set as high as 280 °C, or in vertical processes the immersion time in the solder has to increase by around 50%, leaving more potential for heat induced delamination [5]. The high temperature required for the process also burns off solvent fluxes with the net results being the requirement of strong extraction. Typical solder dipping times are between two and three seconds. The low contact time ensures that the increase in temperature within the solder bath has little impact on the panel integrity.

The mechanism is to wet the appropriately pretreated copper surface with molten tin and then to remove the excess with compressed air knives to maximize coplanarity and open through holes. HASL equipment lines are reliant on compressed air and it is recommended to have a dedicated compressor for this process to guarantee consistent processing.

Water is required for the rinsing but is less significant in this relatively simple process. Waste water after the solder leveling step can be more critical due to the Chemical Oxygen Demand (COD) values from the flux residues. COD is a measure of the capacity of water to consume oxygen during the decomposition of organic matter and the oxidation of inorganic chemicals such as ammonia and nitrite. COD measurements are commonly made on samples of waste water or of natural waters contaminated by domestic or industrial wastes [6].

Removal of residues from the laminate surface after HASL due to ionic contamination by the flux can be required and may need a separate cleaning step. These cleaning steps are typically alkaline, carbonate, or amine based systems.

8.4.1.2 Process Description

The process sequence is fairly rudimentary but has some crucial steps to be considered. A typical process sequence is outlined in Figure 8.10.

Process Step	T°C	Ts	Key Step
Acid Cleaner	40–50	30–60	O
Rinse	RT		
Microetch	20–30	30–60	
Rinse	RT		
Dry			O
Flux Application	n/a	50–70	
Hot Air Solder Levelling	260–280	2–8	O
Post clean	40–50	30–60	
Rinse	RT	30–60	
	n/a	n/a	

Figure 8.10 The typical process flow for hot air solder leveling (HASL).

The cleaning step for lead-free HASL is crucial as the tendency for this medium to de-wet on none optimally cleaned surfaces is greater than in the equivalent lead-containing HASL process. Analysis can be replaced or augmented by frequent solution make-ups, and sufficient feed and bleed rates for the application. This is the step where a filter is recommended to remove debris being carried into the process from previous process steps and storage for work in progress (WIP). Optimal roughness of the copper is a prerequisite to get the best coplanarity of the finished surfaces. This is true whether vertical or horizontal HASL systems are employed.

Drying prior to the flux application is important to avoid diluting the flux. The drying step often uses a combination of sponge rollers and air knives rather than a full-blown drier system.

The performance hot air leveling step, as previously mentioned, is dependent on maintenance. Maintaining the equipment is mandatory but removing the copper from the system is also important for solderability and coverage through wetting as HASL is prone to dissolving copper. After renewing the solder bath there may be evidence of exacerbated copper dissolution. This can be minimized by processing some sacrificial panels in order to introduce some copper into the solder bath.

Figure 8.11 A diagrammatic representation of inadequate solder removal from a plated-through hole (PTH).

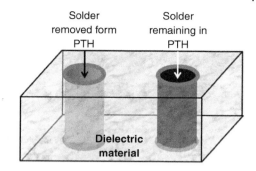

8.4.1.3 Issues and Remedies

8.4.1.3.1 Blocked Plated-Through Holes

Removing excess solder from the plated-through holes (PTHs) is a fundamental requirement for successful HASL production (Figure 8.11). Key to the success of this process are the high pressure air knives. This has been reflected by the degree of complexity C that was allocated to maintenance in Figure 8.9. HASL, however, due to the high temperatures involved, can remove as much as 3 μm of copper from the PWB during processing, which must be taken into account.

8.4.1.4 Summary

HASL is an established process that has developed with time to extend its technical capabilities. HASL is more equipment dependent than the other final finishes in this chapter but it is a surface finish with a long shelf life potential.

8.4.2 High Temperature OSP

OSP systems which apply an organic coating selectively on metals are not true surface finishes but function more like anti-tarnish finishes. The true benefit is cost, small footprint, and process simplicity. The simplicity of the process lends itself to stop-start production. The weak points are that the process is not resistant to corrosive environments and is realistically limited to two reflow cycles in an oxygen reflow atmosphere with more possible in a nitrogen reflow atmosphere.

OSP is commonly used as the solderable finish in SIT applications for mobile devices. When applying SIT, the OSP is combined with a precious metal finish dependent on the wire bonding medium that it is intended to be used with.

8.4.2.1 Process Complexity

The simplicity of OSP is demonstrated in Table 8.7 and Figure 8.12. OSP has the lowest complexity score out of all the final finishes and does not feature any high complexity C scores.

Table 8.7 Complexity score for OSP.

		Complexity A	Complexity B	Complexity C
Maintenance and handling	Analysis		2	
	Dummying			
	Maintenance		2	
Most suitable production environment	Prototype	1		
	Volume	1		
Amenities required	Compressed air			
	Water		2	
	Heating		2	
	Waste water		2	
	Extraction		2	
	Filtration		2	
Process complexity	Score	2	14	0
	Final score		**16**	

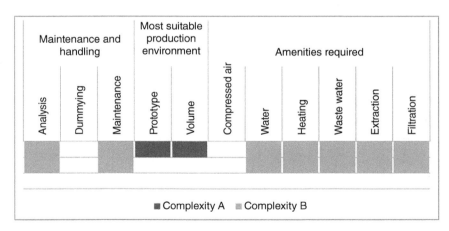

Figure 8.12 A visual breakdown of the complexity scores for OSP.

8.4.2.1.1 *Maintenance and Handling*

OSP is a chemical process and therefore requires chemical analysis. The analysis is straightforward and achievable at most PCB fabricators. Typically the control of the main bath is a UV measurement and the other baths are analyzable by titration.

Because OSP is a selective coating it does not require dummy plating prior to use. This feature means that no material wastage is necessary.

Maintenance can also be considered as low and is mainly centered on the sponge rollers at the entrance of the coating bath. These must be changed regularly and kept clean to ensure an even coverage. Air knives can also be used.

8.4.2.1.2 *Most Suitable Production Environment*
OSP is suited to both high-volume manufacturing and fast turn prototyping. OSP is most suited to horizontal processing.

8.4.2.1.3 *Amenities Required*
The system does not place a high burden on amenities. Waste water may be an issue for waste water treatment as the organics will lead to high CODs. The impact on the waste water will depend on the volume of discharge i.e. production volumes and number of plating lines.

8.4.2.2 Process Description
One-step and two-step OSP processes are used in the field. The example given of a process flow in Figure 8.13 is a two-step OSP process. The purpose of the second step is to minimize the OSP deposition on the precious metal finishes and therefore requires more analysis. For discussion purposes the process sequence in Figure 8.13 is sufficient.

After the OSP coating there is usually a drying or partial drying step in the form of air knives or sponge rollers. This is a key step as it ensures an even coating.

	T°C	Ts	Key Step
Cleaner	35–40	50–70	
Rinse	RT		
Microetch	20–25	50–70	
Rinse	RT		O
Pre-coat	20–25	50–70	
OSP Coating	35–45	50–70	
Rinse	RT		
DI Rinse	25–35		
Dry	n/a	n/a	

Figure 8.13 The typical process flow for OSP.

8.4.2.3 Issues and Remedies

8.4.2.3.1 *Reflow Constraints*

The main limitation for OSP is the reflow atmosphere. Figure 8.14 demonstrates the impact of reflow atmosphere on the wetting angle. A wetting angle of below 20° is considered good for the as-received condition before reflow. The wetting angle will usually deteriorate as the number of reflow cycles increases. For the nitrogen atmosphere the oxygen concentration is kept below 200 ppm.

A nitrogen atmosphere in the reflow oven is recommended to ensure a reliable soldering process.

8.4.2.4 Summary

OSP is a popular final finish due to the low-cost processing that this process offers. The process is also easy to handle.

The key drawback for OSP is that it is not resistant to corrosion and therefore has a short shelf life and requires precise storage considerations. The fragility of the coating also transfers to the board assembly process where board pretreatments/preconditioning steps have to be chosen wisely. In addition to the

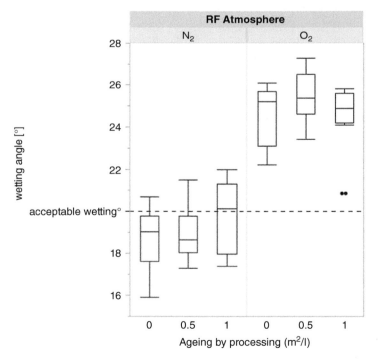

Figure 8.14 A comparison of the wetting angle versus reflow atmosphere for HT OSP at as-received conditions.

pretreatment is the atmosphere in the reflow oven. OSP is best suited to board assembly in nitrogen atmospheres if multiple soldering operations are to be realistic.

OSP is able to fulfill the requirements for high density circuitry because the deposition thickness is only between 0.2 and 0.4 μm.

8.4.3 Immersion Tin

Immersion tin was originally an obvious replacement for eutectic, tin-lead HASL. The metallurgical properties of the finish are similar to HASL and also similar to solder. Immersion tin is a versatile final finish that can be used in the production of double-sided PCBs to IC substrate technology. Immersion tin is one of the preferred finishes for boards used for press-fit connections.

Immersion tin is a good example of regional preferences. While the finish is widely accepted in the European car industry and safety dependent units of the Japanese car industry, Immersion silver is favored at the expense of immersion tin in the United States.

8.4.3.1 Process Complexity

Immersion tin is a robust process that is well suited to high corrosion atmospheres. This is a key aspect that is endorsed by the automotive industry. Because tin is an immersion process it is reasonably easy to handle without the risk of plate out. Plate out can occur in non-immersion electroless process when they become overactive and unstable.

8.4.3.1.1 Maintenance and Handling

Setup analysis is required for immersion tin but horizontal processing usually uses automatic dosing systems to keep the concentration constant after start-up. In addition, the processes are relatively stable with the tin bath itself being most critical.

In some systems auxiliary equipment is used to control the stannic concentration evolution and the increase of copper concentration. These units do require maintenance but the benefits outweigh the negatives.

In vertical plating systems standing time or idle time can result in H_2S gas formation which needs to be purged prior to production. This means that although dummying is not necessary, in vertical systems a start-up procedure may be required after extended periods of down time. Such measures are not required in horizontal processes.

8.4.3.1.2 Most Suitable Production Environment

Immersion tin is ideally suited to high-volume environments as the production equilibrium is more easily achieved and maintained. Stop-start production is also possible as plate out is not an issue and shelf life of the finished panel is approximately one year.

8.4.3.1.3 Amenities Required

Immersion tin is a relatively high consumer of amenities (Table 8.8 and Figure 8.15). Rinse water supply is high after the immersion tin bath because the solution has a high viscosity. Rinsing technology is able to negate the requirement for extreme water feed requirements but immersion tin still requires a large amount of water during the process.

The temperature settings for tin can be considered high but are still lower than those utilized in some of the precious metal plating baths.

Waste water is significant because of the thiourea in the solution. The waste water issue can be offset by the possibility to extend lifetime and so reduce mass waste by needing fewer make-ups. Auxiliary equipment exists to prevent premature aging of the plating bath by copper concentration increase or the build-up of stannic tin. In short, by extending the plating bath lifetime fewer new make-ups are required and the burden on the waste treatment plant is reduced.

Fume extraction is required and not only because of the aroma of thiourea fumes but also as a crucial part of the process to remove H_2S gas from the solution.

Table 8.8 Complexity score for immersion tin.

		Complexity A	Complexity B	Complexity C
Maintenance and handling	Analysis		2	
	Dummying			
	Maintenance		2	
Most suitable production environment	Prototype	1		
	Volume	1		
Amenities required	Compressed air			
	Water			3
	Heating		2	
	Waste water			3
	Extraction			3
	Filtration		2	
Process complexity	Score	2	8	9
	Final score		**19**	

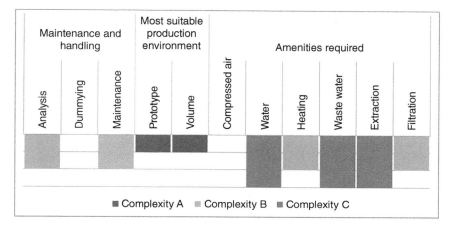

Figure 8.15 A visual breakdown of the complexity scores for immersion tin.

8.4.3.2 Process Description

As previously mentioned, vertical and horizontal processing is possible. Having said that, the process is better suited to horizontal processing as the sump or tank volumes are less and therefore dosing control is more responsive.

The preconditioning is a crucial step for immersion tin (Figure 8.16) as this step prevents copper "mouse bites" and ensures an evenly deposited finish (refer to Section 8.4.3.3).

In some systems the preconditioning step is where the Anti-Whisker Additive (AWA) is concentrated and the step is isolated from the immersion tin step. The AWA is typically silver at a concentration of between 0.5% and 1% w/w of the deposit, while in other systems the AWA is distributed evenly throughout the preconditioning and immersion tin step. In this scenario the preconditioning and immersion tin step are effectively in one tank. The second case lends itself to auxiliary copper control systems where the solution returns to the preconditioning tank cold.

The immersion tin step is the main step in the process and needs to be kept within specification. In horizontal processing there is a possibility to run the line faster to increase throughput. One of the ways this is achieved is by increasing the temperature out of the specification limits. While this will achieve the goal, this action can lead to other issues because the characteristics of the plated layer are changed. This is considered a cause of whiskers as it is generally accepted that stresses in the plating play a role [7].

In addition to directly impacting the plated layer, the other functions of the plating line such as rinsing are also compromised and can have a negative impact on plating. Poor rinsing and drying can result in discoloration and potential degradation of the solderability.

	T°C	Ts	Key Step
Acid Cleaner	35–60	50–70	
Rinse	RT		
Microetch	25–35	70–90	
Rinse	RT		
Pre Conditioning	20–30	50–70	O
Immersion Tin	65–75	480–720	O
Post Immersion Tin	RT		O
Rinse	RT		
Ionic Removal	40–65	60–120	O
Rinse	RT		
Post Dip	RT	50–70	
Dry			O

Figure 8.16 The typical process flow for immersion tin (horizontal processing).

The solder mask type and treatment thereof is also essential for optimum immersion tin plating [8]. Typically a UV bumping machine is placed directly in front of the plating line. This must of course be kept in good working order and the UV lamps checked on a regular basis. The solder mask can also impact the ionic contamination which is limited within the automotive industry to less than or equal to $0.5\,\mu gNaCl\,cm^{-2}$. This value has been tightened from the previous specification of $1.56\,\mu gNaCl\,cm^{-2}$ [9].

Incorrect concentrations or temperature at this stage can also exacerbate discoloration of the tin. This discoloration, although it is usually only cosmetic, is not desirable as on occasions the phenomenon can lead to poor soldering.

Excessive temperature in the drying step can lead to an increased Intermetallic Compound (IMC) formation before storage. Plated panels must, however, be dry prior to vacuum sealing and storage.

Figure 8.17 Diagrammatic representation of "Mouse Bites."

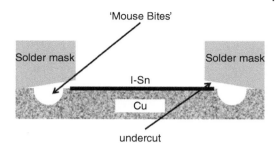

8.4.3.3 Issues and Remedies

8.4.3.3.1 "Mouse Bites"

"Mouse bites" are typically associated with locations where low solution exchange is probable. Solder mask undercuts are an example of this. In these areas replenishing the location with fresh solution is difficult and the tin supply is soon exhausted. The solution matrix, that contains complexors, will subsequently dissolve the copper instead of depositing tin as represented in Figure 8.17.

The conditioning step is designed to form a thin, even coating on the exposed copper surface so that the subsequent main plating step will have an even exchange mechanism instead of a focused exchange mechanism.

Horizontal processing can help remedy this issue because the fluid delivery system can provide better fluid exchange in the micro-feature such as wedges caused by solder mask undercut.

Chemically, a higher tin concentration can be used to improve throwing power in the hole and reduce the impact of poor solution exchange.

8.4.3.3.2 Tin Whiskers

Tin whiskers are a well reported issue in the industry. The recognition of tin whiskers stemmed primarily from electrolytic tin. Pure tin under internal or external stress is required for whisker formation. Tin whiskers are a concern as they have the potential to create shorts because the tin is conductive.

In Figure 8.18 a typical tin copper IMC is depicted. In the diagram the IMC growth is scallop-like in formation, which is considered ideal. The growth of the IMC creates an internal stress that can lead to whisker formation.

Tin whiskers are typically only an issue prior to reflow because after the reflow process most of the pure tin is consumed.

In Figure 8.19, the whiskers appear to be coming from the PCB side but in reality the whiskers are coming from the tin smear originating from the pin.

In the case of press-fit technology, the immersion tin finish is often deemed to be the root cause of whisker generation, whereas, in reality the source of the whisker

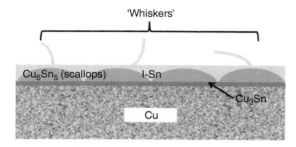

Figure 8.18 The creation of tin whiskers because of internal stress, in this case by the growth of an intermetallic compound (IMC).

Figure 8.19 Tin whiskers in combination with press-fit components.

is from the tin smear left over from the abrasion of the tin coated press-fit pin on the hole. This can be proved as tin whiskers can be found when other final finishes are used [10, 11].

8.4.3.4 Summary

The excellent corrosion resistance of immersion tin makes it the best option for highly corrosive environments typically associated with automotive drive trains. In addition, the reliability of the finish is the preferred option for security systems such as air bags and braking system.

Key to successful immersion tin plating is an awareness of the solder mask used. For multiple soldering operations in board assembly, the solder paste and flux used during assembly and the reflow atmosphere and profile of the reflow oven should be optimized.

Versions of immersion tin can also be used in new fields such as to plate the flanks of singulated quad flat no-leads frames (QFNs/BTCs (bottom termination components)).

8.4.4 Immersion Silver

Immersion silver produces a very coplanar deposit. When stored correctly immersion silver has a long shelf life.

Table 8.9 Complexity score for immersion silver.

		Complexity A	Complexity B	Complexity C
Maintenance and handling	Analysis		2	
	Dummying			
	Maintenance		2	
Production environment	Prototype	1		
	Volume	1		
Amenities required	Compressed air			
	Water		2	
	Heating	1		
	Waste water			3
	Extraction		2	
	Filtration			3
Process complexity	Score	3	8	6
	Final score		**17**	

The excellent conductivity of silver makes it an obvious choice for high frequency applications. This is especially applicable to antenna production and other high frequency or 5G applications.

Immersion silver is susceptible to sulfur-rich environments that can cause creep corrosion.

Immersion silver and immersion tin are perceived as interchangeable, but in reality there are significant differences to these finishes.

8.4.4.1 Process Complexity

Immersion silver is well suited to mass production and is a relatively simple process to control as it is an immersion process and plate out is therefore not a concern (Table 8.9). Plate out is covered in Section 8.4.3.1.

The IPC-4553 standard [12] recommends a thickness in the range of 0.07–0.12 µm for an immersion silver surface finish in general-purpose applications [12].

8.4.4.1.1 Maintenance and Handling

Analysis is straightforward; however, immersion silver systems are sometimes sensitive to light and may need dark covers over the silver processing module. Analysis and physical control of the micro-etch is crucial for optimum performance. The micro-etch needs to be tightly controlled to prevent solder mask undercut which can lead to chemical entrapment which in turn can exacerbate the galvanic corrosion effect, refer to Section 8.4.4.3.

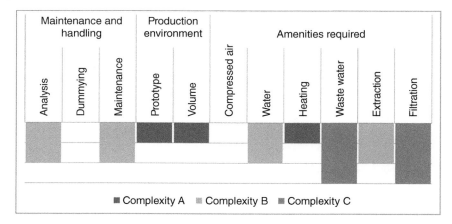

Figure 8.20 A visual breakdown of the complexity scores for immersion silver.

Immersion silver processes require no dummying and little maintenance (Figure 8.20).

8.4.4.1.2 Most Suitable Production Environment

Immersion silver has a small plating equipment footprint and is suitable for mass production. It is slightly more suited to mass production than prototyping as the issues associated with the finish, such as heat related discoloration can be worked out with experience gained through repeat production. However, because there is no risk of plate out it is possible to use immersion silver in a stop-start environment.

8.4.4.1.3 Amenities Required

Immersion silver has no special requirements for rinsing apart from needing a clean deionized water supply. The process is also typically a low temperature process so heating has received a low score. The waste water requires special attention as it is usually nitric acid based. Nitric acid is a problem for waste treatment and is presently a point of concern in China.

Extraction of the working bath is recommended to meet Health and Safety requirements, especially as the immersion silver bath is based on a nitric acid-based matrix. Filtration is highly recommended for all final finish processes as they are functional finishes that need to arrive at the assembly houses in a state ready for soldering or, if applicable, wire bonding.

8.4.4.2 Process Description

The process flow in Figure 8.21 highlights that immersion silver is a low temperature process.

The micro-etch has already been mentioned as a key process step. This is because the etch rate needs to be very low in order to minimize the risk of champagne or micro-voiding and galvanic corrosion at the interface of the solder mask and copper. Champagne or micro-voiding in the soldered joint is predominantly associated with immersion silver processes that contain organics in the matrix. Not all immersion silver processes create champagne voids.

The preconditioning step similar to immersion tin is important to avoid "mouse biting" or excessive copper corrosion where low solution exchange is an issue. Not all immersion silver systems have a preconditioning step. Ultimately the preconditioning step is designed to help establish an even silver deposit in order to prevent exposed copper areas being corroded excessively.

Immersion silver step solutions typically only contain approximately $1 \, g \, l^{-1}$ of silver. One reason for this is the cost of silver. The relatively low concentration of

	T°C	Ts	Key Step
Acid Cleaner	35–60	50–70	
DI Rinse	RT		
Microetch	30–40	60–80	O
Rinse	RT		
Pre Conditioning	40–50	50–70	O
Immersion Silver	40–55	120–240	O
Rinse	RT		
Post Dip	35–60	50–70	
DI Rinse	25–35		
Dry			

Figure 8.21 The typical process flow for immersion silver.

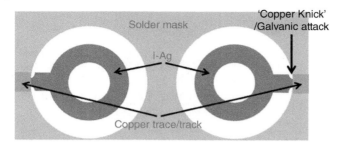

Figure 8.22 Diagrammatic plan view of I-Ag galvanic attack.

silver can result in exhaustion of the reservoir metal where solution exchange is low (refer to Section 8.4.4.3).

8.4.4.3 Issues and Remedies

8.4.4.3.1 Galvanic Attack

Immersion silver is prone to significant galvanic corrosion at the interface between the solder mask and the copper trace (Figure 8.22). As previously outlined this is because the deposition in this location is limited and the silver complexors in the solution matrix preferentially corrode the exposed copper.

In immersion silver systems it is recommended to control the preconditioning step and the etch rate in order to minimize galvanic attack. The galvanic attack between silver and copper is worse than the mouse biting associated with immersion tin because silver is more noble than copper. Tin is less noble than copper and can only be induced to plate on copper by introducing a thiourea. The thiourea changes the potential of the copper so that it becomes less noble than tin.

8.4.4.3.2 Creep Corrosion

Sulfur in a humid atmosphere of around 45–55% relative humidity can form a weak sulfuric acid which can stimulate dendritic growth or creep corrosion originating from the silver. This is shown in Figure 8.23.

Initially creep corrosion can manifest itself as high resistance shorts but as the concentration increases the possibility of a functional short becomes more realistic [14].

Realistically resolving the problem is difficult without creating a hermetic environment. This is one of the root causes that make immersion silver difficult to use in the automotive industry.

8.4.4.3.3 Tarnishing

Immersion silver is easily tarnished by heat treatment. Heat treatment is an analogue for the reflow process. Some of the systems use organics to solve this problem

Figure 8.23 Scanning electron microscopy (SEM) of silver creep corrosion [13].

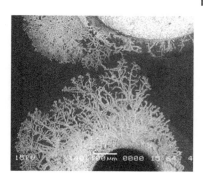

while in some systems increasing the nitric acid concentration during the immersion silver process may help.

8.4.4.4 Summary

Immersion silver has excellent soldering characteristics but is prone to champagne or micro-voiding and tarnishing. As a low safety risk electronic final finish such as higher-end computing immersion silver is a suitable finish. The 5G market will also drive an increase in the use of silver due to the finish's high frequency capability.

8.4.5 Electroless Nickel Immersion Gold (ENIG)

Electroless nickel immersion gold finish is a main finish in the electronics industry and is regarded by many as the best final finish. ENIG is now arguably the most-used finish in the PCB industry since the growth and implementation of the EU RoHS regulation [15].

8.4.5.1 Process Complexity

The previous processes discussed were predominantly horizontal immersion type processes. These types of processes are less complex than the precious metal finishes with autocatalytic steps. Table 8.10 demonstrates the increase in complexity with a score that is six points higher than the next most complex immersion process.

The concept of the ENIG system is to plate an electroless nickel barrier layer as a diffusion barrier between the copper and the gold during soldering or after heat treatments. Migration of copper to the surface has a negative impact on solderability.

8.4.5.1.1 Maintenance and Handling

The ENIG process requires a significant amount of analysis and control (Figure 8.24). The pretreatment needs to prepare the surface but the analysis

Table 8.10 Complexity score for electroless nickel/immersion gold.

		Complexity A	Complexity B	Complexity C
Maintenance and handling	Analysis		2	
	Dummying		2	
	Maintenance			3
Most suitable production environment	Prototype		2	
	Volume	1		
	Compressed air	1		
Amenities required	Water			3
	Heating		2	
	Waste water			3
	Extraction			3
	Filtration			3
Process complexity	Score	2	8	15
	Final score		**25**	

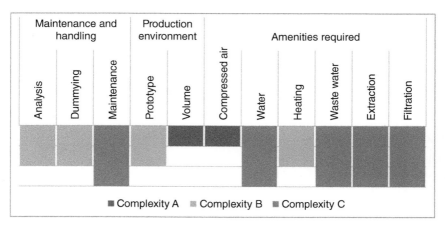

Figure 8.24 A visual breakdown of the complexity scores for electroless nickel/immersion gold.

is similar to the previous processes and the chemistries involved are relatively simple and more importantly stable in terms of concentration fluctuations. The electroless nickel is where the complication starts, as this plating bath is a balance between stability and activity at high temperature. It is highly recommended to use a colorimetric controller with a master slave dosing concept at the very least.

More complicated controllers exist but are commonly installed at high-volume, high end fabricators where process control is critical to ensuring their end users requirements for quality and solderability.

The balance between activity and stability usually makes the need for dummying necessary. In idle times the bath is typically overstabilized and is reinitiated by dummy plating prior to production. The score is only 2 rather than 3 because there are dummy-free ENIG plating systems available in the industry but dummy panels are always available on standby as a precaution. In reality dummy-free electroless nickel processes start slowly and then increase in activity by effectively using the production as dummies.

The electroless nickel bath is a complex mixture of a nickel source, usually sulfate-based, a reducer, usually sodium hypophosphite, complexing agents, stabilizers, and energy, which in this case is heat. This can result in a relative instability in the process bath, particularly during idle times which can result in plate out of the bath onto heaters, tank surfaces and baskets. Bath life is a balance between stability and activity. The impact of plate out is that the nickel bath needs to be transferred to a holding tank or second plating tank to maintain production while the previous tank is stripped of residual nickel. If this procedure is not completely successful, plate out will occur again soon after makeup. The state of the art stripping procedure involves the use of nitric acid. The nitric acid stripping solution must be completely removed in the tank, including all pipework adequately rinsed with water prior to a new makeup, or hyper-corrosion will occur. The nickel will look dark gray if this mistake is made.

8.4.5.1.2 *Most Suitable Production Environment*
ENIG is used in prototype fast turn environments but is labor-intensive and high chemistry consumption can be expected. The best practice for prototype/fast turn production is to batch plate in order to avoid too much inefficiency. Batch plating is where the production is collected until there is a sufficient volume to start the process and be able to achieve somewhere near maximum efficiency. ENIG is best placed in high-volume plating production environments with routine chemical checks and automatic controls in place.

8.4.5.1.3 *Amenities Required*
Due to the large number of process steps the ENIG process places a high burden on amenities. The burden on amenities is exacerbated by the fact that the ENIG process is vertical. Closed horizontal systems require less energy to maintain temperature than open vertical systems.

8.4.5.2 **Process Description**
The typical process as outlined in Figure 8.25 has seven steps. This number can increase if counter-measures to avoid extraneous or excessive nickel plating are included.

	T°C	Ts	Key Step
Cleaner	35–45	180–360	
Rinse	RT	60	
Microetch	25–35	60–120	
Rinse	RT	60	O
Pre Dip	RT	60–180	O
Activator / Catalyst	RT	60–180	O
Rinse	RT	60	
Electroless Nickel	80–90	900–1800	O
Rinse	RT	60	O
Immersion Gold	80–90	420–840	O
Rinse	RT	60	
Dry			

Figure 8.25 The typical process flow for electroless nickel and immersion gold.

Process times are also extended in vertical systems because there are no fluid delivery systems in the plating equipment so the solution exchange is dominated by soak time.

Although, as with all final finishes, the pretreatment is important to prepare the copper surface and includes a cleaner followed by a micro-etch, the key steps in the ENIG process start at the activation stage. Activators or catalysts are susceptible to the drag-in of oxidizing agent from the micro-etch. Also, boards that are heavily oxidized are detrimental to the lifetime of the activation step. This is because copper oxide is easily dissolved in acidic solutions, such as is the case with activator. The rinse after micro-etch and the pre-dip bath are in place to protect the activator from premature aging.

The electroless nickel is, as already mentioned; a balancing act between activity and stability, but the rinse after the electroless nickel step is often overlooked. Nickel is readily oxidized and this can lead to skip plating. The rinse after the electroless nickel step should be kept clean with ample fresh water and the dip time in the rinse should be no longer than one minute. The immersion gold step can be a source for considerable saving if the gold content is controlled well. Ensuring that the immersion gold step is kept in specification can also contribute to improved corrosion resistance of the final finish.

8.4.5.3   Issues and Remedies

8.4.5.3.1   *Excessive/Extraneous Plating*

Nickel foot is the result of excessive/extraneous plating of the electroless nickel process (Figure 8.26). Ultimately the root cause is the hydrolysis of palladium activators which causes precipitation of palladium. This phenomenon can result in shorts through conductive bridging. It can also be assumed that the nickel foot is solderable.

The use of a post-dip between the activator and the nickel can prevent this issue. However, this process step is not usually offered as part of the established "plug and play" ENIG process.

Poor registration and rinsing can also cause this issue.

8.4.5.3.2   *Nickel Corrosion*

Nickel corrosion is a concern in the electronics industry and is the result of the interaction between electroless nickel and immersion gold. Nickel corrosion is believed to contribute to poor solderability. However, this cannot be corroborated.

Corrosion can be significantly improved by the use of high phosphor nickel and/or reduction assisted gold. In Figure 8.27, 1-A represents a case of hyper-corrosion from a cross-sectional perspective and in 1-B the hyper-corrosion is viewed from a planar perspective. In this case the issue is due to the nickel being poorly formed. In 2-A and 2-B the nickel gold interface is very good and corrosion-free. The nickel is well formed but a complementary gold plating system may have been used additionally.

Figure 8.26   SEM of the nickel foot failure and the remedy.

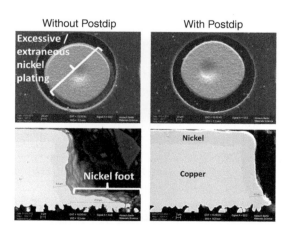

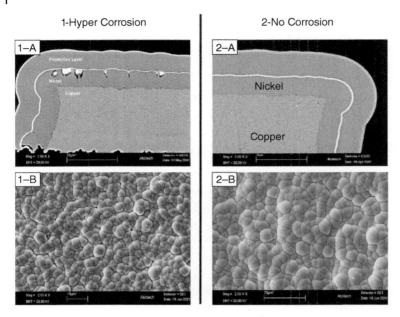

Figure 8.27 SEM investigation of nickel corrosion versus no corrosion in cross-section (top) and from a planar view after gold stripping (bottom).

If an ENIG process is run within supplier specifications, acceptable corrosion can be repeatedly expected. No corrosion at all is difficult to achieve without encountering other issues such as poor adhesion.

Bath age, physical agitation of the nickel and gold processes, and solder mask types can also contribute to nickel corrosion.

8.4.5.4 Summary

Although the ENIG process is most suited to high-volume production situations, it is still used in prototype, small batch, and fast turn-around environments because it is well established and it is possible to perform both lead-free soldering and aluminum wire bonding operations with it as a final finish.

Spike corrosion is a historical issue with ENIG but there is no conclusive correlation between corrosion incidents and soldering failures. Corrosion is unavoidable because the process is an immersion or exchange process. Corrosion can be improved by high phosphor electroless nickel (over 10% phosphor) and/or reduction-assisted immersion gold baths. High phosphor baths were previously attributed to the black pad phenomena as it was theorized that the solder bonded preferentially to the nickel than the nickel phosphor enriched location left when the gold was dissolved into the solder. Further studies showed that the solder may be perfectly wetted to the ENIG finish even with high (up to 12%) P concentration in the nickel layer [16]. High P-content is also beneficial for flex nickel applications, as the co-deposited phosphorus can improve the elasticity of the electroless nickel as measured by Young's Modulus [17].

Until around the year 2000 ENIG was used extensively in server boards. However, server boards are typically back plane panels with a large copper surface area, which translates as a high loading factor in the nickel. Lack of dosing control resulted in black pad and many fabricators moved over to OSP and immersion silver.

Immersion gold is not suitable when gold thicknesses above 100 nm (0.1 μm) are required. Thick gold is required in the ENIG process for gold wire bonding as a substitute for ENEPIG. The communications industry, the medical industry, and the military industry have the highest demand for thick gold application.

Ultimately ENIG has been used successfully for decades.

8.4.6 Electroless Nickel/Electroless Palladium/Immersion Gold (ENEPIG)

ENEPIG was a solution developed to overcome the short comings of the ENIG process. The primary advantage is that palladium acts as a diffusion barrier between the nickel and the immersion gold. The benefits are twofold; nickel corrosion is mostly remedied and gold wire bonding can be performed with relatively thin gold deposits. The soldering performance is similar to ENIG due to the final gold finish. There are two types of palladium used in the industry: palladium phosphor and pure palladium. Qualification is usually the selection decider for this finish, with palladium phosphor dominating the industry. Pure palladium does, however, have some performance benefits over palladium phosphor in that it has a lower hardness value and improves the performance with lower gold thicknesses. Gold thicknesses as low as 40 nm (0.04 μm) are possible with pure palladium.

ENEPIG is a dominant finish for the IC substrate industry.

8.4.6.1 Process Complexity
ENEPIG is inherently more complex than ENIG because there is an additional precious metal plating step involved. The fluctuating value of palladium is a major concern in this process. The process complexity is reflected in Table 8.11.

ENEPIG is more technically capable than ENIG and this has an inevitable impact on the process complexity in terms of control. There is also a drive to use ENEPIG with SIT in an attempt to reduce process costs by selectively plating the precious metal finish. SIT technology does, however, employ additional masking steps than no selective processes, that can also present difficulties such as gold attach or film lift. Potentially, cost saving is achieved at the cost of yield.

Table 8.11 Complexity score for electroless nickel electroless palladium immersion gold.

		Complexity A	Complexity B	Complexity C
Maintenance and handling	Analysis			3
	Dummying		2	
	Maintenance			3
Most suitable production environment	Prototype			3
	Volume	1		
Amenities required	Compressed air	1		
	Water			3
	Heating			3
	Waste water			3
	Extraction			3
	Filtration		2	
Process complexity	Score	2	4	21
	Final score		27	

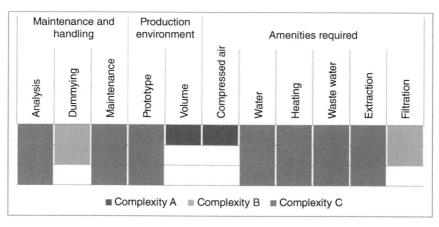

Figure 8.28 A visual breakdown of the complexity scores for electroless nickel/electroless palladium/immersion gold.

8.4.6.1.1 *Maintenance and Handling*

Analysis and control is significant to this process because of the value of the electrolytes and their potential to plate out (Figure 8.28).

Dummying is also required for the electroless palladium and electroless nickel steps. Both processes also require frequent maintenance to prevent plate out. The welded seams of the plating tank are prone to triggering plate out. The reason seams are localized triggers for plate out is that the surface is generally rougher than the tank wall. Once observed, the tank must be cleaned. The cleaning solutions are based on nitric acid. For the nickel tanks an approximate 25% w/w nitric acid mixed with water is utilized, whereas for the palladium tank a nitric solution of between 60% and 65% w/w with sodium chloride is used. In order to be able to maintain production, spare tanks are required which extend the footprint of the line. The nitric acid is also not favorable for waste water treatment and can create an exothermic reaction if not made up in the correct manner. The concentrated acid should be added slowly to water not water to concentrated acid.

8.4.6.1.2 *Most Suitable Production Environment*
ENEPIG is most suited to a mass production environment when down time is limited. It can be implemented when gold wire bonding or another more exotic medium such as silver is required. Typically, ENEPIG and ENIG lines are combined to save space.

8.4.6.1.3 *Amenities Required*
The burden on amenities is significant as electrolytes run at high temperature, the need for extraction is necessary and there are many rinse steps. ENEPIG is a high end finish so that filtration is mandatory.

8.4.6.2 **Process Description**
The complexity of the ENEPIG process is outlined in Figure 8.29. Not only are there many steps, many of them need special attention to ensure that the finish performs successfully. Similarly to ENIG, the activation step requires special attention from the rinse after the micro-etch. Although, as with all final finishes, the pretreatment is important to prepare the surface, the key steps in the ENEPIG process start at the activation stage. Activators or catalysts are susceptible to the drag-in of oxidizing agent from the micro-etch and panels that are allowed to oxidize as the copper content in the activator is the limiting factor for lifetime copper content. Copper oxide is easily dissolved in acidic conditions. The rinse after micro-etch and the pre-dip are in place to protect the activator from premature aging.

The electroless nickel is, as already mentioned; a balancing act between activity and stability, but the rinse after the nickel is often overlooked. Nickel is readily oxidized and this can lead to skip plating. The rinse after nickel should be kept clean and the dip time kept to a minimum. The cleanliness of the rinse after the electroless nickel process is more critical in the ENEPIG process than in the ENIG process. This is because the process that follows the electroless nickel is electroless palladium. Electroless palladium will cover nickel oxide but the complexor in the gold is able to remove it which can potentially initiate peeling, especially at thicker palladium deposits.

	T°C	Ts	Key Step
Cleaner	35–45	180–360	
Rinse	RT	60	
Microetch	25–35	60–120	
Rinse	RT	60	O
Pre Dip	RT	60–180	O
Activator / Catalyst	20–25	60–180	O
Rinse	RT	60	O
Electroless Nickel	80–90	900–1800	O
Rinse	RT	60	O
Electroless Palladium	50–70	300–600	O
Rinse	RT	60	
Immersion Gold	80–85	420–840	O
Rinse	RT	60	
Dry			

Figure 8.29 The typical process flow for electroless nickel electroless palladium immersion gold.

Palladium and gold control is essential to prevent costly plate outs. Even if the metal can be partially reclaimed, the down time and cleaning is costly.

The immersion gold thickness is realistically limited to between 40 and 60 nm. However, applications are requested to be able to achieve up to 100 nm. At this thickness, nickel corrosion is a concern.

When thick gold is requested, semi-autocatalytic or autocatalytic gold is an option. This requirement is requested for gold wire bonding but is not beneficial for solderability.

8.4.6.3 Issues and Remedies

8.4.6.3.1 Excessive/Extraneous Plating

Nickel foot is the result of excessive/extraneous plating of the electroless nickel process. This area has been discussed previously in Section 8.4.5.3.1.

Figure 8.30 SEMs to show in (a) a case of hyper-corrosion to a pin-hole and in (b) a case where the appropriate immersion gold concept can improve/cover the pin-hole and prevent further corrosion.

8.4.6.3.2 Solder Mask Related Pin-Holes

Pin-holes are a recognized feature of ENEPIG processes and are usually located near the solder mask. Pin-holes are found in the electroless nickel deposit and the real concern is whether the pin-holes are further corroded by the subsequent gold process or not.

Having established that the primary concern is the corrosion of the pin-hole by the gold plating step, there are immersion gold concepts that can improve the situation. The main solution to this issue is to apply a reduction-assisted immersion gold bath or, in specialized cases, to employ a semi-autocatalytic gold bath. In Figure 8.30, picture A is an example of a pin-hole showing signs of hyper-corrosion in the location of a pin-hole. In Figure 8.30, picture B the pin-hole has been plated over by the gold plating to ensure that the corrosive exchange process can not cause hyper-corrosion.

8.4.6.4 Summary

ENEPIG brings the possibility of gold wire bonding without higher gold thicknesses. The process is also considered to be a technically improved ENIG. However, the electroless palladium introduces more complexity. In practice the ENEPIG line is often combined with the ENIG line to save on plating equipment footprint.

It has also been mentioned that thick gold is not realistic without the risk of nickel corrosion and that to overcome this autocatalytic gold is an option. The following two sections will cover this option.

8.4.7 Electroless Nickel Autocatalytic Gold (ENAG)

As circuitry becomes denser, the electrolytic nickel electrolytic gold process is no longer capable. The thicknesses are too high and the electrical bussing takes up too much space [18].

The benefit of the ENAG process versus the ENIG process is that the gold is not deposited by an exchange mechanism. This means that thick gold can be achieved without significant nickel corrosion. In this system the classical diffusion barrier scenario is continued by using electroless nickel.

8.4.7.1 Process Complexity

The process complexity is scored at one point higher than ENIG because the process employs semi-autocatalytic gold which typically also needs a strike immersion gold (Table 8.12).

The ENAG process is in essence the same as the ENIG process up until the semi-autocatalytic gold bath, so most of the information on this process is already covered in the ENIG section and will not be discussed here.

8.4.7.1.1 Maintenance and Handling

The ENAG process requires a significant amount of analysis and control (Figure 8.31). The pretreatment needs to prepare the surface but the analysis is similar to the previous processes and the chemistries involved are relatively simple and more importantly stable in terms of concentration fluctuations.

The electroless nickel is where the complication starts as this bath is a balance between stability and activity at high temperature. It is highly recommended to use a colorimetric controller with a master slave dosing concept at the very least. More complicated controllers exist but are commonly installed at higher complexity board fabricators. The balance between activity and stability usually makes the need for dummying necessary. In idle times the bath is typically overstabilized and is reinitiated by dummy plating prior to production.

The dummying score is only Complexity B rather than Complexity C because there are dummy-free systems on the market. However, dummy panels are always

Table 8.12 Complexity score for electroless nickel and semi-autocatalytic gold.

		Complexity A	Complexity B	Complexity C
Maintenance and handling	Analysis			3
	Dummying		2	
	Maintenance			3
Most suitable production environment	Prototype		2	
	Volume	1		
	Compressed air	1		
Amenities required	Water			3
	Heating		2	
	Waste water			3
	Extraction			3
	Filtration			3
Process complexity	Score	2	6	18
	Final score		**26**	

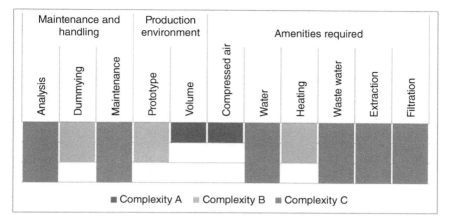

Figure 8.31 A visual breakdown of the complexity scores for electroless nickel semi-autocatalytic gold.

on standby as a precaution. The thickness range of the nickel may reduce the impact of idle time prior to plating. The electroless nickel bath also plates out to some extent as this is the required plating mechanism. The impact of the plate out is that the nickel bath needs to be transferred to a second position and the tank needs to be stripped. The state-of-the-art stripping procedure involves nitric acid.

The gold maintenance can be made easier by using indirect heating and polyvinylidene fluoride (PVDF) plating tank liners. Gold controllers are also recommended to control the gold concentration as closely as possible to the optimum value.

8.4.7.1.2 Most Suitable Production Environment

ENAG is used in prototype fast turn environments but is difficult to control and tends to exhibit high consumption values. The best practice is to batch plate to avoid too much of a loss in efficiency and high consumption. It is best placed in high-volume environments with routine chemical checks and automatic control. Gold controllers are especially recommended to control the gold concentration as closely as possible to the optimum value in order to run as efficiently as possible.

8.4.7.1.3 Amenities Required

Due to the large number of process steps the ENAG process places a high burden on amenities. The burden on amenities is exacerbated by the fact that the ENAG process is vertical.

8.4.7.2 Process Description

The typical process as outlined in Figure 8.32 shows seven steps. This number can increase if counter-measures to avoid extraneous nickel plating are included.

	T°C	Ts	Key Step
Cleaner	35–45	180–300	
Rinse	RT	60	
Microetch	25–35	60–120	
Rinse	RT	60	O
Pre Dip	RT	180–300	O
Activator / Catalyst	RT	60–60	O
Rinse	RT	60	
Electroless Nickel	80–90	900–1800	O
Rinse	RT	60	O
Immersion Gold	80–90	60–120	O
Rinse	RT	60	
Autocatalytic Gold	85–95	2400–3000	O
Rinse	RT	60	
Dry			

Figure 8.32 The typical process flow for electroless nickel/autocatalytic gold.

Process times are also extended in vertical systems because there are no fluid delivery systems so the solution exchange is dominated by soak time. Although, as with all final finishes, the pretreatment is important to prepare the surface, the key steps in the process start at the activation stage. Activators or catalysts are susceptible to the drag-in of oxidizing agent from the micro-etch. Panels that are allowed to oxidize can also be problematic as the copper content in the activator or catalyst baths are the limiting factor for lifetime. Copper oxide is easily dissolved in acidic conditions so in many cases an acidic rinse or acid dip are employed after micro-etch. The function of the pre-dip prior to the activator or catalyst is also to protect the activator or catalyst from premature aging.

The electroless nickel is, as has already been mentioned, a balancing act between activity and stability, but the rinse after the nickel is often overlooked. Nickel is readily oxidized and this can lead to skip plating. The rinse after nickel should be kept clean and the dip time kept to a minimum. The gold can be the source for considerable saving if the gold content is controlled well. Too little gold will make the bath inefficient and may lead to nickel corrosion by the gold complexors.

It is recommended to use indirect heating and tank liners for the semi-autocatalytic gold process.

8.4.7.3 Issues and Remedies
8.4.7.3.1 *Excessive/Extraneous Plating*
Nickel foot is the result of excessive/extraneous plating of the electroless nickel process. This has been referred to previously in Section 8.4.5.3.1.

8.4.7.4 Summary
The ENAG process can overcome nickel corrosion even at gold thickness requirements over 100 μm but it is not suited, as all processes that include nickel, to ultra-fine line applications.

This section of the chapter is here to demonstrate the potential application for semi-autocatalytic gold. The semi-autocatalytic gold could potentially also be applied to electroless nickel and electroless palladium (ENEPAG). The process would simply be more complex.

8.4.8 Electroless Palladium Autocatalytic Gold (EPAG)

The future of high end final finishing will ultimately omit the nickel barrier layer in the final finish. This is because circuit density is increasing and sub 20 μm lines and spaces are inevitable. In addition to high density circuits, low signal loss is required to maximize energy usage and allow high frequency compatibility. Electroless Palladium Autocatalytic Gold (EPAG) is an example of such a finish.

8.4.8.1 Process Complexity
The process complexity is slightly lower than the nickel containing processes as the omission of nickel alone results in less risk of plate out, skipping, or extraneous plating (Table 8.13).

The EPAG process is a high end universal finish so some complexity can be expected. In fact the process is less complex than the other electroless precious metal processes discussed in this chapter.

8.4.8.1.1 *Maintenance and Handling*
Analysis is extensive for this high end process (Figure 8.33). The risk of plate out can be significantly reduced by using indirect heating and PVDF plating tank liners. Analysis is still crucial to ensure that the process stays within the working window to function optimally. No dummy plating is required for the EPAG process. Maintenance is standard but no more challenging than the other final finishes.

Table 8.13 Complexity score for electroless palladium and semi-autocatalytic gold.

		Complexity A	Complexity B	Complexity C
Maintenance and handling	Analysis			3
	Dummying			
	Maintenance		2	
Most suitable production environment	Prototype		2	
	Volume	1		
	Compressed air			
Amenities required	Water			3
	Heating		2	
	Waste water		2	
	Extraction			3
	Filtration			3
Process complexity	Score	1	8	12
	Final score		**21**	

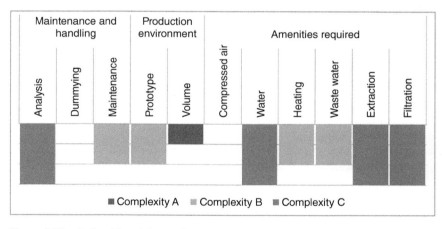

Figure 8.33 A visual breakdown of the complexity scores for electroless palladium/semi-autocatalytic gold.

8.4.8.1.2 *Most Suitable Production Environment*

The EPAG process can be used for either mass production or fast turn prototype work as no dummying is required. The EPAG process is also compatible with SIT processing.

8.4.8.1.3 Amenities Required

The process has many steps so that the water consumption is significant but similar to ENEPIG. Only the semi-autocatalytic gold requires high temperatures at approximately 80 °C so that the heating demand is not as high as other processes. The omission of nickel lessens the burden on waste water treatment. Extraction and filtration are mandatory in this process.

8.4.8.2 Process Description

The process window for EPAG is narrower than some of the other finishes. Therefore there are processes within the system that need to be well controlled (Figure 8.34).

	T°C	Ts	Key Step
Cleaner	35–45	180–300	
Rinse	RT	60	
Microetch	25–35	60–120	O
Rinse	RT	60	O
Pre Dip	RT	50–70	O
Activator / Catalyst	RT	50–70	O
Rinse	RT	60	
Electroless Palladium	80–90	240–360	O
Rinse	RT	60	O
Immersion Gold	80–90	90–150	O
Rinse	RT	60	
Autocatalytic Gold	85–95	900–1800	O
Rinse	RT	60	
Dry			

Figure 8.34 The typical process flow for electroless palladium and semi-autocatalytic gold.

The control of the micro-etch is critical for this process to avoid over-roughening of the copper and potentially exposing granular inconsistencies. This implies regular chemical analysis and regular etch rate monitoring. This is because the subsequent activation step is an immersion process. Similarly to ENEPIG and ENIG the activation system needs to be protected to enable reliability in combination with an acceptable lifetime. This topic has been covered in detail in previous sections of the chapter. The following precious metal steps need to be analyzed regularly and it is recommended to have auto analysis and dosing systems.

8.4.8.3 Issues and Remedies
8.4.8.3.1 Process Novelty
The process is still becoming established and accepted, so there will be some lead time to understand and familiarize with the process.

8.4.8.3.2 Solder Joint Embrittlement
Thicker gold deposits are accepted to perform better with regards to wettability. In this respect thicker gold is beneficial. However, thicker gold can change the characteristics of the solder joint, which can lead to solder joint embrittlement. In Figure 8.35, Sn1.2Ag0.5Cu0.05-0.5Ni lead-free solder balls were used so as to significantly enrich the solder joint with gold. In this case the electroless palladium was maintained at 100 nm (0.1 μm). High Speed Shear Testing was used in this evaluation at a speed of $2\,\mathrm{m\,s^{-1}}$. This speed is much higher than normally used and was selected to enhance the impact of gold embrittlement [19].

The negative impact of increased gold thickness on the solder joint ductility is clearly depicted in Figure 8.35. The implication is that there exists a maximum

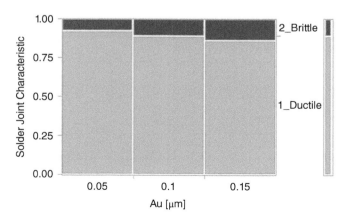

Figure 8.35 The impact of gold thickness on solder joint ductility.

gold thickness where by increasing the gold thickness does not improve the solder joint reliability, nor does increasing the palladium thickness improve wire bondability [20].

8.4.8.4 Summary

Electroless Palladium Autocatalytic Gold (EPAG) is a novel universal finish in that it is solderable and wire bondable. The process has a relatively low complexity but has to be controlled tightly. Equipment is available on the market to aid with control.

The process is suited to high-volume manufacture but can also be used in low volume, fast turn environments.

Unlike ENEPIG the plating line cannot be set up for dual function as this chemistry is significantly different to the classical ENEPIG/ENIG systems.

8.4.9 Electrolytic Nickel Electrolytic Gold

Electrolytic nickel electrolytic gold is an original final finish and it is very reliable. This section will only deal with soft gold. Soft gold is a pure gold (99.9% purity) that is ideal for soldering and wire bonding. This is comparable as a surface finish to the other precious metal finishes in this chapter. Hard gold, which is usually a gold alloy (99.6% purity) that is plated up to 5 μm thick, is used for connectors for which there is still no real substitute. Soldering is not applicable to this application. Electrolytic soft gold was the first IC substrate final finish at the time when IC substrates were ceramic.

The electrolytes for electrolytic nickel electrolytic gold are very stable and operate at low temperature. In addition, the process can achieve thick gold deposits if needed.

For products that are for soldering only, the gold thickness can be reduced but this is only to a minimum of 250 nm (0.25 μm) as opposed to 400 nm (0.4 μm) for wire bonding capabilities.

However, there are disadvantages which limit the process's ability to be used for fine lines and spaces. Most significantly, a bussing system is required to deliver the plating current which ultimately needs to be removed for the final circuitization.

The process is generally speaking easy to handle but there is a high expense aspect to the process which will discussed later.

8.4.9.1 Process Complexity

The process complexity score is similar to the scores that were allocated to the immersion processes (Table 8.14). This is because the electrolytes will not plate out and have long lifetimes.

This process uses the ability to deposit a thick gold layer in order to be both solderable or wire bondable, or in other words to be a universal finish.

Table 8.14 Complexity score for electrolytic nickel electrolytic gold.

		Complexity A	Complexity B	Complexity C
Maintenance and handling	Analysis		2	
	Dummying	1		
	Maintenance		2	
Most suitable production environment	Prototype	1		
	Volume	1		
Amenities required	Compressed air		2	
	Water		2	
	Heating	1		
	Waste water		2	
	Extraction		2	
	Filtration		2	
Process complexity	Score	4	14	0
	Final score		**18**	

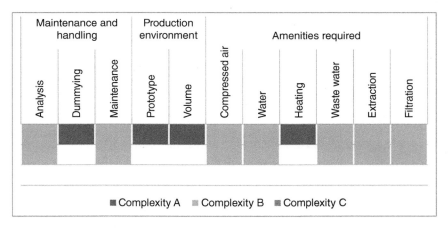

Figure 8.36 A visual breakdown of the complexity scores for electrolytic nickel electrolytic gold.

8.4.9.1.1 Maintenance and Handling

Electrolytic processes are low-maintenance processes (Figure 8.36). Control can be carried out with simple hull cells and in-house titrations.

Dummying is only used as a tool to condition the anodes after down times, while maintenance is centered on contacts and anode maintenance.

8.4.9.1.2 Most Suitable Production Environment

The process is ever ready and suited to both prototype and volume production. Having said this, the process is not suited to fine line applications below 35 μm. New installations of this process plating equipment are rare.

8.4.9.1.3 Amenities Required

The burden on amenities is moderate with rectifier power being the most significant. In all electronic fields cleanliness is paramount to high yields but the electrolytes, in this case, are not viscous and adequate rinsing is easily achieved. However, the process has many rinsing steps. Advanced rinsing technology can potentially ensure that the amount of water consumed is kept to a minimum.

8.4.9.2 Process Description

The process sequence is similar to electrolytic copper processes in that the pretreatment is rudimentary (Figure 8.37).

8.4.9.3 Issues and Remedies

8.4.9.3.1 Cost

Cost is the main issue for this process. The actual cost comparisons with other finishes is discussed in a later section in Figure 8.39.

	T°C	Ts	Key Step
Acid Cleaner	35–45	180–360	
Rinse	RT	60	
Microetch	25–35	60–120	
Rinse	RT	60	
Acid Dip	RT	180–300	
Rinse	RT	60	
Electrolytic Nickel	50–60	As required	
Rinse	RT	60	
Gold Strike	40–60	60	O
Electrolytic Gold	35–50 (hard Au) 60–70 (soft Au)	As required	O
Rinse	RT	60	
Dry		3–6 min	

Figure 8.37 The typical process flow for electrolytic nickel electrolytic gold.

8.4.9.3.2 Requirement for the Bussing System

A bussing system requires extra masking and etching processes that may have an impact on yields which is also a significant factor.

8.4.9.3.3 Gold Embrittlement

The findings from a study on gold embrittlement concluded that the embrittlement and degradation of Sn3Ag0.5Cu lead-free solder due to increasing gold content seemed to be more progressive than abrupt [21].

Figure 8.38 demonstrates that both an increase in palladium and gold content in a 250 μm lead-free solder ball causes the solder joint to become more brittle during high speed shear testing. The black dashed lines in Figure 8.38 are only to indicate the trending increase in brittle failures as the metal content, especially the gold, increases. The study therefore found that gold enrichment in the solder joint is detrimental to the solder joint integrity but does not produce a "line-in-the-sand" result for lead-free Sn3Ag0.5Cu solder [21].

8.4.9.4 Summary

The electrolytic nickel electrolytic gold process is best suited to specialized, corrosion-free, thick gold applications; however, there are alternative solutions for applications that require thick gold. Semi-autocatalytic gold can achieve the same gold thickness requirements without needing a bussing system that is mandatory

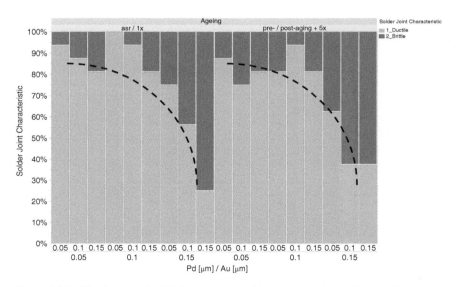

Figure 8.38 The impact of palladium and gold thickness on solder joint embrittlement with regard to 250 μm Sn1.2Ag0.5Cu0.05-0.5Ni solder balls [22].

for electrolytic plating. These finishes have been discussed previously in the sections that cover ENEPIG, ENAG, and EPAG. The key issue with the electrolytic nickel electrolytic gold process is the cost. Gold thickness is typically 400 nm (0.4 μm). Thicker gold deposits have the potential to cause solder joint embrittlement by the creation of gold-tin intermetallic compounds in the solder joint.

Another complexity to take into consideration is that electrolytic processes need a masking step to ensure that plating is restricted to where it is required. Pattern electrolytic processes also need to consider current densities. Isolated features will have relatively higher current densities than areas with a high concentration of features. Isolated features will be more heavily plated than relatively lower current density features. Poor gold distribution can have a significant cost impact.

8.5 Conclusions

The chapter discussed alternates to the conventional electrolytic nickel immersion gold surface finish which have been found to produce similar and sometimes better performance to this industry-accepted benchmark [23].

Ultimately, cost is one of the prime considerations when selecting a final board surface finish. There are also possibilities to combine solderable finishes with precious metal wire bonding. These SIT processes are becoming more popular, especially if cost is a direct driver.

Performance, which was discussed toward the start of the chapter, is another consideration when choosing a final finish. A combination of Figure 8.39 which

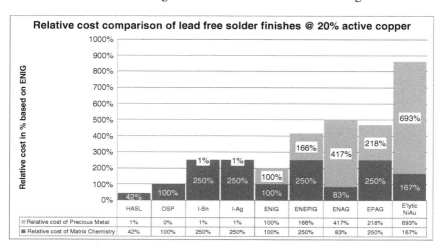

Figure 8.39 The relative cost impact of lead-free solderable final finishes with ENIG as the base line (100%).

focuses on cost and Table 8.2 from earlier in the chapter which focuses on performance should aid in the selection of a final board surface finish.

Selecting a final board surface finish is of significance to the assembly and reliability of the product and it should be carefully considered.

A compromise between cost and performance is the norm with final finish selection. It is not uncommon to have more than one finish at a board fabricator. In any case, specific industries or OEM/EMS usually have a preferred finish or a historical use of a specific finish for their application with good results, which aids in the selection process.

References

1 Vianco, P. (1998). An overview of surface finishes and their role in printed circuit board solderability and solder joint performance. *Circuit World* 25 (1): 6–24.

2 Prismark Partners (2017). The Printed Circuit Report. Prismark Partners Report.

3 IPC 4552A specification. (2017) *Performance specification for electroless nickel/immersion gold (ENIG) plating for printed boards.* IPC International.

4 Sweatman, K. (2009) Hot air solder leveling in the lead-free era. IPC APEX EXPO.

5 Federico, J. (2016) Solderability and tinning: does the industry know the difference. NJMET, Inc., TechTime – Electronics and Technology News, 10 January.

6 The Laboratory People (2009) COD or Chemical Oxygen Demand definition. 16 October.

7 Woodrow, T.A., and Ledbury, E.A. (2005) Evaluation of Conformal Coatings as a Tin Whisker Mitigation Strategy. *IPC/JEDEC 8th International Conference on Lead-Free Electronic Components and Assemblies*, San Jose, CA (18–20 April).

8 Nichols, R., Ramos, G., Nothdurft, L., and Schafsteller, B. (2015) Is it possible to use markers to select the right soldermask to optimize the yield of your selective finish? *SMTA China South 2015 Conference.*

9 IPC specification J-STD-001-D (2005) *Requirements for soldered electrical and electronic assemblies.* IPC International.

10 Tranitz, H.-P. and Dunker, S. (2012) Growth mechanisms of tin whiskers at press-in technology. *IPC APEX EXPO.*

11 He, F., Fung, A., Chan, D. and Yip, D. (2017) Studies of tin whisker growth under high external pressure. *12th International Microsystems, Packaging, Assembly and Circuits Technology Conference (IMPACT).* IEEE.

12 IPC Standard 4553 (2005) *Specification for immersion silver plating for printed circuit boards.* IPC International.

13 Schueller, R. (2007) Creep corrosion on lead free PCBs in high sulfur environments. *SMTA*, Orlando, FL.

14 Wang, W., Choubey, A., Azarian, M.H., and Pecht, M. (2009). An assessment of immersion silver surface finish for lead-free electronics. *Journal of Electronic Materials* 38 (6): 815–827.

15 Wright, A. (2015) Printed circuit board surface finishes – advantages and disadvantages. *SMTA*.

16 Johal, K., Lamprecht, S., Wunderlich, C., and Roberts, H. (2005) Electroless nickel/immersion gold process technology for improved ductility of flex and rigid-flex applications. *SMTA Pan Pacific*.

17 Snugovsky, P., Arrowsmith, P., and Romansky, M. (2001). Electroless Ni/immersion Au interconnects: investigation of black pad in wire bonds and solder joints. *Journal of Electronic Materials* 30 (9): 1262–1270.

18 Huang, Y.H., Hsieh, W.Z., Lee, P.T. et al. (2019). Reaction of Au/Pd/Cu and Au/Pd/Cu multilayers with Sn-Ag-Cu alloy. *Surface and Coating Technology* 358: 753–761.

19 Nichols, R., Ramos, R., and Taylor, R. (2017) The influence of intermetallic compounds on high speed shear testing with a specific interest in electroless palladium and autocatalytic gold. *IEEE CMPT Symposium, Japan (ICSJ)*.

20 Oda, Y., Kiso, M., Kurosaka, S. et al. (2008) Study of suitable palladium and gold thickness in ENEPIG deposits for lead free soldering and gold wire bonding. *41st International Symposium Microelectronics* (2–6 November).

21 Hillman, C., Blattau, N., Arnold, J. et al. (2013) Gold embrittlement in lead-free solder. *SMTAI* (October).

22 Nichols, R., and Heinemann, S. (2017) The impact of deposition thickness on high speed shear test results specifically related to electroless palladium and semi autocatalytic gold. *SMTAI*.

23 Long, E. and Toscano, L. (2011) A study of surface finishes for IC substrates and wire bond applications. *SMTAI*.

9

PCB Laminates (Including High Speed Requirements)

Karl Sauter[1] and Silvio Bertling[2]

[1] *Oracle Corporation, Santa Clara , California, USA*
[2] *Mesa, Arizona, USA*

9.1 Introduction

The use of FR-4 laminate as the dielectric material between imaged copper circuit layers in multilayer printed wiring boards continues to be very cost-effective. In recent years, continuing improvements in laminate material thermal robustness and reliability have contributed to making higher temperature lead-free assembly processing successful. Laminate material improvements have also been made in developing lower loss resin systems to meet higher speed requirements and in meeting other market requirements. Going forward, higher voltage requirements in automotive applications are driving changes in reliability testing and for laminate materials as well. This chapter will address several critical factors relating to laminate materials such as moisture content [1], and describe in some detail why they are important to ensuring that the important finished board performance requirements are met.

9.2 Manufacturing Background

In recent years controlling the rheology of the resin systems being used has become more difficult, due to the increasing use of more complex resin system blends and the use of the higher filler contents for achieving lower loss and other properties. In other words the "operating window" for defect-free press lamination has been getting smaller, which has increased the risk of delamination and other failure modes during subsequent processing and board assembly. Laminate material manufacturing improvements in the treating process are being made

Lead-free Soldering Process Development and Reliability, First Edition. Edited by Jasbir Bath.
© 2020 John Wiley & Sons, Inc. Published 2020 by John Wiley & Sons, Inc.

to ensure better and more consistent wetting of the glass fibers and to minimize debris and contamination when manufacturing prepreg and laminate materials.

Printed circuit board (PCB) fabricators have additional challenges for ensuring that finished board dielectric properties and reliability are achieved, as compared with laminators. To achieve a given finished dielectric thickness, fabricators must use higher resin content materials for providing adequate filling of the adjacent imaged core(s) used in multilayer board constructions. The long-term reliability of multilayer boards critically depends upon having proper resin flow, filling and curing of the prepreg laminate material during press lamination. In contrast, when laminators make core laminate material the resin cure cycle pressures and temperatures between the un-imaged copper sheets are fairly predictable. PCB fabricators, however, are faced with uneven (imaged) copper topography which creates local variations in pressure and which affect resin flow and temperature uniformity during the lamination cycle, making it more difficult to produce a product that will not delaminate during subsequent processing including board assembly.

9.3 PCB Fabrication Design and Laminate Manufacturing Factors Affecting Yield and Reliability

9.3.1 High Frequency Loss

As the industry looks forward to 5G and other future higher speed performance requirements, every aspect of the printed wiring board material characteristics is being evaluated for opportunities for reducing signal loss. Lower loss laminate materials with modified or new resin systems, resin systems with higher filler contents and improved filler materials, lower Dk (dielectric constant), glass reinforcement, more uniform glass weaves, and lower loss copper foils are all being considered. Appropriate desmear processing, appropriate drilling feeds and speeds, appropriate dimensional scaling, and other critical PCB fabrication processes will need to be specifically determined for many of these newer laminate material and copper foil combinations in order to meet the finished product's mechanical and electrical performance requirements.

9.3.2 Mixed Dielectric

Mixed dielectric constructions can affect board yield and the reliability of finished multilayer boards. When using a more expensive very low loss laminate material to meet critical high speed signal requirements, it can be cost-effective to use a compatible lower cost laminate material everywhere else in the board construction stack-up. This type of board construction does involve greater risk to the finished board quality and reliability. The fabricator will have to use the same drill feed and speed rates for both laminate materials, resulting in non-optimum drill smear for at least one of the materials and consequently non-optimum smear-removal processing.

9.3.3   Back-Drilling

Back-drilling can be required on high speed boards in order to not exceed maximum remaining stub length requirements on the higher speed nets. This back-drilling may require both improved laminate material layer-to-layer registration and back-drilling registration in order to avoid having to use the larger back-drill anti-pads needed to ensure long-term reliability. These larger anti-pads required for back-drilling can negatively impact routing density, and board reliability is affected if minimum drill-to-copper clearances are not maintained.

9.3.4   Aspect Ratio

Through-hole via aspect ratio is one of the most well-known PCB fabrication design features affecting yields and long-term reliability. For higher aspect ratio holes either slower lower current density plating, higher agitation and/or more expensive reverse pulse plating is required for sufficiently thick and uniform copper plating thickness on the barrel of the plated-through hole (PTH), especially near the center of the barrel. Consequently, drilling should be optimized for the specific laminate material in order to have a consistently lower and uniform amount of smear for subsequent desmear processing, which helps mitigate chemistry entrapments and improves the reliability of higher aspect ratio PTHs.

9.3.5   PCB Fabrication

There are many ways in which poor processing can affect the characteristics of the laminate material and affect printed wiring board yield and finished product reliability. For example, plasma desmear can create deeper pockets in the drilled hole walls where subsequent processing chemicals can be more difficult to remove. Appropriate processing and controls, and qualification can be done to address most of these issues. One of the more difficult potential problems to address is material and process chemistry residue left on imaged core material prior to press lamination, which in the field can facilitate electrochemical migration (ECM) between the internal copper features and eventual product failure. Therefore appropriate temperature/humidity/bias testing of manufactured test vehicles is recommended for qualifying a new laminate material by a specific PCB fabrication manufacturing facility to ensure that the characteristics of the

Table 9.1 Maximum foil roughness from IPC-4562, metal foil for printed wiring applications.

IPC-4562	Rz (ISO) (µm)
Standard profile	>10.2
Low profile	<10.2
Very low profile	<5.1
TBD (example: HVLP – Hyper very low profile)	<2.5
TBD	<1.25

laminate material are sufficient and have not been compromised by subsequent processing.

9.3.6 Press Lamination

Multilayer board constructions typically have some low-pressure areas during the press lamination cycle of PCB manufacturing which have greater risk of poor bonding and consequently delamination during subsequent high temperature lead-free assembly processing. During press lamination strong bonding is needed between the prepreg's B-staged resin system and the fiberglass reinforcement. Strong bonding should also happen between the B-stage resin system and the exposed fully cured C-stage resin system during press lamination. However, with the increasing use of smoother copper foils that leave a smoother fully cured resin surface on the core dielectric material after etching, there can be less mechanical bond strength developed between the prepreg and core during press lamination (Table 9.1).

Also with smoother copper foils the mechanical bond strength or peel strength between the resin and the copper surface itself can be reduced below 4.0 lb in.$^{-1}$. Accordingly, in order to maintain at least 3.0 lb in.$^{-1}$, improvements in chemical bonding treatments between the resin and copper are being developed to compensate for these new inherent reductions in mechanical bond strength capability.

9.3.7 Moisture Content

The high frequency loss of a finished multilayer board can vary depending upon how much moisture a given PCB fabrication supplier facility's process leaves in the finished multilayer bare board, since the moisture content within these dielectric layers in finished multilayer boards having many internal plane layers is not easily changed. Improved board moisture content controls may therefore be required for high speed products in order to achieve specified high frequency low loss requirements on higher speed nets [2].

Several laminate materials studied have shown that a moisture content increase of about 0.03% by weight correlates with a high speed signal loss increase of about 10%. Although after press lamination a multilayer board is very dry, after subsequent drilling the laminate material dielectric again quickly picks up moisture. Possible remedies include specifying a very short time between drilling and the copper plating of drilled holes, since plated copper on the hole wall is a good barrier against further moisture absorption. If this is not possible, then baking the boards just prior to desmear (plasma and/or permanganate) quickly followed by electroless and electrolytic plating of the drilled holes is recommended to ensure sufficiently low moisture content in finished bare boards for high speed performance applications. Subsequently, in order to mitigate some assembly issues, an additional bake to remove moisture from the external layers prior to solder mask application is advised.

9.3.8   Laminate Material

A recent study identified three classes of internal ECM resistance for laminate materials [3]. Conductive anodic filament (CAF) formation is one type of internal conductive filament formation (CFF) failure that results from a laminate material's poor resistance to internal ECM. Although most THB (temperature/humidity/voltage bias) testing is typically done at closer spacings where the PCB fabrication process more significantly affects the results, testing at larger spacings to accommodate higher voltage testing more clearly shows differences in the internal ECM resistance among various laminate materials. Therefore THB testing based upon the IPC TM-650, 2.6.25 CAF test method [4] is recommended for even standard board designs going through higher temperature lead-free assembly processing in order to screen out the use of the identified lowest class of laminate materials having poor internal ECM resistance.

9.4   Assembly Factors Affecting Yields and Long-Term Reliability for Laminate Materials

9.4.1   Reflow Temperature

The most critical assembly processing factor affecting assembly yields and long-term reliability is of course the reflow temperature. Lead-free alloys, with Sn3Ag0.5Cu being the most common, have a much higher reflow temperature than eutectic tin-lead solder. Simple board designs going through lead-free assembly may require a reflow temperature of only 245 °C peak temperature, but more complex and thicker boards can require a reflow temperature of up to 260 °C. Therefore 6X at 260 °C is typically used to simulate worst-case board assembly and rework preconditioning prior to IST (Interconnect Stress Testing), temperature/humidity/bias testing, and other reliability testing.

9.4.2 Assembly Components

For assemblies having large chip components such as ASICs and large BGAs there may need to be more stringent site-specific board flatness of 4 mils in.$^{-1}$. Boards with long press-fit or greater through-hole pin connector lengths can require lower overall coefficient of thermal expansion (CTE) (higher Tg and/or lower ppm °C^{-1}) in order to mitigate damage from excessive mechanical stress on the laminate material during thermal excursions.

9.4.3 Thermal Stress

Several different types of PTH thermal stress testing are available for evaluating the reliability of PTHs after simulated assembly processing [5]. Oven testing produces a very uniform cyclic strain within the interconnect, resulting in what is considered a more representative acceleration of the particular failure mechanisms that cause product to fail in the field. IST is a more widely used test method, typically testing from ambient or room temperature up to about 150 °C or minimum 10 °C below the laminate material Tg. If the IST test coupon design has a sufficient number of heating elements in the appropriate locations in the separate heating circuits, then a sufficiently uniform cyclic strain within the interconnect can also be achieved with this test method. The Highly Accelerated Thermal Stress (HATS) test method offers a wider thermal cycling range below the laminate material Tg, which avoids the introduction of new non-representative failure mechanism(s).

Regardless of the test method used, test coupons having no failures should always be checked for delamination. Any delamination found within thermal stress test coupons voids the test results, since delamination significantly relieves the thermal stress on PTH walls.

9.5 Copper Foil Trends (by Silvio Bertling)

Copper foil usually has a treatment on the base foil that improves adhesion on either the drum/shiny side (SS), the tooth/matte side (MS), or on both sides. Historically copper foil is most often treated only on the side that will be laminated against the core laminate material, since the opposite side will go through additional PCB fabrication manufacturing processing. This subsequent processing

Table 9.2 Symbols for roughness which were revised.

Type	Symbol of JIS B 0601-1994		Symbol of JIS B 0601-2001
Max. height Roughness	Ry	→	Rz
10 points Mean roughness	Rz	→	Rz(JIS)

includes etching and another treatment which is applied on the exposed copper features of cores prior to press lamination when making a multilayer board.

The following are definitions for each abbreviation used in the following sections:

- **Rz (10 points mean roughness)** – Some symbols for roughness were revised in accordance (Table 9.2) with the ISO standard [6] from JIS (Japan Industrial Standard) B 0601-2001 version [7]. The 10 points mean roughness was eliminated from the JIS 2001 version, but it still remains as a Rz JIS reference, since it was popular in Japan.
- **Rz** (ISO) is obtained from the distance in micron meter between the highest peak and the lowest valley in the range of sampled reference length to the direction of mean line of the roughness curve.
- **Rz** (JIS) is obtained from the total in micron meter of the mean value of each distance between the mean line and five peaks from the highest one, and the mean value of each distance between the mean line and the five valleys from the lowest one, of the roughness curve in the range of sampled reference length.
- **Ra** is calculated as the roughness average of a surface measured with microscopic peaks and valleys.
- **Rq** is the root mean square average of the roughness profile ordinates. Rq is also called RMS (root mean squared).
- **Sz** is defined as the sum of the largest peak height value and the largest pit depth value within the defined area. White light interferometry-based optical (non-contact) profilometers are typically used for measuring and calculating these area (S) surface roughness values.
- **Sa** is the arithmetical mean height of a line to a surface. It expresses, as an absolute value, the difference in height of each point compared to the arithmetical mean of the surface.
- **Sq** is the root mean square value of ordinate values within the definition area. It is equivalent to the standard deviation of heights.

Based upon IPC-4562 [8], the industry standard for metal foil in printed wiring board applications, there are three different categories of copper foil surface roughness based upon the copper foil's usually rougher tooth/matte side peak-to-valley height (Rz surface roughness):

1) Standard profile or standard high tensile elongation (HTE) typically has Rz = 8–11 μm
2) Low profile (LP) copper foil typically has Rz = 6–8 μm
3) Very low profile (VLP) has Rz = 5–7 μm
4) TBD (example: HVLP (Hyper Very Low Profile)) has Rz = 2.5–4.5 μm
5) TBD has Rz = 1.25–2.0 μm (instead of Rz the Sz value is a better choice)
6) TBD has Rz = <1 μm (instead of Rz the Sz value is a better choice)

It should be mentioned that 2 μm surface roughness is the lower limit for mechanical stylus surface roughness measurements, However, stylus measurements for profiles <5 μm could increase this uncertainty factor significantly since, depending upon the profile shapes and structure, an accurate value may be difficult to generate. Laser profilometers are now being used to measure the Rz surface roughness of smoother copper foils. These Rz values are typically reported on a copper foil manufacturer's data sheet, except that Ra values may be reported for the surface roughness on the drum side (SS). Rz and Rq values are typically not reported for the shiny side of the copper foil. Some copper foil manufacturers not only report the IPC required R values but also include S values for the matte and shiny side of the copper foil.

Currently many copper foil names are used in the industry, such as VLP2 (Very Low Profile <2 μm), HVLP, SVLP (Super Very Low Profile), VLP-1.5 (Very Low Profile [>1.5 μm Rz]), HVLP-2 (Very Low Profile [>2 μm Rz]), etc. These names can be misleading because there is not consistent information about what the actual profile height is and which value type it represents. Within the IPC efforts are being made to find an agreement between some of the above names and what type of profile type and range should be associated to each name. At this time it would be better to refer to a specific profile height such as Rz (ISO) or Sz, even though the names appear to be more convenient to use. It should be noted that if Rz (JIS) values are reported they are typically about 20% lower than Rz (ISO) values.

The industry would like to correlate copper foil surface roughness, especially for profile heights Rz <2.5 μm with actual high frequency loss, but so far any attempt to do so has not been successful. There are several reasons why such efforts have not provided the desired results. For example, copper foil suppliers provide surface roughness information for the foil only meaning on the treated side and the non-treated side as mentioned earlier. However, after a PCB is manufactured the copper foil untreated side goes through a surface roughening process which will

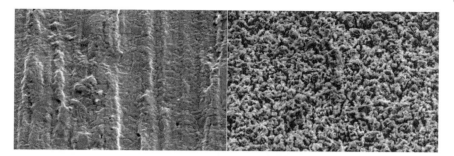

Figure 9.1 Smoother foil before fabricator treatment (left). Typical surface after fabricator treatment (right).

change the copper surface structure by mostly removing some copper from the surface to better obtain adhesion to the prepreg for manufacturing a multilayer board.

The board fabricator's surface roughening process leaves a very different topography behind (Figure 9.1) and therefore is not comparable with the profile treatment process applied at a copper foil supplier. Thus when measuring the surface roughness after the fabricator deployed their roughening treatment/roughening process it is important to understand two major issues. Firstly the new surface structure can have a significantly different topography, and secondly when the roughness measurements are performed they can be higher in value than the copper foil surface prior to the board fabricator's treatment. This could especially be an issue when copper foil profile heights of below 2.5 µm are desired.

The IPC 3-12a committee has been working on developing a non-contact optical test method (IPC TM-650, 2.2.22) [9] and copper foil surface roughness categories based upon Sq/Sa measurements that are expected to correlate better with actual high frequency loss. These measurements are more useful in selecting VLP and smoother copper foils for use in higher frequency products that require lower copper surface roughness.

Commonly used standard electro-deposited Grade 3 base copper foils are as follows:

1) Standard HTE foil with the reverse treat foil (RTF) type treated on the shiny side (SS) which can be used with most resin systems.
2) Standard HTE foil with the RTF type treated on the shiny side (SS) for use with polytetrafluoroethylene (PTFE) dielectric material and other materials used in radio frequency (RF) applications (>20 kHz).
3) Flex foil (some are only Grade 2 at this time).
4) Battery foil, no treatment (this expensive lowest loss copper foil is used for lithium-ion batteries).

5) Battery foil with treatment is used for high speed low loss and RF applications. It comes as standard matte side treatment and as a RTF type on the shiny side treated versions.
6) Carrier foil is a very thin functional foil that is placed on the separation layer of a copper carrier. This copper foil is considered Grade 1 because it cannot be tested based upon the IPC-4562 specification [8]. IPC-4562 specifies three classes of foils in Section 1.3 Quality/Performance Classification as Grade 1, Grade 2, and Grade 3.

   Each grade reflects a specific minimum quality standard.

   Carrier foil is therefore used for high end HDI (High Density Interconnect) applications often in connection with MSAP (Modified Semi Additive Process).
7) Rolled annealed (RA) foil.

The selection of the type of copper foil to be used depends upon the application (high speed frequency, data rates, power, trace width, trace thickness, etc.).

The power requirements on printed wiring boards are increasing, particularly for some automotive applications. Therefore some copper foil manufacturers are working on improving the measurement and control of their copper foil volume resistivity. Comparative copper foil volume resistivity testing should always be done at the same temperature, typically 20 °C. Manufacturers of copper foil for these applications may need to determine more accurately the volume resistivity for different base foils, including measuring the impact that the copper treatment has on the overall volume resistivity value of the foil. For power applications the measurement of the thermal conductivity of the copper foil may also require new data, because different base foils and different copper treatments could potentially impact the thermal conductivity of the foil.

9.6 High Frequency/High Speed and Other Trends Affecting Laminate Materials

9.6.1 High Speed Standards

PCI Express 4.0 standard and similar and future high speed serial computer expansion bus standards (5G) will continue to drive the use of lower loss laminate materials and copper foils [10]. Good high speed signal integrity simulation requires accurate dielectric thickness as well as accurate dielectric constant (Dk) and accurate dissipation factor (tan δ or Df) signal loss values which vary with both the resin content of the dielectric material and the frequency of the signal [11]. Tighter dielectric thickness requirements than the typical ±10% for core laminate materials may be needed. Lower loss laminate materials are more expensive, so in most cases the laminate material selected should be what is just good enough for the specific application [12].

9.6.2   Adhesion Treatment (Prior to Press Lamination)

As a potentially cost-effective alternative to using a lower loss laminate material, improved retention of the copper surface smoothness when using lower loss copper foils may be accomplished by using a primarily chemical bonding type of adhesion treatment that does not roughen the exposed copper surface of imaged core material prior to press lamination. Some of the currently more commonly used alternative oxide treatments applied prior to press lamination can significantly roughen the exposed copper surfaces and negatively impact the signal integrity of high speed inner layer signal nets. Also, some of the currently used oxide treatments contain small amounts of metals that can reduce the conductivity on the exposed copper surfaces.

9.6.3   Laminate Material Filler Content

Dielectric or resin fillers are increasingly being used to reduce the laminate material loss characteristics. Fillers can also be used to improve other laminate material properties such as reducing Z-axis expansion. The critical resin flow/rheology during press lamination is strongly affected by both the filler content and the resin content of the dielectric. Filler contents approaching 30% typically have greater risk of delamination, poor rheology for good wetting and insufficient resin filling during press lamination. If the filler used has larger particle sizes (lower cost), then these larger particles can even significantly roughen the adjacent copper foil during press lamination and potentially affect signal loss.

9.6.4   Glass Weave Effect

The construction of the laminate material is known to contribute to what is called the glass weave effect on higher speed signal traces. When one trace in a differential pair is routed over the woven glass crossover points (or "knuckles") in the adjacent glass weave and the other trace happens to be routed over relatively resin rich areas, this can cause skew and affect signal integrity. For this reason in these applications it is recommended that long segments routed along the X or Y axis be avoided. To a lesser extent and depending upon the pitch of the differential pair and pitch of the adjacent glass reinforcement the length of segments at a 45° angle may need to be limited as well.

9.6.5 Halogen-Free

The bromine-based TBBPA (tetrabromobisphenol A) flame retardant system used in most FR-4 laminate materials has been very effective in mitigating the fire risk of printed wiring boards, and has not been confirmed as a significant environmental or health hazard in these applications. In general, the alternate more "halogen-free" flame retardant systems used in laminate materials often have some degradation in electrical/mechanical performance, and reliability may be affected by a premature activation of the alternate flame retardant mechanism. However, for consumer products where long-term life in the field is not a critical requirement, the use of laminate materials having more "halogen-free" flame retardant systems continues to increase.

9.7 Conclusions

The development of new laminate materials has been proceeding at a more rapid pace in recent years, responding to market demands for improved "halogen-free" laminate materials, for lower loss laminate materials, for lower cost laminate materials and for more thermally robust laminate materials. Although resin systems are consequently becoming more complex, the use of dielectric materials between imaged copper circuit layers in multilayer printed wiring boards continues to be very cost-effective. Laminate material improvements are expected to continue to meet higher speed requirements and going forward even some higher voltage requirements in automotive applications. This chapter has addressed several critical factors relating to laminate materials, and their importance.

Over the last 20 years the increase of high speed digital and RF performance requirements has resulted in many copper foil suppliers developing lower profile PCB copper foils. The impact of the skin effect at high frequency on standard High Temperature Elongation copper foil features having a surface roughness ~10 μm Rz resulted in excessive high frequency signal loss, therefore alternative lower profile foils are being required. PCB copper foil requirements with regard to mechanical bond strength or peel strength have also been changing to meet the product needs associated with the lower loss laminate materials and smoother copper foils. Lower profile copper foil for high speed, low loss and RF applications have become mandatory. For more accurately measuring the functional profile height and shape of profiles, the optical measurements that generate S values such as Sa, Sz, and Sq seem to be more suitable for profile heights below Rz 2.5 μm. The demand for high speed low loss laminates and copper foils with below 2.5 μm Rz surface roughness is expected to increase further as more applications will require higher data rate capabilities and/or operate at higher frequencies.

References

1 Coombs, C.F. Jr. (2001). Water and moisture absorption. In: *Printed Circuits Handbook*, 5e, 131–132. McGraw-Hill.

2 Sauter, K. (2019) Effect of moisture on high speed loss. *IPC APEX Expo 2019*, S-21.

3 Biehl, E. (2017) The necessity of temperature humidity bias testing with high voltages </= 1000 V. SMTA International, September 2017.

4 IPC TM-650, 2.6.25. *Conductive anodic filament (CAF) resistance test: X-Y axis test method*. IPC International.

5 Sauter, K. (2007). PCB laminates. In: *Lead-Free Soldering* (ed. J. Bath), 199–220. Springer Science+Business Media, LLC.

6 ISO 4287 (1997) *Geometrical Product Specifications (GPS) – Surface Texture: Profile Method – Terms, Definitions and Surface Texture Parameters*. International Organization for Standardization.

7 JIS B 0601-2001 (2001) *Geometrical Product Specifications (GPS) – Surface Texture: Profile Method – Terms, Definitions and Surface Texture Parameters*. Japanese Industrial Standards.

8 IPC-4562A (2008) *Metal foil for printed wiring applications*. IPC International.

9 IPC TM-650, 2.2.22. *Non-contact metallic foil surface topography/texture*. IPC International.

10 Hornig, C.F. (2012) Signal integrity and PCB physical parameters. Printed Circuit Design and Fabrication/Circuits Assembly, February.

11 Bogatin, E., DeGroot, D., Huray, P. and Shlepnev, Y. (2013) Which one is better? Comparing options to describe frequency dependent losses. *DesignCon* (January).

12 Smetana, J., and Sauter, K. (2015) High frequency loss test methods for laminate materials comparison. *IPC Apex-Expo*.

10

Underfills and Encapsulants Used in Lead-Free Electronic Assembly

Brian J. Toleno

Microsoft, Mountain View, California, USA

10.1   Introduction

Adhesives play an important role in the manufacturing of high density surface mount assemblies. With the fabrication of lead-free electronic assemblies there are two adhesive applications used widely for increased reliability: underfills and encapsulants. This chapter will discuss some of the important physical properties related to these adhesive materials such as rheology, glass transition temperature (T_g), coefficient of thermal expansion (CTE), modulus, and methods of curing. Once there is a clear understanding of these physical properties there will be a more detailed discussion of each of these important applications.

When choosing an adhesive system to use within an application, several factors need to be considered. Firstly, the properties of the uncured material, such as shelf life, pot life, and rheology. Secondly, the method by which the adhesive is processed, dispensed or applied relates back to the rheology, but the method of curing must also be considered (e.g. heat or light cure). Finally, the properties of the final cured material. Typically, someone begins with the final item and works their way backwards, but all three play an important role in selecting the correct material for an application. This chapter will discuss these properties in detail in order to provide a better understanding of adhesives. Then, different applications using adhesive materials will be reviewed. Knowledge of these properties and understanding the applications will aid design and production engineers in selecting the right material for their product.

Lead-free Soldering Process Development and Reliability, First Edition. Edited by Jasbir Bath.

10.2 Rheology

Rheology is an often misunderstood, but important property for adhesives used in electronics assembly [1–4]. Rheology is the science of the deformation and flow of matter. The whole concept of rheology and flow behavior and the terms involved can be summarized in Figure 10.1.

Consider an amount of liquid of thickness x between two parallel plates (for clarity, Figure 10.1 shows this as a two-dimensional area). The bottom plate is kept stationary, while the top plate is forced to move. The liquid can be considered to be made up of layers, which move relative to each other. This relative movement of layers is known as shearing. To characterize this process only three parameters can be measured.

1. *Dimensions of the material.* This is trivial and is governed by the geometry of the particular system. Units = m.
2. *Force applied.* This is how hard the upper plate is pushed. Units = N.
   The force required to move the upper plate is obviously related to the area of the plate. It is therefore conventional to divide the force by the area to give the Stress, τ.

$$\text{STRESS} = \tau = \frac{\textbf{FORCE(N)}}{\textbf{AREA(m}^2\textbf{)}} \quad \textbf{Units} = \textbf{Nm}^{-2} = \textbf{Pa}$$

3. *Velocity.* The speed of the upper plate relative to the lower. Units = ms^{-1}.
   The layered structure of the material is analogous to a pack of cards. The thicker the pack, the less each card moves relative to its neighbors for a given movement of the upper plate, and the lower the shearing. Thus, we can define a shear rate, D (or $\dot{\gamma}$), as

$$\textbf{SHEAR RATE} = \textbf{D} = \frac{\textbf{VELOCITY(ms}^{-1}\textbf{)}}{\textbf{THICKNESS(m)}} \quad \textbf{Units} = \textbf{s}^{-1}$$

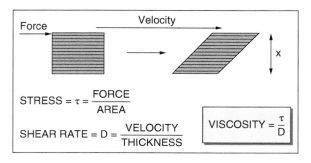

Figure 10.1 Basic rheology concepts.

The shear rate is obviously related to the shear stress in some manner. We define the ratio of these two to be the viscosity, η.

$$\text{VISCOSITY} = \eta = \tau/D \quad \text{Units} = \text{Pa/s}^{-1} = \text{Pas}$$

So, it can be seen that the viscosity is not a fundamental, measurable property of a material in the same way that mass is. It is merely the constant of proportionality between shear stress and shear rate. This is a commonly misunderstood concept. Since it is not an intrinsic property of the material, the measurement method will affect the value obtained. Therefore, when comparing viscosities, it is important to determine the method used and compare quantities determined using the same method [1–4].

Some typical viscosities are shown in Table 10.1. Although the formal unit is Pas, it is common to see viscosity represented in mPas, since water has a viscosity of 1 mPas. Note that air has a viscosity of approximately one hundredth of that of water (0.01 mPas).

Typical shear rates of some common processes are shown in Table 10.2. Processes which have very small sample thicknesses, such as brushing or rubbing, have extremely high shear rates, up to 10^6 s^{-1}. At the other extreme, sedimentation involves very small relative movement of layers, and therefore is still a shearing process, with a typical value of 10^{-6} s^{-1}.

10.2.1 Rheological Response and Behavior

If a sample is sheared at different shear rates, and the corresponding stress required is measured, the data can be plotted to form a flow curve (see Figure 10.2). This is basically a spectrum that can be used to characterize the flow behavior of a material. The simplest response is termed Newtonian. This is where the shear stress

Table 10.1 Typical viscosities.

Material	Viscosity/Pas
Bitumen	100 000 000
Polymer melt	1 000
Golden syrup	100
Liquid honey	10
Glycerol	1
Olive oil	0.01
Water	0.001
Air	0.000 01

Table 10.2 Typical shear rates.

Process	Shear rate s^{-1}
Sedimentation	0.000 001–0.000 1
Leveling	0.01–0.1
Draining	0.1–10
Chewing	1–100
Brushing/Pumping	1–1 000
Rubbing	10 000–1,000,000

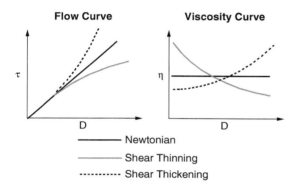

Figure 10.2 Simple rheological response.

varies linearly with shear rate over the entire range. If the corresponding viscosity is calculated and plotted, the resulting graph is termed a viscosity curve. For Newtonian materials the viscosity is constant at all shear rates. Examples of Newtonian fluids include water, ethylene glycol, simple oils, and cyanoacrylate monomers.

The majority of fluids are not Newtonian. Most materials require less stress than might be expected at higher shear rates. The viscosity will therefore decrease with increasing shear rate. Such materials are described as being shear thinning or pseudoplastic. Examples include polymer melts.

A wide variety of molecular arrangements can give rise to shear thinning behavior. Some of these are illustrated in Figure 10.3. A common example occurs when irregularly shaped particles line up in the direction of the flowing liquid. They present a smaller surface area, where there is less viscous drag and consequently less shear stress is required than might be expected. As soon as the shearing is stopped, Brownian motion ensures that the particles become randomly oriented once again.

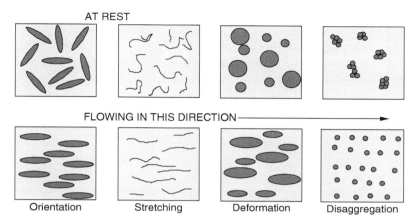

AT REST

FLOWING IN THIS DIRECTION

| Orientation | Stretching | Deformation | Disaggregation |

Figure 10.3 Shear thinning behavior of dispersions.

Other processes which can lead to shear thinning behavior are also shown in Figure 10.3. These include stretching of long chain polymer molecules, deformation of soft pliable particles, and the breaking up of agglomerates (disaggregation).

10.2.1.1 Thixotropy

The shear thinning behavior described above is not time-dependent. This means that the same viscosity is obtained at a particular shear rate, no matter how long the sample is sheared at this rate. In other words, there is no hysteresis in the viscosity curves for increasing or decreasing shear rates. The situation where there *is* time dependency is called thixotropy. This is illustrated in Figure 10.4. A material whose viscosity drops with time when held at a constant shear rate is termed thixotropic.

The effect on such a material of cycling the shear rate is shown in Figure 10.4. Increasing the shear rate (*"up curve"*) gives rise to the normal shear thinning behavior. However, on subsequently decreasing the shear rate (*"down curve"*) much less stress is required. The result is a hysteresis or thixotropic loop between the two curves. An alternative way of detecting thixotropy is by plotting the viscosity at a constant shear rate as a function of time. A Newtonian or simple shear thinning product will show no change, but a thixotropic material will show

Figure 10.4 Rheological response of a thixotropic fluid.

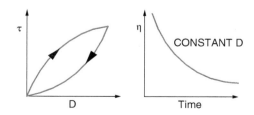

a decrease in viscosity with time. Another characteristic of thixotropy is that a sample will regain its original structure when left unsheared for a while.

Thixotropy is a very useful property in some adhesive products. Whenever there is a need to pump a high viscosity paste through a nozzle, thixotropy is of great benefit. For example, epoxy chip bonders have very high viscosities, but they are capable of being pumped through syringe needles of 400 μm or less. They restructure rapidly to prevent slumping of the product on the circuit board. This is also an important characteristic for underfill materials and encapsulants (see Section 10.7).

The most common raw material which imparts thixotropy (a "thixotrope") is silica. This consists of roughly spherical silicon dioxide particles with diameters in the region of 100 nm. These do not exist individually, but form aggregates. These in turn clump together to form agglomerates. When a dispersion of such agglomerates is left to stand, a three-dimensional network builds up as shown in Figure 10.5. This gives the product (which is now a "gel") its structure, very much like reinforcements in concrete. The attractive forces between the agglomerates which drive this structuring can be either Hydrogen bonding or Van der Waals' forces.

When the gel is sheared, the network breaks down and flow is facilitated. The cycle is reversible but time-dependent. The longer the shearing, the more the structure breaks down. At rest the material gradually regains its structure until the entire network is rebuilt.

One common way to describe this behavior is to measure and report a thixotropic index. This is the ratio of viscosities at low speed and high speed. The most common method for obtaining the thixotropic index is to divide the viscosity at low speed by the viscosity at high speed, where the high speed is 10× the low speed. For example, one can measure the viscosity of an underfill at 2 rpm (slow speed) and get a value of 12 000 cP, measure the same material at 20 rpm and get a value of 10 000 cP. This material would then have a thixotropic index of 1.2. Newtonian fluids would have a thixotropic index of 1.

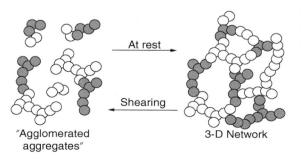

At rest

Shearing

"Agglomerated aggregates"

3-D Network

Figure 10.5 Particle interactions in a silica dispersion.

Since underfills need to be dispensed then flow via capillary action under area array components these materials are typically slightly thixotropic. They have thixotropic indexes in the order of 1.1 up to 2 or more for highly filled systems. Encapsulants, on the other hand, tend to have higher thixotropic indexes due to their need to be dispensed easily, yet form a dome and minimize slumping when on the printed circuit board (PCB) covering a component. These materials typically have a thixotropic index of 1.5 up to as high as 4 or even 5. Although there are some encapsulants that are just high viscosity with little thixotropic character, these are difficult to precisely dispense onto PCBs with fine pitch components.

10.2.2 Measuring Rheology

When comparing and evaluating materials one parameter to consider is rheology. It is important to understand the different measurement methods can produce different values for the same material. When comparing values it is critical to be sure that the method and conditions (e.g. shear rate) are consistent between the values being compared [2–4]. In this section different methods of measuring rheology will be discussed.

10.2.2.1 Spindle Type Viscometry

One piece of equipment used to measure viscosity is the spindle type viscometer, shown schematically in Figure 10.6. This consists of a disk, or spindle, which is placed in a beaker of the product (400 ml) in a water bath at 25 °C. For the majority of instruments, the spindle can rotate at a limited number of defined speeds, and the instrument gives a reading of "viscosity." This is suitable for Newtonian products, but measurement of shear thinning materials presents more problems.

This is shown by an examination of Figure 10.6. The two parameters required for the determination of viscosity are shear stress, τ, and shear rate, D. The stress required to keep the spindle rotating at a particular rpm is measured accurately by a torque spring on the instrument. The shear rate, however, is more problematic. Recall that this is defined as the velocity, v, divided by the sample thickness, x.

Figure 10.6 Spindle type viscometer.

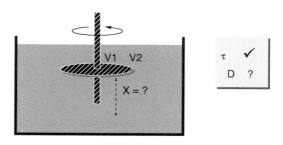

Neither of these parameters can be accurately measured. A rotating disk has a wide range of velocities, the outside of the disk moving much faster than the inside. The thickness of sample which is undergoing shear, x, is undefined. Hence it is impossible for the shear rate, v/x, to be determined.

This is not a problem for Newtonian liquids, since the viscosity is independent of the shear rate. However, for shear thinning materials an undefined shear rate means that the viscosity cannot be defined. This is particularly relevant at low shear rates where the viscosity declines rapidly.

Therefore any measurement of spindle viscosity for non-Newtonian fluids is not an absolute value. It is only a relative number, which depends on the particular spindle used, the spindle speed, and the time taken. As discussed earlier, the ratio of differences between viscosities at different spindle speeds is used to calculate thixotropic index.

10.2.2.2 Cone and Plate Rheometry

Many of the problems with the spindle type viscosity equipment outlined above can be overcome using a Cone and Plate system. Figure 10.7 shows a schematic diagram of the configuration. The sample (1–2 ml) is placed between a cone and plate, which is thermostatically controlled to ±0.1 °C. The plate is stationary, but the cone (which can be 2–5 cm in diameter, with an angle of 1°–4°) rotates at controlled, programmable rates. As in the case of the spindle type viscometer, the stress required to turn the cone can be measured accurately.

Analysis of the shear rate measurement is shown in Figure 10.7. As before, the velocity along the cone is greatest at the outside – i.e. $v_2 > v_1$. However, this time the thickness of the sample is well-defined and greater at the outside of the cone than the inside (i.e. $x_2 > x_1$). The angle of the cone is designed such that $v_2/x_2 = v_1/x_1$. In other words, the shear rate is precisely defined and constant throughout the whole of the sample. The viscosity, τ/D, can therefore be calculated accurately and can be compared with the results obtained from any system which uses a well-defined shear rate.

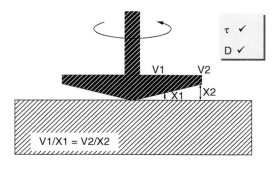

Figure 10.7 Cone and plate configuration.

There are a number of advantages of the Cone and Plate system.

- As outlined above, the shear rate, and therefore the viscosity, is precisely defined.
- A very small sample size is used.
- The temperature of the sample is controlled very precisely and temperature equilibrium is reached within a matter of minutes.
- A wide range of shear rates is possible – up to $1000\,s^{-1}$. The shear rates are programmable, so a continuous "spectrum" can be obtained.
- It is possible with the Cone and Plate system to do a number of sophisticated measurements, such as yield point determination, pipe flow prediction and temperature profile.

An example of the importance of using this method with a non-Newtonian fluid is the type of output obtained from the Cone and Plate rheometer of a surface mount adhesive material in Figure 10.8. Clearly this is a dramatically shear thinning product, so the question "What is the viscosity of a shear thinning fluid?" is meaningless, unless the shear rate is specified. The spindle type viscometer will give a number, but there are questions as to what the number means. The shear rates involved in the spindle type viscometer are low (c. 1–$10\,s^{-1}$). This is where the curve is most steeply sloped and therefore the greatest error occurs.

Cone and Plate would be the preferred method for measuring the rheology of underfill and encapsulant materials since, as discussed, they are typically thixotropic in nature. Underfills, in particular, can have a wide range of viscosity values. From the low viscosity room temperature flow materials that can be as low as 2000 cPs up to highly filled flip-chip underfills with viscosities in the order of 25 000 cPs.

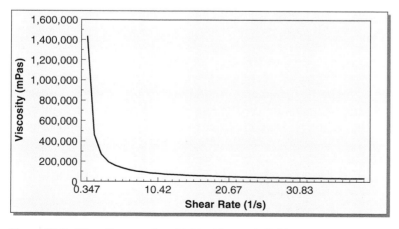

Figure 10.8 Viscosity curve for a highly thixotropic fluid.

Encapsulant materials typically range from 4000 cPs up to 100 000 cPs based on the application need, base chemistry, and filler load. In addition, since these are more thixotropic in nature Cone and Plate measurements are better suited to provide a more accurate representation of their rheological behavior.

10.3   Curing of Adhesive Systems

The adhesive and polymer systems discussed in this chapter all start as a liquid or a paste that must be cured in order to achieve the final properties presented in their respective technical data sheets. Polymers can cure via a number of mechanisms. The most common cure mechanisms for the materials used in the assembly of high density electronic devices are thermal, ultraviolet (UV), room temperature vulcanization (RTV), and catalyzed (two-part). Also, the path and/or conditions of the cure may affect the final properties of the material. Since underfill and encapsulant materials cure primarily by either thermal or UV cure, these will be the methods focused on.

10.3.1   Thermal Cure

Many polymer systems used in electronic assemblies are cured via thermal energy. Most often this curing occurs using a batch oven (e.g. underfill materials), or curing is done in seconds on a hot plate or curing station (e.g. snap cure systems). Also, some curing reactions that will progress at room temperature can be accelerated through the application of thermal energy. The data sheets supplied with these materials often contain the optimum cure profile. Deviations from that optimum cure profile will result in the final properties of the material being different than if the material was cured at the optimum profile. The most notable effect is on the glass transition temperature (T_g, see Section 10.4).

The most common heat-cured system are pre-mixed frozen epoxies (e.g. underfills). These materials have already begun to cure once they are mixed, but in order to obtain the optimum properties (and also to achieve reasonable cycle time) heat cure cycles are used to complete the curing. The higher the temperature, the more rapid the cure time, but there is a trade-off. Higher temperatures result in a larger stress and can quickly evaporate any volatile materials (if present), leading to voids. On the other hand, too low a temperature may not achieve sufficient temperature to "kick-off" the reaction, leading to long cycle times and unfavorable final properties.

Most underfills are one-part, epoxy-based, heat cure materials which are liquid at ambient temperature but when exposed to relatively high temperatures will harden to a tough, glass-like polymer within minutes. The active raw materials in

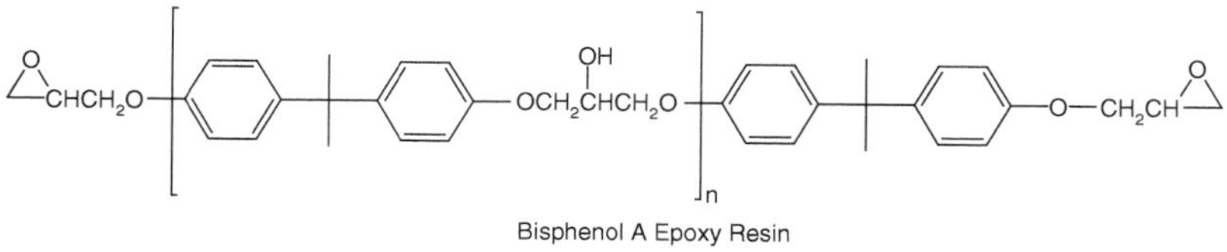

Figure 10.9 Bisphenol a diglycidyl ether epoxy resin.

the adhesive (from the point of view of curing) are the epoxy resins (Figure 10.9) and the curing agent (latent hardener) or catalyst/curative blend.

The curing agent and/or catalyst and/or latent hardener – hereafter referred to as the curing agent – is normally dispersed as a very fine powder or liquid into a specific blend of various epoxy resins of different viscosity (chosen to give an over-all target viscosity suited to the rheological requirements of the particular mode of application). Under ambient temperature conditions minimal chemical reaction occurs between the curing agent and the epoxy groups. In order to effect cure the adhesive must be heated to elevated temperature i.e. sufficient energy must be applied in order for a solid hardener to undergo a melting process/phase change or activate the catalyst and curing agent chemically such that it can react by ring-opening of the epoxy as illustrated in Figure 10.10.

The chemistry in the curing agent usually possesses more than one active site capable of reacting with an epoxy group. A particular epoxy resin chain also usually possesses more than one epoxy group, for example, bisphenol A diglycidyl ether is difunctional. Hence, the adhesive initially consists of a blend of low viscosity, multifunctional epoxy resins and a dispersed curing agent. No chemical reaction takes place between the curing agent and the epoxy groups under the recommended storage conditions. Depending on the type of curing system used,

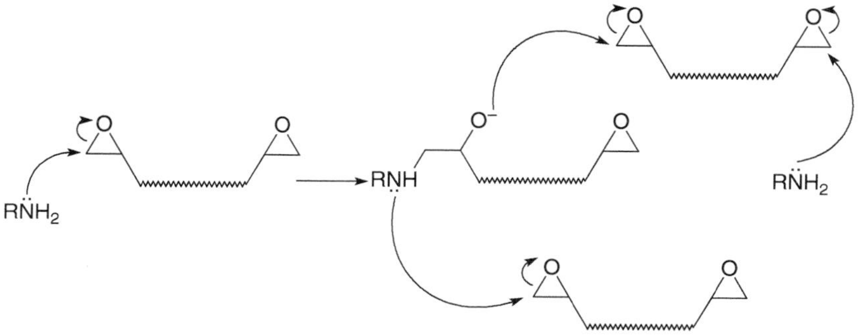

Figure 10.10 Ring opening of epoxy by curing agent RNH_2.

the storage temperature can range from as low as −40 °C up to 8 °C. As heat is applied to the adhesive the curing agent activates and the reactive sites on the curing agent start to chemically react with the epoxy groups on the individual epoxy resin chains so that these chains become increasingly chemically linked. This chain growth process is accompanied by an increase in viscosity.

Once ring-opened by chemical reaction with the hardener, the epoxy group is converted to a new chemical entity known as an alkoxide (-O⁻). Such alkoxide groups are themselves reactive toward other epoxy groups (if heated to sufficiently high temperatures) and this results in the formation of new ether linkages between adjacent polymer chains. This process produces a highly cross-linked three-dimensional network. As the number of cross-linked chains increases the viscosity of the adhesive increases until the adhesive eventually becomes a solid polymer when cure is complete. The cure process is illustrated in Figure 10.11. As the adhesive approaches 100% cure it assumes 100% of its cured properties. Incomplete cure will reduce the cured properties of the adhesive.

A typical oven temperature profile for heat curing the epoxy underfills is shown in Figure 10.12. This shows a time of 65 seconds at 150 °C and 130 seconds at 125 °C with a peak of 155 °C – this is more than adequate to fully cure the adhesive. The

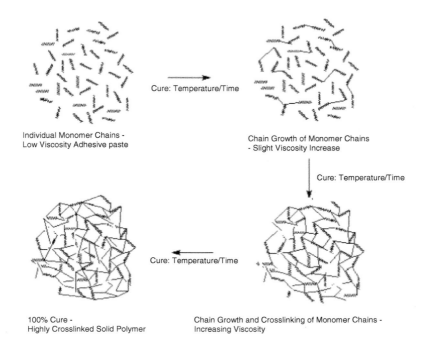

Individual Monomer Chains - Low Viscosity Adhesive paste

Cure: Temperature/Time

Chain Growth of Monomer Chains - Slight Viscosity Increase

Cure: Temperature/Time

Cure: Temperature/Time

100% Cure - Highly Crosslinked Solid Polymer

Chain Growth and Crosslinking of Monomer Chains - Increasing Viscosity

Figure 10.11 Illustration of the curing process.

BONDLINE TEMPERATURE, °C

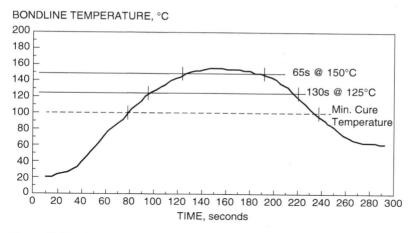

Figure 10.12 Typical oven profile for curing an underfill.

cure profile can be further optimized to reduce dwell time in the oven and peak temperature by monitoring the effect on bond strength.

The degree of cure and the rate of cure is most often modeled using differential scanning calorimetry (DSC). An isothermal DSC curve gives a measure of the time required to achieve 100% cure when heated at a constant temperature. In this test the adhesive sample is heated up at a rapid rate ($100\,°C\,min^{-1}$) to a target temperature (e.g. 125 °C) and then held at this temperature over a period of several minutes. The degree of conversion of the adhesive from the uncured, liquid state to the cured, solid state can then be expressed as a percentage degree of cure for different time periods of, for example, three minutes, four minutes, five minutes, etc. A typical series of isothermal DSC curve is shown in Figure 10.13 as a graph of percent conversion (%) against time (minutes).

The isothermal curve indicates how long the adhesive should be heated at a certain fixed temperature in order to achieve full cure. The curing oven must be set and maintained to give a temperature-time profile that matches or exceeds the heat input required by the adhesive. Although the heater panels of a typical conveyor oven may be set to attain a certain target temperature, the actual temperature that the adhesive experiences will depend on several factors. These include the heat capacity of the PCB and the individual components, the component population density on the PCB surface and the conveyor belt speed (which depends on the total number of boards to pass through the oven).

Consequently, the actual temperature achieved on the board should be measured against time to produce a cure profile. The temperature-time profile of the cure station should be checked regularly to confirm adequate curing conditions. It is important to measure the temperature of the adhesive at the bondline of a

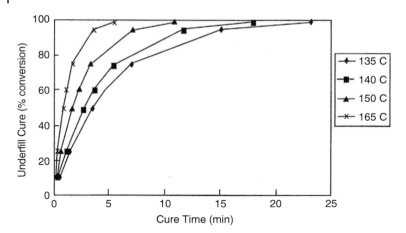

Figure 10.13 DSC curves illustrating the effect of temperature on degree of cure on an underfill material.

fully populated PCB so as to take into account all thermal transfer effects. This is important since oven set-point temperature is not going to be exactly the same as the temperature at the adhesive, and if the temperature is lower – then the material can be under-cured, leading to possible field failures due to non-ideal physical properties. The bondline temperature of components located in regions of the board adjacent to large heat sinks (such as large PLCC components) should be checked to confirm that they are receiving the minimum cure requirement.

DSC can also be run in a dynamic mode to measure changes in the heat flow characteristics of an adhesive when the material is heated under precisely controlled conditions. It measures the heat of reaction generated when a liquid adhesive product is converted to a cured or solid state. In a dynamic DSC run, the temperature of a liquid adhesive sample is ramped up at a rate of (typically) $10\,°C\,min^{-1}$, from room temperature to $250\,°C$, and the heat of reaction measured. There are three key pieces of information available from dynamic DSC tests:

- The temperature at which the adhesive starts to undergo cure or polymerize (T_{onset})
- The temperature at which the maximum rate of reaction is achieved (T_{peak})
- The total amount of heat evolved in the polymerization reaction (ΔH, joules per gram of product)

A typical dynamic DSC curve is shown in Figure 10.14.

The dynamic DSC trace illustrates the reaction profile of a typical surface mount adhesive and it indicates that a certain minimum threshold temperature is required to initiate the cure mechanism.

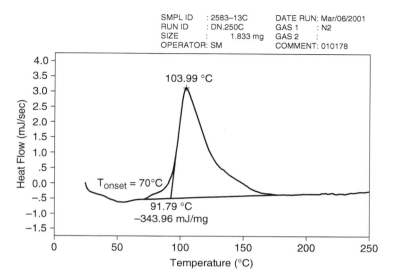

SMPL ID : 2583–13C   DATE RUN: Mar/06/2001
RUN ID : DN.250C   GAS 1 : N2
SIZE : 1.833 mg   GAS 2 :
OPERATOR: SM   COMMENT: 010178

Figure 10.14 Typical dynamic DSC trace.

Ideally, the recommended cure conditions should be followed, with the understanding that the true bondline temperature of the adhesive may not necessarily be identical to the oven set temperature. If a cure profile other than the recommended profile is needed, DSC studies can be used to determine the appropriate cure schedule and the final physical properties of the cured material should be determined with the cure schedule being utilized.

10.3.2 Ultraviolet (UV) Light Curing

An UV light curing material has a curing mechanism that is initiated with exposure to UV and/or visible light. Within the electromagnetic spectrum, UV light refers to the wavelengths in the range of ~200–400 nm while visible light encompasses those from 400 to 750 nm, Figure 10.15. The wavelength of light is inversely proportional to energy. Therefore, shorter wavelengths (e.g. UV) possess higher energy.

Polymerization begins when the energy from light of a specific wavelength contacts the photo initiators present in the resin. The free radicals that form as a result initiate the polymerization reaction between the monomers of the resin, leading to varying degrees of cross-linked polymer chains.

The cure begins at the surface and continues as far as the wavelengths can penetrate into the material. Curing starts instantly and can occur over a time span as short as one or two seconds depending on the intensity of light and volume of

The Electromagnetic Spectrum

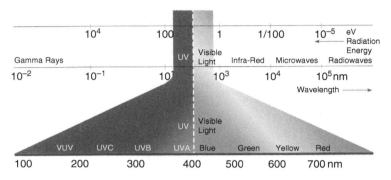

Figure 10.15 The electromagnetic spectrum.

material. The chemical type of these adhesives includes acrylates, modified acrylates, acrylated urethanes, and modified acrylic esters along with many others. Since the material must be exposed to the light in order to cure, this method is not used for underfills, but is common for many encapsulants.

Two important process demands that must be addressed when using a UV cure material are depth of cure and surface cure. When a deeper cure of the UV material is needed, high-intensity light ranging between 300 and 400 nm is required. The longer wavelengths are able to penetrate farther into the material, yielding a higher cure depth. It is also important to note that if the UV material is colored, this will significantly inhibit the depth of cure due to a boundary layer of dark, cured material forming on the surface blocking any further light penetration.

On the other hand, if a hard, tack-free surface is required, it may be necessary to expose the material to high-intensity light in the wave bands below 300 nm. The shorter wavelengths will provide the energy needed to polymerize the surface of the material. This is typically the case with systems that cure via a free radical propagation. This free radical reaction is retarded by the oxygen at the surface (some of the radicals react with the oxygen at the surface), and therefore can lead to a tacky surface.

When using a UV cure material it is important to choose a UV curing system capable of delivering both the proper intensity and wavelength needed to cure the material in question (for examples of UV curing systems see Figure 10.16). Most UV curing materials cure in the order of seconds at an intensity of mW cm^{-2} (given a thickness in the order of 0.1 mm) at the photo-initiator wavelength. Specifications on a UV curing system may list an intensity in the order of hundreds of watts per square cm. At first this appears to be overly sufficient, several magnitudes above what is needed. This output is measured over the entire spectral output of

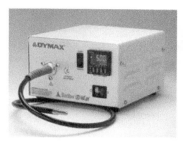

Figure 10.16   Example UV curing systems.

the bulb. Therefore before choosing a UV curing oven it is important to know the spectral output (what wavelengths are emitted by the bulb) and the intensities at the wavelengths of choice.

One common observation with certain UV-cured materials is that the surface may be tacky to the touch after irradiating with UV light. Some UV photo initiators experience what is termed "oxygen inhibition". This is caused by the oxygen molecules in the atmosphere interacting and "using up" the cure initiator at the surface of the material being cured. This can be alleviated in three ways. First would be to blanket the material in an oxygen-reduced atmosphere (e.g. nitrogen). This can be costly to set up and to operate with large batches of materials. Second would be to have a secondary cure mechanism available (e.g. heat, moisture, etc.) that can complete the curing on the surface. Finally, sometimes this can be overcome through the use of a higher intensity and/or a higher energy light source.

Some materials may be cured by both UV and other mechanisms, which are termed dual-cure systems. These materials, while first partially cured using UV light, are fully cured using another mechanism. These secondary cure mechanisms include heat, activators, ambient moisture, and resin/hardener mixtures. In some cases photo initiators are added to an already existing technology, for example silicones or epoxies, in order to achieve dual-cure mechanisms.

Multiple curing mechanisms are important for many reasons. The most important of these is the need to cure the material even in areas where the light cannot penetrate, or shadowed areas. In these cases, the UV light is used to immobilize the material initially, thus enabling the assembly to be moved to another stage in the process. This increases production throughput preventing a bottle neck in the system.

Another benefit that is gained by adding photo initiators to other materials revolves around the fact that one is able to take advantage of the inherent properties of these materials. Epoxies, silicones, and cyanoacrylates are all examples of already existing technologies that have been fused with UV technology in order to gain the benefits of both.

Another approach to curing epoxies with UV light is by means of a technology referred to as cationic UV curing. These are typically one-part epoxies that exhibit extremely low outgassing during cure. Once cured these cationic UV epoxies can provide high temperature bonds that demonstrates toughness, high peel, and high impact resistance.

The cure of these epoxies is initiated by UV light. However, the cationic curing mechanism differs from typical UV cure in the fact that once the cure is initiated it will continue even if the UV light is removed. These products also require less UV energy to cure than other UV-curable materials, therefore it is not difficult to find a UV light source that will provide enough energy to initiate cure.

10.3.3 Moisture Cure

Another method by which polymers can be cured is via a reaction with atmospheric moisture. These materials can come as two-part or one-part systems. The two most common base chemistries for this cure system are urethanes and silicones.

In the urethane system the common reaction is where the isocyanate system reacts with the atmospheric moisture to produce an amine (with carbon dioxide as a by-product), which then reacts with additional isocyanate to form the polyurea. This moisture reaction is also the reason that urethane (e.g. isocyanates) foams or form bubbles in the presence of moisture. The evolution of carbon dioxide produces the bubbling often observed.

There are three main types of moisture reactive silicones: acetoxy, methoxy, and oxime. These are the silicone chemistries that are often referred to as room temperature vulcanizing (or RTV). The acetoxy systems are very common in household applications such as bathroom caulk but are not suitable for electronic applications due to the by-product of acetic acid, which would be corrosive on an electronic assembly over time. The other two systems have by-products that are not harmful for electronic assemblies and are therefore suitable for electronic applications.

Since these systems cure via a reaction with moisture, simply heating them will not accelerate the polymerization reaction as is the case of most other chemical reactions, since this additional heat will actually drive off moisture. If an acceleration of this type of reaction is required, using moist heat is the method of choice (e.g. a humidity cabinet). One advantage with this type of system is that no additional energy is required such as heat or UV radiation. The other is that in hybrid systems, this reaction can occur over time after the rest of the material is cured via another mechanism (most commonly UV).

10.4 Glass Transition Temperature

The glass transition temperature (T_g) of an adhesive indicates the temperature at which the adhesive changes from being a glassy solid to a "rubbery" material. This is not a sharp point such as the melting point of a solid, but rather a broad transition. There is often a change in the physical properties after transitioning through the T_g [5–10]. Just because some of the physical properties change this does not mean that materials cannot be used above their T_g. For example, the glass transition temperatures of silicones are typically below 0 °C, yet they perform very well even at elevated temperatures. Even within the same sample, the glass transition occurs within a range of temperatures and not as a single point. Factors such as intrachain stiffness, polar forces, and comonomer compatibility can affect the size of the glass transition region.

DSC is the quickest and simplest method to measure T_g. The method requires small samples (typically 5–20 mg) that require no special preparation and material from components on processed boards can be utilized. The method consists of heating the sample in a closely calibrated thermocell where the temperature of the sample is compared to the temperature of a blank reference point within the same cell. Thermodynamic transitions such as melting points and reaction exotherms are easily measured, and the change in heat capacity at the T_g is seen as a shift in the baseline for cured materials as shown in Figure 10.17.

Unfortunately, this fast and convenient method is not universally applicable to all materials. High filler loadings, high crosslink densities, and other thermo-molecular processes can mask the shift due to the T_g and make the transition difficult or impossible to identify.

Figure 10.17 Example of DSC to measure T_g.

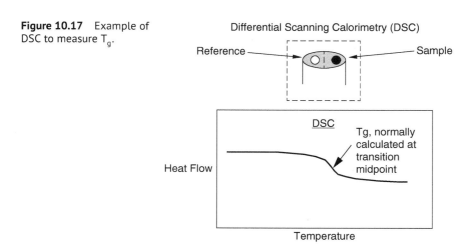

Thermomechanical analysis (TMA) is the most common test used to determine thermal expansion coefficients. Since there is a shift to a higher thermal expansion coefficient above the T_g due to changes in molecular free volume, the method can also be used to measure the glass transition temperature as shown in Figure 10.18.

The technique simply consists of heating the sample upon an expansion-calibrated platform, and measuring the dimensional change of the sample with an instrumented probe. The method will also easily follow cure-stress relaxations in and around the glass transition region, which sometimes leads to ambiguity in the assignment of a specific T_g, and can yield a different value for the same specimen if measured at a different point (for instance the cure stress may be significantly different when measured near the edge of a sample versus its center).

Dynamic mechanical analysis (DMA) consists of oscillating flexure energy applied to a rectangular bar of the cured material. The stress that is transferred through the specimen is measured as a function of temperature. Components of material stiffness are separated into a complex modulus and an elastic modulus. The technique is highly accurate and reproducible, although the Tg can be defined in different ways which will have different values, as illustrated in Figure 10.19. Large samples that must be accurately machined, coupled with longer setup times, make this test relatively expensive to run.

Each of these methods will produce different data for the same material. For example, a single specimen of a developmental epoxy encapsulant material was cast and cured for two hours at 145 °C after gelling at 100 °C for one hour. It was then machined into specimens for DSC, TMA, and DMA analysis. The results show T_gs ranging from 130 °C for the TMA, to 146 °C for the DMA measurement as shown in Table 10.3.

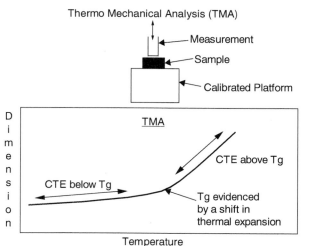

Thermo Mechanical Analysis (TMA)

Figure 10.18 Thermomechanical analysis used to measure T_g.

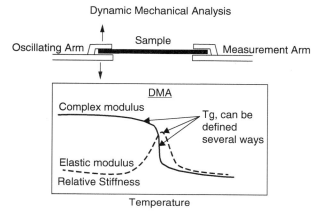

Figure 10.19 Dynamic mechanical analysis used to measure T_g.

Table 10.3 Instrument effect on T_g.

Instrument	T_g in °C
DSC	142
TMA	130
DMA	137 (G″) 146 (tan delta)

10.5 Coefficient of Thermal Expansion (CTE)

The CTE is a measure of the fractional change in dimension (usually thickness) per degree rise in temperature. For microelectronics encapsulants, it is often quoted in "ppm/°C" (value $\times 10^{-6}$/°C). Chemical composition, filler loading, and cure cycles all affect the value. For typical materials that have non-linear expansion, the specified temperature range will also have an effect on the data, with measurements closer to the T_g yielding higher values than those quoted across lower temperatures, as shown in Figure 10.20.

The CTE of adhesives used in microelectronics assembly is often a critical parameter [5, 8–10]. The different polymer systems used in electronics have vastly different CTEs for the base system. These CTEs can be modified from this base level through the addition of fillers, changes in the backbone chemistry, and changes in the hardener system. For example, most underfill materials are used to mitigate the CTE mismatch between silicon die (1.8 ppm °C^{-1}) or component package (4–10 ppm °C^{-1}) and the FR-4 substrate (15 ppm °C^{-1}). Therefore, the

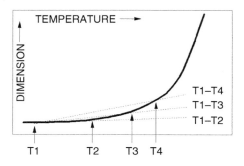

Figure 10.20 Effect of temperature on CTE.

CTE of the base epoxy is often manipulated with the addition of fillers. This effect can be observed by examining an example matrix consisting of a standard epoxy anhydride which is loaded with a proprietary mixture of low expansion fillers. All samples were cured for two hours at 145 °C after gelling for one hour at 100 °C. CTEs decrease rapidly and the viscosity increases (see Table 10.4).

Also, the CTE often increases (as much as threefold in epoxies) after passing through the glass transition temperature. These values are often defined as α_1, below T_g, and α_2, above T_g. When looking at a material and the amount of stress it may impinge on a system or the amount it can absorb it is good to examine the CTE, T_g, and modulus together when choosing a system.

Deciding the best physical properties needed for the material in the application is always a trade-off between the properties (performance) and application. Since the application of underfills and encapsulants are driven by a reliability enhancement need, those reliability conditions drive the material properties needed.

From an underfill perspective, one major differentiator in performance requirements is based on a flip-chip versus chip scale package (CSP) application. The difference in performance needs is illustrated in Table 10.5.

This does not necessarily tell the whole story in PCBs, where there is a high degree of reliability required (e.g. military/aerospace, automotive, etc.)

Table 10.4 Effect of filler on CTE.

Filler loading (%)	CTE (TMA method) (ppm °C^{-1})
60	34.5
65	28.8
70	24.1
75	20.0

Table 10.5 Different performance requirements for flip chip and chip scale package underfill.

Flip chip underfill	CSP underfill

• Absorbs stresses caused by CTE mismatch between chip and board
• Critical characteristics:
  • Low ionic content
  • High adhesion
  • Low CTE
  • High Tg
  • No voids during capillary flow
	• Primarily used to improve shock resistance (drop test reliability)
	• Necessary characteristics:
	  • High adhesion
	• Less important:
	  • Low ionic content
	  • Low CTE, High Tg
	  • No voids

where most underfills and encapsulants used are similar to the flip-chip underfill materials with a low CTE (below 45 ppm $°C^{-1}$), high modulus materials (> 4 MPa), and high $T_g > 125\,°C$. For other applications where the area array devices are designed more around consumer electronics and drop/shock is a larger concern, these underfills are often unfilled and have a CTE > 50 ppm $°C^{-1}$, but with a low modulus (<3 MPa), and the T_g can range from below $0\,°C$ to $100\,°C$. For any application the underfill needs to be tested to the reliability conditions required. Encapsulants follow a similar pattern, with high reliability applications typically utilizing low CTE, high T_g materials (typically thermal cure) and other applications typically utilizing low T_g, flexible (low modulus) materials. Once again, the reliability requirements typically drive the physical property requirements.

10.6 Young's Modulus (E)

Young's modulus is the same as tensile modulus. Tensile modulus is the ratio of stress to strain within the elastic region of the stress – strain curve (prior to the yield point). Young's modulus characterizes the elastic properties of the material under tension or compression irrespective of the sample geometry. Typically, the lower the modulus the more elastic the material. Thus, a material with a low modulus can be considered very rubbery and can absorb more stress before fracturing [6, 10].

To measure tensile modulus, the stress is gradually increased on a sample then the measure of the elongation the sample undergoes at each stress level is recorded. This is continued until the sample breaks. Then a plot of stress versus elongation is produced, as in Figure 10.21.

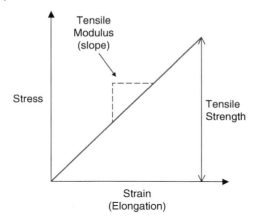

Figure 10.21 An example of a strain-stress graph for determining modulus.

10.7 Applications

10.7.1 Underfills

Underfill materials are polymer systems (filled or unfilled) that are used to increase the reliability of a variety of area array packages used in high density electronic assemblies [11–25]. Underfill systems are typically epoxy-based chemistries (e.g. bisphenol epoxies and cycloaliphatic epoxies) that sometimes have a filler added. There are two main reasons for using an underfill material.

Firstly, to relieve stress from the large CTE mismatch between a component (flip-chip or wafer level CSP) and the substrate it is bonded to (e.g. FR-4). Secondly, to increase the reliability of the component, such as a CSP or ball grid array (BGA), with respect to physical shock and vibration. With the advent of lead-free electronics, assemblers have found these materials are more brittle in nature than lead-containing materials, and therefore there is an increased need for underfill in these applications due to an increased risk in drop and shock conditions [24].

Most manufacturers that use flip-chip on board (FCOB) use underfill due to the large CTE mismatch between the silicon die and the FR4 substrates typically used in manufacturing these products. The other packages may only be underfilled if there is either a perceived risk to the products (e.g. a cell phone may undergo frequent drops) or for a high reliability application (e.g. avionics). The CTE matching of the underfill materials is typically achieved through the addition of silica- or alumina-based fillers. This filler addition lowers the CTE from ~80 ppm °C^{-1}, to as low as ~20 ppm °C^{-1} for highly filled systems in order to provide a gradient between the silicon chip and the substrate. The need for a low CTE value in an underfill is an often misunderstood concept. Not all underfills necessarily need to have a low CTE to provide reliability enhancement for all devices. The balance between CTE and modulus should be examined in order to provide the highest reliability enhancement possible. In general, low modulus materials (which tend to have higher CTE) are better for drop/shock reliability enhancement, and lower CTE materials (which tend to have higher modulus) are better for enhancing reliability when failures are occurring during thermal cycling. A graphic showing the relationship between modulus, CTE, and two common failure modes is shown in Figure 10.22. As this figure illustrates understanding the failure mode of the device is critical in choosing the material with the right properties. Also, note that using a material with a low CTE does not guarantee an increase in reliability if the device is susceptible to fatigue cracking. Figures 10.23 and 10.24 illustrate the reliability of various underfill materials under drop testing (Figure 10.23) and thermal shock (Figure 10.24). The materials tested were two fast flow snap cures (FFSC) and two reworkable underfills.

Table 10.6 shows the different reliability and processing requirements for flip chip underfill and CSP underfill.

10.7.1.1 Capillary Underfill

These low viscosity liquids are designed to flow under a component by capillary action, wetting to the chip and substrate surfaces and encapsulating the solder joints. Underfills designed for assembly-level flip chip components generally cure in five minutes or less at 165 °C to form a hard seal with high adhesion to both the component (flip chip or CSP) and the substrate. More recently, the trend has been for underfill materials to cure at temperatures of below 100 °C (usually curing under 30 minutes). Underfills designed for package-level assemblies (e.g.

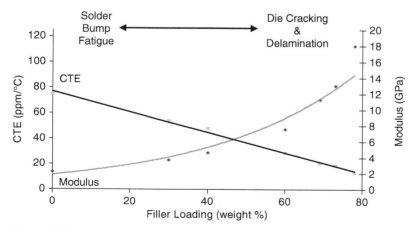

Figure 10.22 Graph showing the relationship between CTE, modulus, and failure modes.

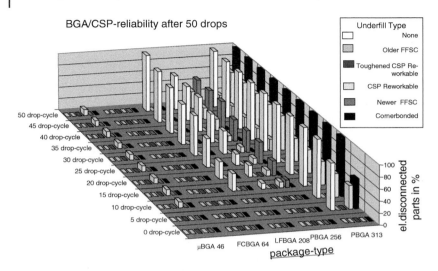

Figure 10.23 Reliability of various underfill materials with respect to drop tests.

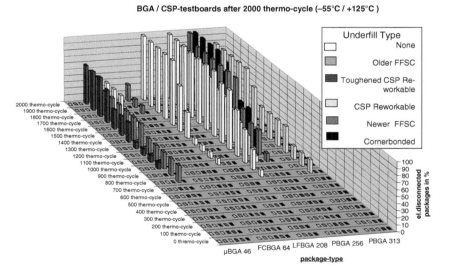

Figure 10.24 Underfill reliability with respect to thermal cycling.

SiP (System in Package)) offer improved assembly reliability but generally require more time to flow under the die and to cure. Faulty or skewed components must be detected prior to underfill cure, since mostly all capillary flow underfills are permanent (see Section 10.7.1.3 for reworkable underfills).

Table 10.6 Different reliability and processing requirements for flip chip underfill and CSP underfill.

Flip chip underfill	CSP underfill

• Absorbs stresses caused by CTE mismatch between chip and board
 • Critical characteristics:
   • Ionic content <25 ppm
   • High adhesion
   • CTE <40 ppm/°C
   • Tg >125 °C
   • No voids during capillary flow
   • Low moisture uptake
 • Process:
   • Heated substrate (for flow)
   • Cure temp >120 °C
   • Cure time >30 min, step cure
	• Primarily used to improve shock resistance (drop test reliability)
	 • Critical characteristics:
	   • High adhesion
	   • Modulus <4 MPa
	 • Less important:
	   • Low ionic content
	   • CTE
	   • Tg
	 • Process:
	   – Room temp flow
	   – Cure temp <120 °C
	   – Cure time <10 min

Underfill is applied close to the edge of a flip chip or CSP to enable capillary forces to encapsulate the gap between the component and the board. Dispensing of capillary underfill materials requires specialized equipment to achieve the accuracy and precision required for high-volume assembly. At a minimum, the dispenser must reproducibly position for successive assemblies and apply a predetermined volume of underfill to the edge of the component. Previously, heating the substrate was a common secondary requirement that accelerates capillary flow. Over the past 10 years many high-volume consumer electronic applications have driven the process toward room temperature flow materials. Cure is usually accomplished in belt-style reflow or curing ovens.

Several dispensing patterns are used for applying underfill to die and packages. The simplest pattern involves applying underfill to a single side and maintaining a reservoir as material wicks between the package and substrate. Previously, the underfill was applied to the remaining three sides using a "U" pattern to form a fillet. With newer generation materials the filleting step can be eliminated. The dispensing of a single line or "I" pattern is the most common for smaller packages.

Another common flow pattern recommended for fast processing of smaller or simpler devices involves dispensing a "L" of material (two sides), waiting for flow out and then dispensing an inverted "L" to complete the fillet, if needed.

The underfill flow rate or underfill time is one of the key attributes of an underfill encapsulant. Underfill suppliers typically use smooth or frosted glass slides with a predetermined gap established by shim stock to evaluate underfill times. In some cases, a fixture is used to maintain a stable gap and provide a more efficient (more than one slide assembly can be prepared at a time) means of running the test.

The fixture is placed on the hot plate and monitored for the selected underfill temperature (room temperature up to 100 °C). When the selected temperature is reached, underfill material is dispensed at the gap area. The flow front is timed to determine the time required to reach 1 cm. Information regarding underfill time and voiding predisposition can be obtained.

Other methods, such as glass slides set up with a variable gap that runs from ~0.13 mm to zero (creates a wedge shape), can be used to expeditiously determine the compatibility of an underfill with various gap sizes. Another approach, that may be utilized more by IC manufacturers, is the use of a quartz bumped die that is consistent with the IC size to be used in the application. Different combinations of substrates with slides such as FR-4, silicon, etc. can be evaluated. The predisposition to voiding can also be assessed before and after cure. The glass slide/quartz die tests are not designed to reflect underfill times in specific applications, since solder masks, chip size, chip passivation coating, bump count, bump spacing, bump location and surface cleanliness will all have a dramatic effect on underfill times. This test may be used to compare alternative encapsulant suppliers, alternative underfill temperatures, and shelf life/pot life of underfill encapsulants.

10.7.1.2 Fluxing (No-Flow) Underfill

Fluxing, or so-called "no-flow" underfills are an attempt to make underfill processing more compatible with conventional surface mount assembly processes by eliminating the dedicated oven required for cure. A fluxing function incorporated into the underfill combines the component attachment and underfill cure processes [16, 19–21].

Fluxing underfill is either applied directly to the attach site on the substrate prior to component placement or the component is dipped into the material and then placed onto a substrate. During reflow, the underfill acts as the flux, enabling interconnect formation and self-centering prior to curing of the underfill. Ideally, underfill cure is completed in the reflow oven. Otherwise, subsequent reflow or deliberate heat treatment is required to complete underfill cure.

Fluxing underfills are different from capillary flow materials in several ways. Viscosities and thixotropic index are typically much higher than that of capillary underfills. For example, viscosities of fluxing underfills can be in the order of 8000 cps or higher, versus less than 3000 cps for most room temperature flow underfills. In addition, the thixotropic index for printed or dippable materials can be greater than 2. Since inorganic fillers in the underfill will generally impede essential contact between the CSP solder balls and the attach pads, fluxing underfills are unfilled, which results in a higher CTE, which could be a detriment to reliability if thermal cycling performance is the primary reliability failure mode. The entire board must be discarded if defective or skewed components are detected, if the underfill is not removable/reworkable.

10.7.1.3 Removable/Reworkable Underfill

Occasionally malfunctioning flip chip or CSP components are discovered in the factory or in the field. Conventional underfills are not removable once cured, requiring that a malfunctioning assembly be discarded, sometimes at considerable cost. Removable/reworkable underfill enables repair of the assembly [26]. The use of reworkable underfill materials has increased significantly in recent years. One of the factors for this increase is the higher density circuitry on PCBs, raising the value of the assembly, leading to more interest in using a reworkable underfill for these devices. The goal of using the reworkable underfill material is to reduce the amount of scrap generated. These savings can easily be negated if the rework process is overly complex or the PCB is destroyed in the rework process.

Removable/reworkable underfill is typically dispensed and processed similar to other capillary underfills. Extra functionality enables them to soften or degrade when sufficiently heated. The assembly repair process is similar to that of other soldered surface mount technology (SMT) components with the exception of an extra step or extra time required to remove residual underfill. Figure 10.25 shows an example of what a rework site would look like. The rework process for underfilled components is significantly more complex than rework of components that are just soldered. When reworking underfill careful process control is critical [25, 26]. The opportunity for failures such as solder mask damage, pad pull-out, and solder extrusion/bridging on adjacent devices during rework are increased.

10.7.1.4 Staking or Corner Bond Underfill

An alternative approach to complete underfilling of CSP devices for increased reliability with respect to mechanical shock and drop is to bond the corners or edges of a CSP (Figure 10.26). Some materials can be applied after the device is soldered to the substrate and some materials can be dispensed with the solder paste prior to reflow. The advantage with these materials is that you can reduce the capital equipment expense by eliminating a curing oven and a separate dispenser and reduce cost by using less material, with four to eight dots of underfill versus the entire

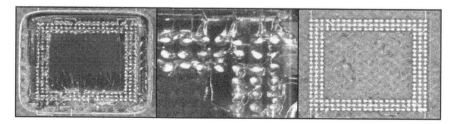

Figure 10.25 Images of pads during rework process (from left to right: after component removal; close up of pads; pads after flux and solder removal). Source: Images courtesy of Henkel.

(a)

(b)

(c)

Figure 10.26 Corner bonded CSP. Source: (a) Image courtesy of Dymax. (b and c) Images courtesy of Henkel.

surface area under the package. There is a limit to the mass of the component whose reliability can be enhanced (if the mass of the component is too great the adhesive may not hold it in place). Also, when dispensing corner dots with packages that have corner bumps, care must be taken not to interfere with the reflow soldering process [21, 26].

10.7.2 Encapsulant Materials

Another type of polymeric material that has seen increased usage in lead-free electronic devices in recent years is encapsulation materials. These materials are typically used to provide environmental protection of some sub-section or portion of components on an electronic assembly. Whereas conformal coatings or potting materials are used over the entire assembly, encapsulants are designed to be dispensed and cured in very discrete locations. Similar to underfill materials, there are very different physical property requirements based on usage.

10.7.2.1 Glob Top

One common form of encapsulation that has been in use since the 1990s on PCBs is glob top encapsulation [27, 28]. These materials are used for FCOB or chip on board (wire-bonded) application and consist of epoxy-based, low-ionic, high Tg, low CTE materials. These are also usually heat-cured and very rigid. The other type of glob top materials in use are silicone-based materials. These are used for their high flexibility and high heat resistance. These materials are very common for LEDs and other optical components. In this case the silicone material can be made optically clear as well as resist the heat generated by these optical devices [29]. Overall these glob top materials need to have a balance of viscosity and thixotropy since in many applications they need to flow through wire bonds (without causing the wires to touch, aka wire bond sweep) or under a bumped package, while not being so low in viscosity that they flow outward too far. Common viscosity ranges are 20 000 cPs up to 200 000 cPs. This balance of viscosity typically results in a domed appearance as shown in Figure 10.27.

10.7.2.2 Component Encapsulation

The more recent trends have been in the area of small area and discrete component encapsulation as shown in Figure 10.28. These materials are typically different than traditional glob top materials in that they are designed to encapsulate small areas and be processed relatively quickly (i.e. no long cure times). These types of materials are used primarily to enhance reliability in two ways. The first is environmental protection. Many consumer products and wearable electronics are now

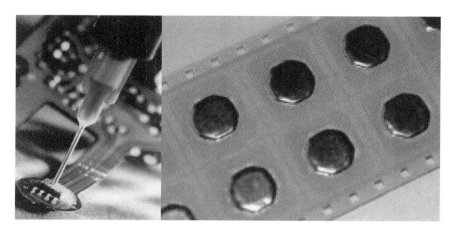

Figure 10.27 Glob top.

(a)

(b)

(c)

Figure 10.28 (a) Encapsulant. (b and c) Encapsulant with UV-indicating dye (with and without black-light illumination). Source: Images courtesy of Henkel.

designed to be waterproof to some extent, defined by the IEC 60529 specification (Degrees of protection provided by enclosures (IP Code)).

There are a number of ways to achieve the desired level of waterproofing, with component encapsulation one common tool. In order to protect the sensitive components, it is critical that the material fully encapsulates the components and does not leave any exposed edges or corners. The edges of area array packages and discrete components can be difficult to wet and any material selection for this type of application needs to consider the ability of the material to coat these edges and corners. The other common reliability enhancement is to protect small discrete components (0402 sized components and smaller) against damage due to handling. As designers use every possible space on a PCB and more and more functionality is condensed into these small spaces, the components get smaller and closer to the edges. In a manufacturing environment the small components can be dislodged easily, especially due to the more brittle nature of lead-free solder versus tin-lead solders. These encapsulants help to protect these small discrete components and increase the reliability during handling.

Common chemistry platforms for these applications are typically acrylated urethanes and UV-activated epoxies [30–32]. As stated earlier, these encapsulants are primarily designed to coat and encapsulate discrete components, so the key characteristics are the ability to coat over edges and corners and not have a high degree of overflow. This combination of properties results in materials with a lower viscosity (2000 cPs up to 50 000 cPs) than what is typically seen in the glob top materials described earlier, but with a higher thixotropic index (2–4).

10.7.2.3 Application

The method of application is typically dependent on two main attributes, the rheology of the material and the density of the PCB. Typically, for higher viscosity materials (>250 000 cPs) precision dispensing is done with an auger valve [33], although time/pressure dispensers can also be used. However, the shot to shot accuracy is typically not as good although the speed is faster. These methods (auger and time-pressure) are most common for glob top type of materials. For the low to medium viscosity fluids (and with the proper setup, even with higher viscosity fluids) and very dense assemblies non-contact dispensing, commonly called jetting [34, 35] is most common. This is often the application of choice for component encapsulants.

With respect to the curing process, thermally cured materials are still common (see Section 10.3.1), and the UV cure and UV dual-cure systems (see Section 10.3.2) are more common. The same cautions and process controls still need to be enforced here as well such as thermally profiling components to ensure the material is reaching the time and temperature required. For UV systems ensuring the proper light dosage is received at the adhesive is important, with

Figure 10.29 Low-pressure molding. Source: Images courtesy of LPMS-USA.

the UV-cured component encapsulants since there is likely to be material under components which is shadowed from the UV lamp or wand. To ensure the highest level of reliability a dual-cure system is recommended.

10.7.2.4 Low-Pressure Molding

A more recent method of encapsulation for electronic devices is using a polyamide or poly-olefin material with a low-pressure molding process [36, 37]. Unlike the underfills and encapsulants described previously, this method uses thermoplastic materials to encapsulate the components and/or the PCB as shown in Figure 10.29. By using a thermoplastic instead of a thermoset material, the curing step is eliminated, since the material just needs to be melted and re-solidified. The low viscosity of these resins when melted allows a low-pressure molding process. At their melt temperature the viscosity is in the order of 3000–20 000 cPs at 190–220 °C. This allows the molding pressure to be as low as 50 psi up to 200 psi. The material will flow around components without the stress of the typical high pressure injection process. Even though the process temperature is close to the melt point of lead-free solder as the material is injected into the mold, it cools quickly and the solder joints remain un-disturbed [38].

These materials and the process has been utilized in a number of automotive and consumer electronic sensors and connectors. The primary reliability enhancement here is waterproofing as well as dust and particulate protection. More recently, this technology is being applied to a wider range of electronic devices and new standards have been published regarding this technology [39].

10.8 Conclusions

The proper selection and application of underfills and/or encapsulants can significantly enhance the reliability of lead-free electronic devices. As with any design element there are always competing attributes that must be carefully balanced and evaluated to ensure the maximum performance enhancement. The chapter has provided an overview of the most critical performance and process parameters that should be evaluated and understood before assembly in order to minimize unintended consequences. The addition of these polymers, when carefully chosen and properly applied and cured, will enable the production of robust systems.

References

1 Schramn, G. (1994). *A Practical Approach to Rheology and Rheometry*. Haake GmbH.

2 Barnes, H.A., Hutton, J.F., and Walters, K. (1989). *An Introduction to Rheology*. Elsevier.

3 Ferguson, J. and Kemblowski, Z. (1991). *Applied Fluid Rheology*. Elsevier.

4 Fay, H.G. (1997) Rheology for the Bewildered. Loctite Process Technology Document.

5 Konarski, M. (1998) Cure parameter effects on the Tg and CTE of flip chip encapsulants. Circuits Assembly.

6 Voloshin, A.S. and Tsao, P.H. (1993) Analysis of the stresses in the chip's coating. Proceedings of the 43rd Electronic Component and Technology Conference.

7 Roller, M.B. (1982). The glass transition: What's the point? *Journal of Coatings Technology* 54 (691): 33–40.

8 Fahmy, A.A. and Raga, A.N. (1970). Thermal-expansion behavior of two-phase solids. *Journal of Applied Physics* 41 (13): 5108.

9 Goldstein, M. (1963). Some thermodynamic aspects of the glass transition: free volume, entropy and enthalpy theories. *Journal of Chemical Physics* 39 (12): 3369.

10 Brinson, H.F. (ed.) (1990). *ASM Engineered Materials Handbook Volume 3 Adhesives and Sealants*. ASM Publications.

11 Lau, J.H. and Ricky Lee, S.W. (1999). *Chip Scale Package*. McGraw-Hill.

12 Lau, J.H. (1996). *Flip Chip Technologies*. McGraw-Hill.

13 Buchwalter, S.L., Edwards, M.E., Gamota, D. et al. (2001). Underfill: the enabling technology for flip-chip packaging. In: *Area Array Interconnection Handbook* (eds. K. Puttlitz and P.A. Totta), 452–499. Kluwer Academic Publishers.

14 Babiarz, A., Quinones, H. (2001) Advances in fast underfill of flip chips. ICEP/Microelectronics, Tokyo, Japan, February 2001.

15 Liu, J., Johnson, R.W., Yaeger, E. et al. (2002) CSP underfill, processing and reliability. Proceedings from the Technical Program, IPC APEX 2002.

16 Previti, M.A. (2001) No-flow underfill: a reliability and failure mode analysis. Proceedings from the Technical Program, IPC APEX 2001.

17 Carson, G. and Edwards, M.E. (2001) Factors affecting voiding in underfilled flip chip assemblies. Proceedings from the Technical Program, SMTA International, 2001.

18 Yaeger, E., Szczepaniak, Z.A., Konarski, M. et al. (2002) Underfill materials, processing and reliability for fine pitch flip chip on laminate assembly. Proceedings from the Technical Program, IPC APEX 2002.

19 Houston, P.N., Smith, B.A., and Baldwin, D.F. (2001) Implementation of no flow underfill material for low cost flip-chip assembly. SMTA International, 2001.

20 Poole, N., Vasquez, E., and Toleno, B.J. (2015) Epoxy flux material and process for enhancing electrical interconnections. SMTA International, 2015.

21 Toleno, B. and Schnieder, J. (2003) Processing and reliability of corner bonded CSPs. Proceedings from the IEMT, SEMICON West, San Jose, 2003.

22 Tu, P., Chan, Y.C., and Hung, K.C. (2001). Reliability of microBGA assembly using no-flow underfill. *Microelectronics Reliability* 41 (12): 1993–2000.

23 Radhakrishnan, J. (2017) Application methods and thermal mechanical reliability of polymeric solder joint encapsulation materials (SJEM) on SnAgCu solder joints. SMTA China, 2017.

24 Zhang, S., Chang, F., Yee, S., and Shiah, A.C. (2003). An investigation on the reliability of CSP solder joints with numerous underfill materials. *Journal of Surface Mount Technology* 16 (3).

25 England, J., Toleno, B., and Israel, J. (2014) Development of an underfill rework process and evaluation of underfill reworkability. SMTA International.

26 Xie, F., Wu, H., Hodge, K. et al. (2016). Evaluation of reworkable edge bond and corner bond adhesives for BGA applications. *Journal of Surface Mount Technology* 29 (3).

27 Burkhart, A. (1991) New epoxies for advanced surface mount applications. Surface Mount International Conference and Exposition, San Jose, CA, August 1991.

28 Goodrich, B. (1995). New generation encapsulants. *International Journal of Micro Circuits and Electronic Packaging* 18 (2): 133–137.

29 Lei, I., Lai, D., Don, T. et al. (2014). Silicone hybrid materials useful for the encapsulation of light-emitting diodes. *Materials Chemistry and Physics* 144 (s1–2): 41–48.

30 Dymax. Encapsulation materials. https://dymax.com/application-areas/electronics-assembly/encapsulants (accessed 4 February 2020).

31 Panacol USA. Encapsulants. www.panacol-usa.com/products/encapsulants.html (accessed 13 August 2019).

32 Adhesives and Sealants Magazine (2005) HENKEL: UV-Cure Encapsulants. 27 January. www.adhesivesmag.com/articles/85817-henkel-uv-cure-encapsulants (accessed 4 February 2020).

33 EFD (2005) Augur valve dispensing. https://smtnet.com/library/files/upload/EFD_-_Auger_Valve_Dispensing.pdf (accessed 4 February 2020).

34 Babiarz, A. (2006) Advances in jetting small dots of high viscosity fluids for electronic and semiconductor packaging. Pan Pacific Symposium, 2006.

35 Quinones, H., Babiarz, A., and Fang, L. (2002) Jetting technology for microelectronics. IMAPS Nordic, Stockholm, Sweden, September 2002.

36 Pierce, M. and Hansen, R. (2006) Manufacturing improvements enabled by new low-pressure molding process. SMTA International, 2006.

37 LPMS USA. https://www.lpms-usa.com (accessed 4 February 2020).

38 Brandenburg, S. (2007) Evaluation of overmolded electronic assembly packaging using thermoset and thermoplastic molding. Advancing Microelectronics, January/February 2007.

39 IPC-7621 (2018) *Guideline for design, material selection and general application of encapsulation of electronic circuit assembly by low-pressure molding with thermoplastics.* IPC International.

11

Thermal Cycling and General Reliability Considerations
Maxim Serebreni

Department of Mechanical Engineering, University of Maryland, College Park, MD, USA

11.1   Introduction to Thermal Cycling of Electronics

Solder materials are the most widely used second-level interconnects in electronic packaging. Solder interconnects, often referred to as solder joints, provide the mechanical and electrical connection between electronic packages and printed circuit boards (PCBs). Electronic packages are assembled onto PCBs using a soldering process. This soldering process results in the formation of solder joints, which vary in shape and size, based on package style, stencil thickness, and board pad size. A large percentage of electronic failures are due to thermally induced stresses and strains that are caused by differences in the coefficient of thermal expansion (CTE) mismatch between the components and boards. A study by the United States Air Force on field failures, related to operating environments, shows that about 55% of the failures are due to high temperatures and temperature cycling, 20% of the failures are attributed to vibration and shock, and, finally, 20% are caused by humidity [1]. Under fluctuating temperature conditions, the large CTE mismatch can cause cyclic fatigue failure to occur in solder joints. Due to the geometry of the surface mount component, thermomechanical fatigue is induced by the accumulation of shear stress on the solder joints, as shown in Figure 11.1. The distance from the center of the package to the solder joint is called the distance to the neutral point (DNP). In general, components with larger DNP will be more affected by CTE mismatch, which means that the farthest joint from the package center will experience the highest shear strains.

Some operating service conditions subject components to high temperatures during the dwell time. This allows creep to become the dominant inelastic deformation mechanism. Table 11.1 lists the temperature range experienced during field conditions for several electronic product categories.

Lead-free Soldering Process Development and Reliability, First Edition. Edited by Jasbir Bath.
© 2020 John Wiley & Sons, Inc. Published 2020 by John Wiley & Sons, Inc.

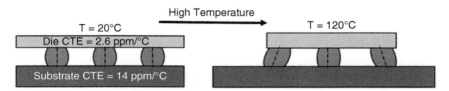

Figure 11.1 CTE mismatch in electronic assemblies at elevated temperatures.

Table 11.1 Field conditions for various industries [2].

	Temp range (°C)	Cycles per year	Service time (yr)	Failure rate (%)
Consumer	0 to 60	365	1	1
Computer	15 to 60	1460	5	0.1
Telecom	−40 to 85	365	7–20	0.01
Aircraft	−55 to 95	365	20	0.001
Automotive	−55 to 95	100	10	0.1

Field conditions differ based on device usage and application, even for the same products and components. Specific applications have detailed specifications that define (i) the maximum and minimum temperature range, and (ii) the dwell time at each temperature extreme. In some products there are different requirements for each life phase. In situations where the given CTE mismatch is large enough and the temperature range and cyclic frequency are such that solder joints will experience plastic or creep deformation, low cycle fatigue failure can occur as shown in Figure 11.2. In this figure, 2512 chip resistors were assembled on a 1.6 mm thick PCB and subjected to accelerated thermal cycling with a temperature range of −40 to 125 °C according to IPC sta ndard 9701A [3]. Often, the accelerated thermal cycling tests are used to qualify the performance and reliability of surface mount solder attachments of electronic assemblies. During the accelerated thermal cycling tests, creep strain in the solder joint is induced due to CTE mismatch between the PCB and the component. For the case illustrated in Figure 11.2, solder joints in the chip resistors predominantly experience shear strains due to the rigid alumina component and PCB with a CTE of 7.8 and 17 ppm °C^{-1}, respectively.

After a sufficiently long number of cycles, a fatigue crack will nucleate and start to propagate through the solder joint resulting in a mechanical failure of the joint. The interruption in the physical connection of the solder joint disrupts the electrical signal path, which renders the component or device inoperable.

Determining time to failure under thermal cycling most often involves resistance monitoring of the test vehicle on which electronic components are soldered.

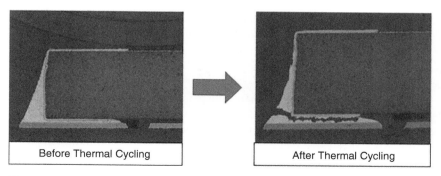

| Before Thermal Cycling | After Thermal Cycling |

Figure 11.2 Fatigue cracks in solder joints of 2512 chip resistors after thermal cycling from −40 to 125 °C.

A Kink in wire
B Crack in PTH solder due to wire pull
C Cracking in internal copper barrels
D Short between two joints
E Pad cratering on PCB or component
F Failure of joint near mounting point

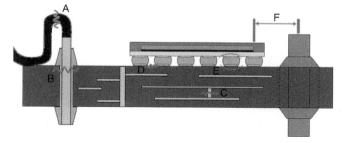

Figure 11.3 Potential failure modes in thermal cycling test boards.

Board layout and test configuration could inadvertently induce additional failure modes other than solder fatigue cracking that could be detected and misinterpreted as fatigue failures. Such failure modes are shown in Figure 11.3 and range from failure of the monitoring wire to internal cracking of copper traces or vias. For test boards that utilize hand soldering wires to plated-through holes, adequate strain relief should be provided to avoid wear-out and kinking of the wiring during thermal expansion and contraction. Hanging test boards using monitoring wires as support should be avoided as it can cause fatigue in the PTH solder joint of the wires. Failure internal to the board could occur in copper traces or vias due to epoxy/resin CTE mismatch with copper and board bending. It is recommended to design stitching vias around component lands in routing layers to distribute thermal loads in the boards. Multiple ground wires should be soldered for redundancy for each component or test board depending on desired electrical monitoring configuration. Cleanliness of boards prior to thermal cycling is often overlooked in many accelerated tests. Fine pitch components and features are more susceptible to contamination from flux residues that could induce electrochemical migration or corrosion failures.

Proximity of components to the mounting point could cause accelerated fatigue failure of solder interconnects due to the elevated board strains near the vicinity of constrained areas. Similar effects have been found to occur from over constraining of components using heat sinks as well. In general, any configuration that could impede the characteristic thermal expansion of components and boards should be avoided when the fatigue endurance of a particular component is desired.

11.1.1 Influence of Solder Alloy Composition and Microstructure on Thermal Cycling Reliability

The microstructure of solder joints is influenced by a number of factors such as the reflow profile, solder composition, and dissolution of metallization from the board or components. It is virtually impossible to control or replicate the microstructure of solder joints from one assembly to another. Even within a single ball grid array (BGA) package, several different microstructures can form [4]. The microstructure is characterized by a small numbers of grains that form across the joint or a single grain that comprises the majority of the joint. Such a low number of grains with high misorientation provides different properties for every joint. Studies on the mechanical properties of single grain solder joints demonstrate the anisotropic nature of tin [5, 6].

Anisotropy of solder joints is most evident in the intrinsic properties that have influence on the mechanical degradation under thermal and mechanical loading. Solder alloy Young's modulus has been previously found to possess a linear relationship with strain rate and temperature. The yield stress of solder alloys is found to have a near linear relationship at cold temperature and slowly progresses to a non-linear dependence with increasing temperature and decreasing strain rate [7]. A similar behavior is observed in eutectic tin-lead solder with the exception of anisotropic behavior due to differences in microstructural features and their response to temperature cycling [8].

An example on the effect of silver content on the solder alloy is demonstrated by the following study that investigated the thermal cycling performance of doped Sn1Ag0.5Cu solder [9]. The alloys investigated were Sn-1Ag-0.5Cu-0.02Ce (SACC) and Sn-1Ag-0.5Cu-0.05Mn (SACM). The 244 I/O BGA packages were assembled with five solder alloys and subjected to drop and thermal cycling testing. The low silver Sn1Ag0.5Cu alloys demonstrate drop test results similar to SnPb without sacrificing thermal cycle performance, as shown in Figure 11.4, which remains as good as Sn3Ag0.5Cu for a BGA device. The trace additions of Mn and Ce were

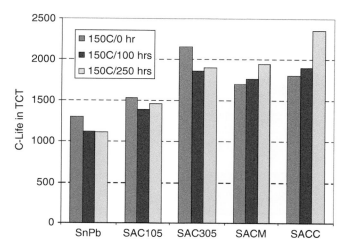

Figure 11.4 Characteristic life when thermal cycling (−40 °C/125 °C) for a 244 I/O BGA (0.5 mm pitch). Before and after preconditioning as shown. Sn-1Ag-0.5Cu-0.05Mn (SACM) and Sn-1Ag-0.5Cu-0.02Ce (SACC) performed similar to Sn3Ag0.5Cu [9].

credited with suppressing intermetallic growth (especially on Ni) and reducing coarsening of the precipitates in the matrix – which in turn also helped prevent grain growth.

11.2 Influence of Package Type and Thermal Cycling Profile

Conventional solder fatigue theory dictates four major contributing factors influencing the strain range solder joints experience in electronic packages under temperature cycling. These factors are CTE mismatch between component and PCB, solder joint height, distance to neutral, and the temperature range. The classical distance to neutral effect is not a dominant factor in thermal cycling fatigue for certain electronic components as originally thought. For BGA packages, the ratio of die to package size parameter can be seen to dominate the characteristic life among BGA packages, where larger die to package ratio results in lower characteristic life [10]. Quad flat no-leads/bottom termination component (QFN/BTC) components can be seen on the other end of the spectrum. The thermal fatigue life of BTC components assembled using stencil printing of solder paste are more susceptible to variation in final solder joint geometry that could result in significantly lower characteristic life than expected [11]. Wafer-level chip scale packages (WLCSPs) are found on the lower end of the spectrum similar to QFN/BTC components due

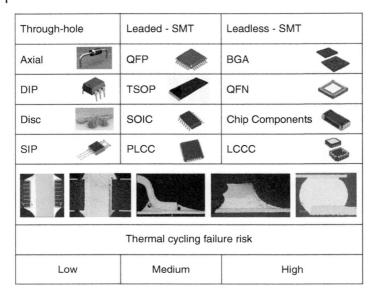

Figure 11.5 Thermal cycling failure risk for various through-hole and surface mount electronic components.

to the higher CTE mismatch solder joints experience between the package and PCB. Failure in the WLCSP under thermal cycling has been found to occur in solder joints and the redistribution layers and characteristic life of such experiments should be adjusted to account for components that failed from solder fatigue to avoid a biased result that can be skewed by early failure in the redistribution layers. Figure 11.5 shows various electronic components for which thermal cycling failure risk in solder joints is classified into three categories: Low, Medium, and High. Through-hole and leaded surface mount components are grouped into the low to medium risk of failure. The higher lead compliance provides a strain relief for the solder joints under thermal expansion enabling the lead to absorb the brunt of mechanically induced strain due to CTE mismatch between components and board. Leadless surface mount components are placed under the high-risk category specifically due to their higher susceptibility to failure under thermal cycling. This behavior is true for all leadless surface mount components, from passive resistors to complex integrated circuits. This understanding is often useful for solder qualification studies. Selection of the most critical components or a package that is known to fail early can identify the weakest link in an electronic assembly as well as shorten the qualification program duration.

To demonstrate the criticality of package type on thermal cycling failure risk the following data is reviewed. Data was gathered for Sn0.7CuNi lead-free solder

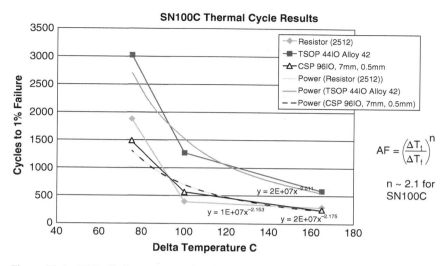

Figure 11.6 Weibull plots were used to determine 1% failure at three different delta T values. Regardless of component type, the exponent n in the acceleration factor was near 2.1 for Sn0.7CuNi lead-free solder.

at three temperature ranges, 25 to 100 °C, 25 to 125 °C, and −40 to 125 °C. This was done for three component types (chip scale package [CSP], resistor, and thin small outline package [TSOP]) and the results plotted in Figure 11.6. Calculations can be made by testing the alloy to different delta T values, determining the $N_{1\%}$ at each, and fitting the curve to find the acceleration factor which is shown in Eq. (11.1).

$$AF = \left(\frac{\Delta T_t}{\Delta T_f}\right)^n \tag{11.1}$$

where ΔT_t is the temperature range under the test environment and ΔT_f is the temperature range of the thermal cycle under field conditions and n is an exponent.

Temperature cycling itself contributes to a significant portion of damaged solder joints experienced in electronic components. It is well understood that the failure behavior of SnAgCu solder is dependent on the dwell time of the temperature profile. Figure 11.7 summarizes three data sets reporting on the influence of dwell time greater than 30 minutes on 0–100 °C temperature cycle on a chip resistor, ceramic BGA, and a plastic BGA. Using this experimental data set and finite element modeling, an extrapolation of the dwell time effect on failure rate was performed. Data shown in Figure 11.7 was normalized by a dwell time of 10 minutes.

The dwell time dependence shown in Figure 11.7 demonstrate the relationship between CTE mismatch and dwell time on failure behavior between the different

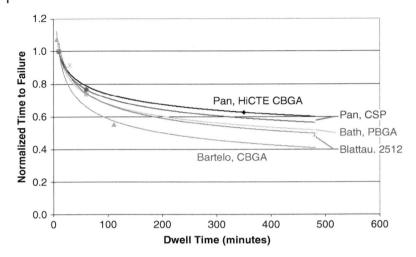

Figure 11.7 Normalized time to failure as a function of dwell time at maximum temperature for SnAgCu solder.

surface mount components. Under a temperature difference of 100 °C a ceramic BGA could experience 60% lower time to failure between a field and test environment with an eight-hour compared to a 10-minute dwell time. However, plastic BGA components show a smaller percent difference in time to failure in similar temperature ranges. This indicates that the component characteristics that control CTE mismatch will have a different correlation between field conditions and accelerated tests.

11.2.1 Influence of Board and Pad Design

PCBs are the backbone of every electronic system. The transition to lead-free solder has meant that boards are exposed to higher reflow temperatures that result in higher board expansion. To address these challenges manufacturers have looked to adjust the resin/glass of the board to achieve lower CTE laminates to compensate for the increased temperature of the lead-free reflow process. Increasing the glass transition temperature also assists with reducing board warpage to minimize manufacturing defects during reflow. Decisions on board stack-up are made to ensure signal integrity and electrical performance. The final mechanical properties set by the board such as the CTE and modulus could have significant influence on board-level reliability of electronic components that are soldered to the boards.

In addition to influencing solder fatigue, selection of epoxy and glass style have been shown to influence other failure mechanisms such as conductive anodic

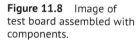

Figure 11.8 Image of test board assembled with components.

filament (CAF) formation that results in internal shorts. This failure mechanism is characterized and driven by an electrochemical reaction that forms between the degraded glass/resin interface [12]. Boards with fine pitch applications are at higher risk of CAF formation.

The following study illustrates the impact of changing glass style and pad size on the reliability of surface mount components under thermal cycling [13]. Thermal cycling (-40 to $125\,°C$) tests were done on two PCB glass styles (1080 and 7628). The in-plane and out-of-plane CTE were measured and found to be about 3–$4\,ppm\,°C^{-1}$ different in the X-Y directions of the boards. To facilitate the reliability assessment of surface mount technology (SMT) components four 0-Ω resistor sizes (2512, 1206, 0602, and 0402) were mounted on each board as shown in Figure 11.8. Each board had 20 resistors with multiple orientations. The overall sample size consisted of 32 resistors of each size per glass style. In addition, the PCBs had two pad sizes.

To illustrate the difference in the layers between the two PCBs, the glass weave was exposed by first placing the boards in a furnace at $550\,°C$ to burn off the epoxy, followed by a soak in ferrous chloride acid to remove the copper traces and through holes. Figure 11.9 demonstrates the cross-section of the two glass styles along with the exposed glass. A drastic difference in the weave density in the warp and fill direction of each glass style can be seen. In addition, the layer count in the 1080 glass style is notably higher than the 7628. The larger spacing of the 1080 glass style allows for a higher volume fraction of epoxy resin to be impregnated during the lamination process. The higher volume fraction of epoxy to glass increases the effective CTE of the PCB. The test boards in this study were designed with four copper layers. The two internal copper layers facilitated routing between the front and back sides of the boards. It is important to note that multilayer boards with heavy copper fill per layer will influence the effective CTE of PCBs.

Figure 11.10 shows cross-sections of the difference in the pad dimensions and the resultant bond line thickness for the solder joints for comparison. Solder

(a) (b)

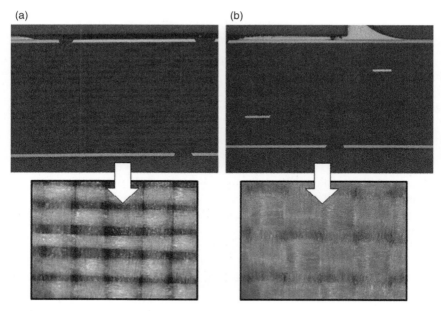

Figure 11.9 FR-4 PCB with (a) 19 layers of 1080 and (b) 7 layers of 7628 glass styles.

Small Solder pads
Pad size: 3.2x0.847 mm
Solder height: 22-41 μm

Large Solder pads
Pad size: 3.2x3.2 mm
Solder height: 42-46 μm

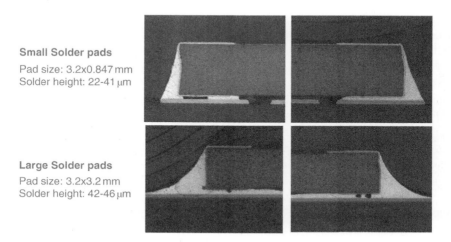

Figure 11.10 Cross-sections of assembled 2512 resistors with small and large PCB pads.

joint height between the two pad sizes are found to be consistent as shown in Figure 11.10. Solder joint height is determined by the space underneath the component that is soldered to the PCB. The main differences observed between the two pad dimensions is shown to be in the size of the meniscus or solder joint fillet. Larger solder joint fillets were present in the larger 3.2 mm PCB pads. It was found that variation in solder joint height across the different chip sizes could vary from 5 to 70 μm. This variation is a direct result of the solder paste printing process that involves the pick and place step as well as the component weight itself. The solder joint meniscus remain consistent between pad sizes and pick and place SMT procedure. In all cases, the solder fillet size was consistent between components with smaller solder joint height.

The Weibull data shown in Figure 11.11 shows the comparative results for the 2512 and 1206 size components on the two different size pads. The plots indicate that the 2512 resistors behaved better on the 7628 glass with the larger pads while the 1206 parts behaved better on the 1080 glass style with the larger pads.

Test boards were subjected to a total number of 4000 thermal cycles, although failure was recorded only for 2512 and 1206 resistor sizes according to IPC SM-785 [14]. Failure analysis was performed to validate failure mode and location. It is important to note that no particular differences were found between chip resistors mounted in the warp and fill direction of the boards and were aggregated into the same Weibull distribution of Figure 11.11. A reason for this could be the small CTE mismatch between the two orientations used in these test boards.

Cross-sections and optical microscopy were used to determine extent of cracking after thermal cycling experiments. Figure 11.12 shows the impact on the solder joints for the resistors using optical inspection on two resistor sizes. Cracking of 1206 size resistors were found to be consistent with the images shown in Figure 11.12. Fatigue cracking in 2512 resistors were found to have a lift-off effect when cracks fully propagated through the joints. This behavior was not seen in the smaller sized resistors. Cracking along the fillet of the joints was found for smaller sizes of resistors on the 1080 glass style PCBs only. This observation indicates a higher contribution of global CTE mismatch between the resistor and the effective PCB properties, which would be driving the damage in solder joints in this test.

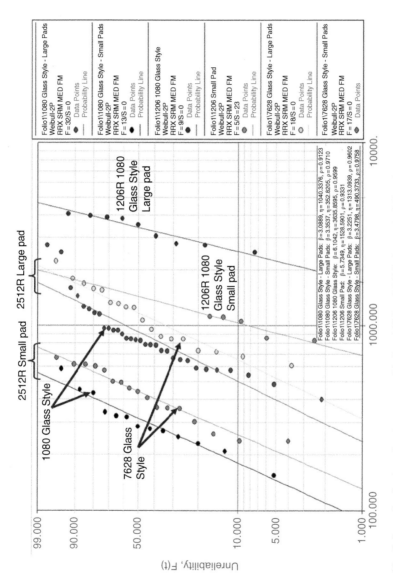

Figure 11.11 Failure distribution of resistors assembled on 1080 and 7628 glass style PCBs with two pad sizes.

Figure 11.12 Optical image showing cracking in solder joints of (Left) 0402 resistor (Right) 2512 resistor.

11.3 Fatigue Life Prediction Models

11.3.1 Empirical Models and Acceleration Factors

One of the most widely used empirical models for reliability assessment of solder interconnects in electronic packaging is the Norris-Landzberg (NL) model shown in Eq. (11.2) [15]. The NL model was originally developed in the 1960s to predict fatigue failure of SnPb solder joints in flip-chip devices. The NL model assumes proportionality between plastic strain range and temperature range in the following relationship.

$$AF = \frac{N_1}{N_2} = \left(\frac{f_1}{f_2}\right)^{-m} \left(\frac{\Delta T_1}{\Delta T_2}\right)^{-n} e^{\frac{E_a}{K}\left(\frac{1}{T_{max,1}} - \frac{1}{T_{max,2}}\right)} \tag{11.2}$$

Where N_1, and N_2 are fatigue life at two different thermal environments, f corresponds to temperature cycling frequency, m and n are constants, ΔT is the temperature range, E_a is the activation energy for the solder, K is the Boltzmann's constant, and T_{max} is the maximum temperature at each thermal environment. Once the model constants are determined from experimental data the projected reliability of other thermal environments can be predicted. The NL model has been found to provide good correlation when the model has been calibrated for a specific package style, solder alloy, and use conditions [16]. Other researchers have identified weaknesses in the NL model between various package types and lead-free solder alloys [17]. They indicated that damage accumulation in solder joints occurs throughout the duration of the thermal cycle and varies for each segment of the cycle and that a simple acceleration factor cannot be taken as a damage indicator for the complex loading conditions solder joints experience during thermal cycling. Semi-empirical fatigue life prediction models such as the Engelmaier model implemented first-order analytical equations to correlate the plastic strain in solder joints as a damage indicator for empirical Coffin-Manson type fatigue equations [18]. Engelmaier approximated the plastic strain in the solder joint using the first-order approximation due to temperature range, solder joint height and CTE mismatch and distance to neutral point

$$\Delta\gamma = F\frac{D\Delta\alpha\Delta T}{2h} \tag{11.3}$$

F is an empirical factor accounting for second-order effects, D is the distance to neutral point, $\Delta\alpha$ is the CTE mismatch between component and PCB, ΔT is the temperature range, and h is the solder joint height. After the plastic shear strain is determined, fatigue life prediction is made using the following relationship

$$N_f = \frac{1}{2}\left(\frac{\Delta\gamma}{2\varepsilon_f'}\right)^{\frac{1}{c}} \tag{11.4}$$

where ε_f' is the fatigue life ductility coefficient and c is the fatigue ductility exponents given by the following equation:

$$c = -0.442 - 6x10^{-4}\overline{T} + 1.74x10^{-2}\ln(1+f) \tag{11.5}$$

where $\overline{T}$ is the mean cyclic temperature and f is the cyclic frequency. The popularity of the Engelmaier model by the electronics industry enabled its adoption into industry standards such as IPC-D-279 and IPC SM-785 [14, 19]. Although the Engelmaier model has been found to follow fatigue life trends of certain package types, it did not provide sufficient accuracy for leadless packages where the solder joint height could be substantially lower than leaded packages [20].

11.3.2 Semi-empirical Models

Semi-empirical fatigue life prediction models incorporate an analytical expression of strain with the empirically calibrated damage model. Such models improve over the Engelmaier model by introducing package specific attributes to describe the shear strain solder joint experience. One such model developed by Suhir is based on the bi-metallic strip model. Shear and axial strain in underfill BGA packages can be calculated by the following equations from Suhir's model [21]:

$$\gamma_{zx}(x) = k\frac{3\Delta\alpha\Delta T sinhkx}{3G_2\lambda coshkl} \tag{11.6}$$

x distance from chip center
$2l$ chip length
t_i thickness of layer i
$G_i = E_i/2(1+v_i)$ shear modulus of layer
$D_i = E_i t_i^3/12(1+v_i^2)$ Flexural rigidity i
$D = D_1 + D_2 + D_3$ Flexural rigidity

$t = t_1 + t_2 + t_3$ Assembly thickness

$\lambda = \frac{(1-v_1)}{E_1 t_1} + \frac{(1-v_3)}{E_3 t_3} + \frac{t^2}{4D}$ Axial compliance

$K = \frac{t_1}{3G_1} + \frac{2t_2}{3G_2} + \frac{t_3}{3G_3}$ interfacial compliance

$k = \sqrt{\lambda/k}$

The axial strain can be calculated with the following equation

$$\varepsilon_z = \left((\alpha_{uf} - \alpha_{sold}) + (\alpha_{uf} - \alpha_{eff})\frac{2v}{(1-v)} \right) \Delta T \tag{11.7}$$

where α_{uf} is the thermal expansion of the underfill, α_{sold} is the thermal expansion of the solder, and α_{eff} is an "effective" thermal expansion reflecting the in-plane forced deformation of the underfill layer in between the die and substrate. The shear and normal strain components are then used to calculate the effective plastic strain

$$d\varepsilon_p = \frac{\sqrt{2}}{3}\sqrt{d\varepsilon^2 + \frac{3}{2}d\gamma^2} \tag{11.8}$$

$$d\varepsilon^2 = (d\varepsilon_x - d\varepsilon_y)^2 + (d\varepsilon_y - d\varepsilon_z)^2 + (d\varepsilon_z - d\varepsilon_x)^2 \tag{11.9}$$

$$d\gamma^2 = d\gamma_{xy}^2 + d\gamma_{yz}^2 + d\gamma_{zx}^2 \tag{11.10}$$

The cycle to failure can then be calculated using Solomon's equation [22], which provides a relationship between the number of cycles to failure and equivalent plastic strain as shown in Eq. (11.11):

$$N_f = C(d\varepsilon_p)^n \tag{11.11}$$

where C and n are constants.

Regardless of the complexity of semi-empirical models to predict the plastic strain in solder joints, certain limitations toward package type, loading environments, solder alloys will always be present. Accelerated testing between SnPb and Sn3Ag0.5Cu solders have shown Sn3Ag0.5Cu solder as more reliable than eutectic Sn37Pb solder alloy at more benign thermal environments but proves to be less reliable at more aggressive temperature ranges [23]. Capturing such complex behavior under thermal cycling requires a more accurate modeling of material behavior and geometry in addition to identifying a more suitable damage indicator to correlate with cycles to failure using empirical data.

11.3.3 Finite Element Analysis (FEA) Based Fatigue Life Predictions

Numerical models provide a quantitative approach to predict failure in solder interconnects in electronic packages. Finite element analysis (FEA) methods has been extensively used to investigate and optimize thermal and mechanical performance of electronic packages [24]. Throughout the decades many simulation techniques and model types have been used to analyze complex microelectronic packages. Some researchers have found that thermal cycling simulations of flip-chip packages using simplified axisymmetric, 2D or strip models can provide comparable fatigue life predictions with various 2-D and 3-D FEA models used in the industry [25].

In addition to modeling techniques, constitutive material models of the solder alloy itself can account for a significant role on the accuracy of the prediction [26]. Fatigue life models implement a damage indicator that correlates with the number of cycles to failure. FEA simulations are used to approximate the value of the damage indicator that will be substituted into the fatigue equation. Coefficients in fatigue equations are calibrated using a combination of empirical data and simulation results. Therefore, experimental data of failure is necessary for any fatigue life prediction methodology and is not exclusive to semi-empirical models that were previously discussed.

One of the first energy-based models was developed by Morrow, and predicted fatigue life by taking the plastic strain energy density as a damage indicator [27]. The plastic strain energy density is calculated by the area enclosed in the hysteresis loop of stress and strain. Energy density is then related to cycles to failure in the following power law relationship

$$N_f^m W_p = C \tag{11.12}$$

where m is the fatigue exponent and C is the material ductility coefficient.

The plastic energy density W_p is obtained from the observed area of the hysteresis loop during cyclic loading. Some of the popular energy-based fatigue models for SnAgCu solder in microelectronics were proposed by Schubert and Syed [28, 29].

The following two equations are based on the Morrow model and have been adapted for solder fatigue life prediction. Schubert's model is shown in Eq. (11.13) and Syed's model is shown in Eq. (11.14). Both of these models utilize total dissipated creep energy density obtained from finite element models.

$$N_f = 345 W_{cr}^{-1.02} \tag{11.13}$$

$$N_f = (0.0019 W_{cr})^{-1} \tag{11.14}$$

In both models the coefficients and exponents were determined by combining FEA simulation results and real test data. A notable difference between the two models is that Schubert's model was calibrated using failure data of flip-chip packages with underfill while Syed's data was calibrated using data from chip scale packages. Both models used accumulated creep strain energy density as a damage indicator. Other models partition the elastic, plastic, and creep component responsible for accumulation of energy density as developed by Dasgupta et al. [30].

$$1/N_f = 1/N_{fe} + 1/N_{fp} + 1/N_{fc} \tag{11.15}$$

This model predicts the creep-fatigue damage from the deviatoric energy densities of the elastic, plastic, and creep components. The motivation behind energy partitioning is that the plastic and creep deformation mechanism result in different types of material damage. Solder joints experience creep at elevated temperature and plastic deformation at colder temperatures. Dasgupta's model was calibrated using separate sets of cyclic tests on small-scale test specimens at various temperatures and strain rates to account for creep and plastic deformation regimes [31].

Alternatively, Darveaux correlated measured crack growth data and correlated inelastic strain energy density in solder [32]. This method yielded in two models that predict the number of cycles to crack initiation and crack propagation.

$$N_0 = k_1 \Delta W_{ave}^{k_2} \tag{11.16}$$

$$\frac{da}{dN} = k_3 \Delta W_{ave}^{k_4} \tag{11.17}$$

Where ΔW_{ave} is the volume averaged inelastic strain energy density per stabilized cycle at the interface layer. The volume averaging calculation of energy density per element is carried out using the following equation.

$$\Delta W_{ave} = \frac{\sum \Delta W V}{\sum V} \tag{11.18}$$

Where ΔW is the inelastic energy density accumulated per cycle and V is the volume of each element in the layer of interest. This model provided a set of empirical constants k_1 through k_4 for various modeling methodologies of geometry and solder constitutive model. The characteristic life can be predicted knowing the area of the solder joints through which the crack will propagate to cause complete failure.

$$\eta_w = N_0 + \frac{a}{da/dN} \tag{11.19}$$

Where a is the solder joint diameter at the interface and η_w is the characteristic life. Since its inception, the Darveaux model was adopted for various BGA package types and solder alloys by calibrating the model constants using experimental data [33–36]. Each of the models discussed here can have advantages and limitations depending on the model methodologies, assumption, and accuracy of the empirical data used for calibration of model parameters.

Comparison of select fatigue life prediction models have been performed to determine the most accurate model for a specific package type. Jiang et al. [36] compared the prediction using the Darveaux model and the classic Engelmaier model for a BGA package with Sn3Ag0.5Cu solder under thermal cycling. Their results indicate that the Engelmaier model provided more conservative predictions than the Darveaux model with respect to characteristic life. Towashiraporn et al. [37] suggested that much of the error in empirical models may be attributed to inaccuracies of the modeling at the time of model generation or application of the model to a different package type than that used for model calibration. Their study suggests that there exists a length scale dependence in empirical models developed using empirical data due to the characteristics of the microstructure, geometry, and stress state that are specific to electronic packages. Such an observation reinforces the notion that the influence of assembly process parameters on thermal cycling reliability cannot be ignored. Therefore, the selection of a fatigue life prediction approach whether empirical or numerical must incorporate a combination of geometric and material dependencies. An example is an empirical model that incorporated these dependencies for QFN/BTC packages subjected to thermal cycling [38]. Such a model incorporates the change of mold compound, temperature dependent properties on the stress experienced by solder joints for a single package style and is not intended to be used for all variations of QFN/BTC packages. The associative relationship between temperature and creep deformation utilizes a modified Arrhenius relationship to correlate linear elastic and creep strains. Although empirical in nature, such models have been shown to provide more realistic fatigue life predictions compared with finite element-based models [39].

Nonetheless, thermal cycling life prediction provides a deterministic component to a failure mechanism that possesses a statistical variation in its nature due to the factors previously discussed in this chapter. Most solder joint failure due to thermal cycling fatigue can be characterized by a failure rate that can fit the Weibull distribution as demonstrated in Figure 11.11. The characteristic life of the Weibull distribution corresponds to the time or number of cycles at which 63.2% of the units in a population are expected to fail. This value provides a useful metric that incorporates the various differences in solder microstructure and assembly quality between components. Differences between the prediction techniques such as analytical equations and numerical models are attributed to the applicability of the data used to calibrate the model to the one being assessed. When choosing an approach for reliability assessment, a rule of thumb is to select a model that was qualified for the component or package style of interest by other investigators and calibrated using a similar solder alloy and loading regime. This is particularly important so as to isolate a model that is calibrated under the same failure mechanism as the one under assessment.

11.4 Conclusions

This chapter provides an overview of thermal cycling reliability pertaining to solder joints. The reliability of solder interconnects was demonstrated to be influenced by all aspects of the electronic assembly ranging from the package style, circuit board construction, and the solder composition. Industry standards are widely adopted to define the time and stress components should be exposed to without exhibiting failure. As new electronic components are introduced, they are often expected to meet or exceed existing accelerated life tests defined for older components. Selection of accelerated life profiles can often be more extreme than the field environment, which can result in activating a failure mechanism that cannot occur in the more benign field environment characterized by lower temperatures. Such failures can shift from solder joints to other aspects of the board or components. Therefore, when testing the reliability of solder joints it is of critical importance to consider the type of electronic packaging, circuit board construction and solder alloy used in the desired temperature profile and carefully leverage that knowledge with the desired test outcomes.

A robust risk assessment for evaluating the reliability of solder interconnects under thermal cycling should utilize a combination of FEA and analytical models. The implementation of the analytical models is to guide results from simulations into a trend that closely matches observed failure rates. Acceleration factors obtained from FEA can provide correlation to expected failure in field environments that cannot be easily verified without testing. Moreover, simulations can provide a more accurate prediction for the reliability assessment under more complex thermal loading conditions that cannot be easily explored experimentally. The largest driver behind virtual qualification of electronics is the cost saving associated with eliminating or reducing the scale of test programs to determine risk of solder joint failure. As more complex electronic components are introduced in harsh use environments, the more important it is to predict the fatigue life of solder joints. Reliability test data will always remain an integral component of predicting solder joint fatigue and is not going to diminish with the increased implementation of modeling techniques to predict the reliability of electronic components under thermal cycling.

References

1 Steinberg, D.S. (2000). *Vibration Analysis for Electronic Equipment*, xviii. Hoboken, NJ: Wiley.
2 Sharon, G. (2015) Thermal Cycling and Fatigue. White Paper. DfR Solutions.
3 IPC-9701 (2002) *Performance test methods and qualification requirements for surface mount solder attachments*. IPC International.
4 Lehman, L.P., Kinyanjui, R.K., Wang, J. et al. (2005). Microstructure and damage evolution in Sn-Ag-Cu solder joints. In: *Proceedings Electronic Components and Technology*. ECTC'05, 674–681. IEEE.

5 Bieler, T.R., Jiang, H., Lehman, L.P. et al. (2008). Influence of Sn grain size and orientation on the thermomechanical response and reliability of Pb-free solder joints. *IEEE Transactions on Components and Packaging Technologies* 31 (2): 370–381.

6 Matin, M.A., Coenen, E.W.C., Vellinga, W.P., and Geers, M.G.D. (2005). Correlation between thermal fatigue and thermal anisotropy in a Pb-free solder alloy. *Scripta Materialia* 53 (8): 927–932.

7 Shi, X.Q., Zhou, W., Pang, H.L.J., and Wang, Z.P. (1999). Effect of temperature and strain rate on mechanical properties of 63Sn/37Pb solder alloy. *Journal of Electronic Packaging* 121 (3): 179–185.

8 Liang, Y.C., Lin, H.W., Chen, H.P. et al. (2013). Anisotropic grain growth and crack propagation in eutectic microstructure under cyclic temperature annealing in flip-chip SnPb composite solder joints. *Scripta Materialia* 69 (1): 25–28.

9 Schueller, R., Blattau, N., Arnold, J., and Hillman, C. (2010). Second generation Pb-free alloys. *Journal of Surface Mount Technology* 23 (1): 18–26.

10 Lee, T.-K., Bieler, T.R., Kim, C.-U., and Ma, H. (2015). *Fundamentals of Lead-Free Solder Interconnect Technology*. New York, NY: Springer.

11 Heinrich, S.M., Elkouh, A.F., Nigro, N.J., and Lee, P.S. (1990). Solder joint formation in surface mount technology—Part I: Analysis. *Journal of Electronic Packaging* 112 (3): 210–218.

12 Sood, B. and Pecht, M. (2018). The effect of epoxy/glass interfaces on CAF failures in printed circuit boards. *Microelectronics Reliability* 82: 235–243.

13 Serebrini, M. and Caswell, G. (2018). The impact of glass style and orientation on the reliability of SMT components. *International Symposium on Microelectronics: Falll 2018* 2018 (1): 000699–000706.

14 IPC SM-785 (1992) *Guidelines for accelerated reliability testing of surface mount attachments*. IPC International.

15 Norris, K.C. and Landzberg, A.H. (1969). Reliability of controlled collapse interconnections. *IBM Journal of Research and Development* 13 (3): 266–271.

16 Vasudevan, V. and Fan, X. (2008). An acceleration model for lead-free (SAC) solder joint reliability under thermal cycling. In: *58th Electronic Components and Technology Conference, ECTC 2008*, 139–145. IEEE.

17 Syed, A. (2010). Limitations of Norris-Landzberg equation and application of damage accumulation based methodology for estimating acceleration factors for Pb free solders. In: *11th International Conference on Thermal, Mechanical & Multi-Physics Simulation, and Experiments in Microelectronics and Microsystems (EuroSimE)*, 1–11. IEEE.

18 Engelmaier, W. (1983). Fatigue life of leadless chip carrier solder joints during power cycling. *IEEE Transactions on Components, Hybrids, and Manufacturing Technology* 6 (3): 232–237.

19 IPC-D-279 (1996) *Design guidelines for reliable surface mount technology printed board assemblies*. IPC International.

20 Chauhan, P., Osterman, M., Lee, S.W.R., and Pecht, M. (2009). Critical review of the Engelmaier model for solder joint creep fatigue reliability. *IEEE Transactions on Components and Packaging Technologies* 32 (3): 693–700.

21 Suhir, E. (1987)). Die attachment design and its influence on thermal stresses in the die and the attachment. In: *Proceedings of 37th Electronic Components Conference*, 508–517.

22 Solomon, H. (1986). Fatigue of 60/40 solder. *IEEE Transactions on Components, Hybrids, and Manufacturing Technology* 9 (4): 423–432.

23 Osterman, M. and Dasgupta, A. (2007). Life expectancies of Pb-free SAC solder interconnects in electronic hardware. *Journal of Materials Science: Materials in Electronics* 18 (1–3): 229–236.

24 Kelly, G. (2012). *The Simulation of Thermomechanically Induced Stress in Plastic Encapsulated IC Packages*. Springer Science & Business Media.

25 Che, F.X. and Pang, J.H.L. (2013). Fatigue reliability analysis of Sn-Ag-Cu solder joints subject to thermal cycling. *IEEE Transactions on Device and Materials Reliability* 13 (1): 36–49.

26 Ng, H.S., Tee, T.Y., Goh, K.Y. et al. (2005). Absolute and relative fatigue life prediction methodology for virtual qualification and design enhancement of lead-free BGA. In: *Proceedings of the 55th Electronic Components and Technology Conference*, 1282–1291. IEEE.

27 Morrow, J.D. (1964) ASTM STP 378, 45. Philadelphia, PA: ASTM.

28 Schubert, A., Dudek, R., Auerswald, E. et al. (2003). Fatigue life models for SnAgCu and SnPb solder joints evaluated by experiments and simulation. In: *Proceedings of the 53rd Electronic Components and Technology Conference*, 603–610. IEEE.

29 Syed, A. (2004). Accumulated creep strain and energy density based thermal fatigue life prediction models for SnAgCu solder joints. In: *Proceedings of the 54th Electronic Components and Technology Conference.*, vol. 1, 737–746. IEEE.

30 Dasgupta, A., Oyan, C., Barker, D., and Pecht, M. (1992). Solder creep-fatigue analysis by an energy-partitioning approach. *Journal of Electronic Packaging* 114 (2): 152–160.

31 Zhang, Q., Dasgupta, A., and Haswell, P. (2005). Isothermal mechanical durability of three selected Pb-free solders: Sn3.9Ag0.6Cu, Sn3.5Ag and Sn0.7Cu. *Journal of Electronic Packaging* 127 (4): 512–522.

32 Darveaux, R. (2002). Effect of simulation methodology on solder joint crack growth correlation and fatigue life prediction. *Journal of Electronic Packaging* 124 (3): 147–154.

33 Zhang, L., Patwardhan, V., Nguyen, L. et al. (2003). Solder joint reliability model with modified Darveaux's equations for the micro SMD wafer level-chip scale package family. In: *Electronic Components and Technology Conference*, 572–577. IEEE.

34 Lall, P., Islam, M.N., Singh, N. et al. (2004). Model for BGA and CSP reliability in automotive underhood applications. *IEEE Transactions on Components and Packaging Technologies* 27 (3): 585–593.

35 Chaparala, S.C., Roggeman, B.D., Pitarresi, J.M. et al. (2005). Effect of geometry and temperature cycle on the reliability of WLCSP solder joints. *IEEE Transactions on Components and Packaging Technologies* 28 (3): 441–448.

36 Jiang, L., Zhu, W., and He, H. (2017). Comparison of Darveaux model and Coffin-Manson model for fatigue life prediction of BGA solder joints. In: *18th International Conference on Electronic Packaging Technology (ICEPT)*, 1474–1477. IEEE.

37 Towashiraporn, P., Subbarayan, G., McIlvanie, B. et al. (2004). The effect of model building on the accuracy of fatigue life predictions in electronic packages. *Microelectronics Reliability* 44 (1): 115–127.

38 Serebreni, M., Blattau, N., Sharon, G. et al. (2017). Semi-analytical fatigue life model for reliability assessment of solder joints in QFN packages under thermal cycling. In: *Proceeding, International Conference for Soldering and Reliability, SMTA*, 6–8.

39 Serebreni, M., McCluskey, P., Ferris, T. et al. (2018). Modeling the influence of mold compound and temperature profile on board level reliability of single die QFN packages. In: *19th International Conference on Thermal, Mechanical and Multi-Physics Simulation and Experiments in Microelectronics and Microsystems (EuroSimE)*, 1–8. IEEE.

12

Intermetallic Compounds

Alyssa Yaeger[1], Travis Dale[2], Elizabeth McClamrock[1], Ganesh Subbarayan[2], and Carol Handwerker[1]

[1] School of Materials Engineering, Purdue University, West Lafayette, IN, USA
[2] School of Mechanical Engineering, Purdue University, West Lafayette, IN, USA

12.1 Introduction

Intermetallic compounds (IMCs) are intrinsic in solder joints. They form between tin (Sn) and other components in the system (typically Cu, Ni, Au, Ag, and Pd) and can have profound effects on solder joint reliability in thermal cycling, drop/shock, and in other environments depending on whether the device is used for consumer electronics or military systems. IMCs are present in many forms and locations within solder joints: in the bulk during solidification of the solder from the melt and at the interface by reaction with the metallizations/surface finishes on circuit boards during soldering and subsequent aging. IMCs are, therefore, found dispersed in specific microstructure patterns within the solder: inside tin grains as the tin solidifies, between tin dendrite arms as the composition approaches the eutectic, and concentrated along the interfaces with the metallizations and surface finishes.

As the assemblies age, the IMCs can coarsen within the solder matrix, by both an increase in the size of the average IMC particle and by precipitation on interfacial IMCs. Along the solder/surface finish and solder/metallization interfaces, the IMCs can continue to become thicker due to interdiffusion between the tin from the solder and the remaining surface finish and metallization layers. In this chapter we examine IMC formation and evolution for different solder alloy/surface finish combinations from the point of view of their impact on solder joint reliability. We present an overview of the mechanical properties of lead-free solders as formed and as they age as a function of composition, i.e. characterizing solders by the types and amounts of IMCs they contain, with a focus primarily

Lead-free Soldering Process Development and Reliability, First Edition. Edited by Jasbir Bath.
© 2020 John Wiley & Sons, Inc. Published 2020 by John Wiley & Sons, Inc.

on the reliability and mechanical properties of the resulting composite joint as influenced by IMCs. By "composite" joint we consider that its behavior is determined by its multi-phase structure, its geometry, and thermal history. These include the IMC-containing solder, the interfacial IMC layers, the remaining surface finishes, and any additional reaction products, processing effects, aging effects, and geometrical effects on the microstructure and bonding in the joint.

These combine to determine solder joint reliability, not just in the mechanical properties of the materials in the joint but also void formation, gold embrittlement, Ag_3Sn plate formation, IMC spalling, and poor wetting, such as from black pad defects. These can cause either widespread damage or localized defect formation, undermining solder joint performance. This chapter has been constructed to address several common questions regarding intermetallics in solder. The chapter will discuss:

- The roles that IMCs play in determining solder joint reliability, and how those roles change as a result of aging or damage induced by thermal cycling
- The performance of common lead-free solder alloys in combination with metallizations and surface finishes to understand what to expect in these specific systems
- The problems that may arise when combining new solder alloys and surface finishes/metallizations and the methodologies that can be used to separate out the different possible root causes.

12.1.1 Solders

As in other metal alloy systems, intermetallic particles in solders cause alloy hardening, making the alloy more creep resistant and raising the yield stress. The magnitude of these effects depends on IMC volume fraction, size, morphology, and location within the joint, making it important to understand IMC formation and evolution to predict the joint behavior. Figures 12.1–12.4 show some effects of composition on IMC formation. Figure 12.1 shows Cu_6Sn_5 and Ag_3Sn in the bulk of a Sn-3.8Ag-0.7Cu solder joint. Large primary Ag_3Sn particles form first during cooling, and finer particles of Ag_3Sn and Cu_6Sn_5 form together at a lower temperature (Figure 12.17). In contrast, Figure 12.2 shows large primary Cu_6Sn_5 particles, with no fine Sn-Cu IMCs apparent from solidification of the eutectic.

12.1.2 Interaction with Substrates

Interfacial IMCs can contain many elements, including Sn, Cu, Ni, and Au, but Ag will only form Ag_3Sn in the bulk. The solder in Figures 12.1 and 12.3 are of the same composition and are expected to contain three phases: Sn, Ag_3Sn, and

Figure 12.1 SEM image of bulk SnAg3.8Cu0.7 solder joint. Sn-rich, Cu_6Sn_5, and Ag_3Sn phases are visible in this microstructure. Source: Adapted from [1].

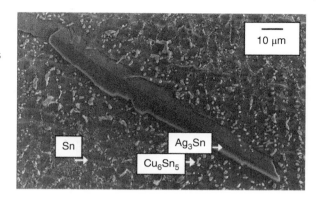

Figure 12.2 Sn-2.8Cu solder joint with Cu_6Sn_5 intermetallics in the bulk. Source: Adapted from [2].

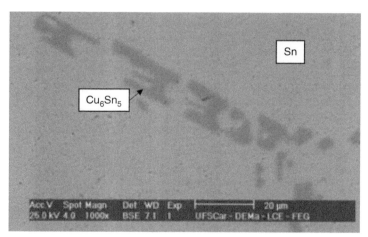

Figure 12.3 SEM image of Sn-Ag3.8 –Cu0.7/Ni-P solder joint interfaces after ageing 440 hours at 80 °C. Source: Adapted from [3].

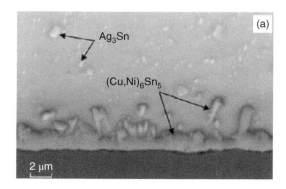

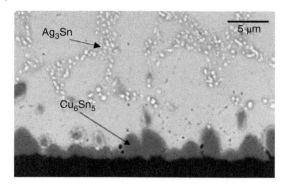

Figure 12.4 Cu_6Sn_5 IMC layer between SAC-305 and OSP (Cu) pad finish. Source: Adapted from [4].

Cu_6Sn_5. However, due to the interaction with the Ni-P substrate after annealing in Figure 12.3, the solder is composed of Sn and Ag_3Sn, as the Cu_6Sn_5 IMCs have formed $(Cu,Ni)_6Sn_5$ at the interface. Due to the length of time and limited volume of Cu in the system, the $(Cu,Ni)_6Sn_5$ intermetallics have started to spall into the bulk of the solder. However, the interface between the $(Cu,Ni)_6Sn_5$ and the Ni-P has remained intact. The dark layer in Figure 12.3 between the IMCs and Ni-P indicates a P-rich layer formed due to the depletion of Ni from the Ni-P surface finish. This P-rich layer can be weakly bonded to the interfacial intermetallics and can cause a brittle fracture such as seen in Figure 12.7.

The interaction between Sn-3.0Ag-0.5Cu solder and a Cu substrate, coated with organic solderability preservative (OSP) is shown in Figure 12.4. (Note: The following naming convention is used throughout this chapter: Sn-3.0Ag-0.5Cu is called "SAC305"). The Cu_6Sn_5 from the solder has migrated to the interface, leaving only fine Ag_3Sn precipitates in the bulk. The morphology of the interfacial intermetallics can be changed between a binary Sn-Cu and ternary Sn-Cu-Ni IMC, as can be seen in Figures 12.3 and 12.4. Tables 12.1 and 12.2 provide a guide to this chapter, based on the solder/substrate system or the reliability issues faced. Per the European Union RoHS (Restriction of Hazardous Substances) legislation, this chapter will mainly focus on lead-free solders, although Section 12.4 will focus on high-lead, RoHS-exempt solder.

12.2 Setting the Stage

This section describes the basics of how solder joints behave under different loading conditions. In this section, we attempt to bridge the gap between the mechanical and materials engineering views of solder joints. We discuss their behavior in the context of the solder alloy, the surface finish composition and geometry, and

Table 12.1 Relevant sections of this chapter, organized by solder alloy.

Solder alloy	Substrate	Relevant sections
SnAg	Cu	12.3. Common Pb-Free Alloys
		12.3.1.2. Sn-Ag and Sn-Ag-Cu solder alloys on Cu
	ENIG	12.3. Common Pb-Free Alloys
		12.3.2.1. Ni-Sn
		12.3.3. Au-Sn
		12.3.2.2. Sn-Ag solder alloys on Ni
		12.3.2.3. Spalling
		12.3.2.4. Effects of P concentration
SnAgCu	Cu	12.3. Common Pb-Free Alloys
		12.3.1.2. Sn-Ag and Sn-Ag-Cu solder alloys on Cu
	Cu and ENIG	12.3. Common Pb-Free Alloys
		12.3.3. Au-Sn
		12.3.3. Sn-Ag-Cu on Cu and Ni
		12.3.2.2.1. Sn-Ag-Cu and Sn-Cu solders on Ni, and Ni-Cu-Sn intermetallics
		12.3.2.3. Spalling
		12.3.2.4. Effects of P concentration
	ENIG	12.3. Common Pb-Free Alloys
		12.3.3. Au-Sn
		12.3.2.2.1. Sn-Ag-Cu and Sn-Cu solders on Ni, and Ni-Cu-Sn intermetallics
		12.3.2.3. Spalling
		12.3.2.4. Effects of P concentration
SnCu	Cu	12.3. Common Pb-Free Alloys
		12.3.1.1. Sn-Cu
	ENIG	12.3. Common Pb-Free Alloys
		12.3.3. Au-Sn
		12.3.2.2.1. Sn-Ag-Cu and Sn-Cu solders on Ni, and Ni-Cu-Sn intermetallics
		12.3.2.3. Spalling
		12.3.2.4. Effects of P concentration
High Pb	All	12.4. High Pb – Exempt

the internal loading conditions as the IMCs coarsen and grow. In the next section, these principles are applied in describing how IMC formation occurs and affects reliability in Sn-Ag-Cu alloys (SAC), Sn-Cu alloys, and Sn-Ag alloys in contact with Cu, electroless Ni-immersion Au (ENIG), and Au surface finishes, metallizations, or pads/leads.

Table 12.2 Issue-based guide to the chapter.

Issue	Relevant sections
Brittle fracture	12.2.1. Mechanical and Thermomechanical Response of Solder Joints
	12.3.2.4. Effects of Phosphorus concentration in ENIG on solder joint reliability
Gold embrittlement	12.3.3. Au-Sn
Spalling	12.3.2.3. Spalling
Phosphorus effects in ENIG on solder joint reliability	12.3.2.4. Effects of P concentration in ENIG on solder joint reliability
Fast IMC growth	12.3. Common Pb-Free Alloys

12.2.1 Mechanical and Thermomechanical Response of Solder Joints

Solder is used to form electronic and mechanical bonds between electrical and electronic components and circuit boards with the reflow process. This process consists of flux activation, melting of the solder alloy, wetting of the pads and component terminations by solder, spreading/redistribution of the solder on to the components leads/terminations, and printed circuit board (PCB) pads, dissolution of the surface finish into the solder alloy, intermetallic formation at the interfaces between the solder and surface finishes, and finally solidification of the alloy to form solid solder joints. Solder joints have a wide range of configurations and sizes.

Areas wetted by the solder may be fixed as in area array joints or may be undefined as in peripheral array solder joints. Area array joints have fixed interfacial areas of contact, such as in ball grid arrays (BGAs), with the standoff height depending on the total solder volume, the pad areas, and the weight of the component. Peripheral array joints are less constrained in terms of wetted area as the solder wets the leads on the circuit boards and bottom terminations or leads and wicks up the remainder of the component metallization or lead. The final distribution of the solder and standoff height is determined by the solder volume, the wettability of lead and pad surfaces and the weight of the component. Examples of area array joints (chip scale package- CSP) and peripheral array joints (quad flat no-leads joint (QFN)) are shown after thermal cycling in Figure 12.5, and serve to illustrate crack formation in solder joints which can ultimately lead to failure.

Solder joints typically fail under thermal and mechanical loads in two possible ways: ductile creep-fatigue fracture along the IMC interface (seen in Figures 12.5 and 12.6) and brittle fracture through the IMC or between the IMC and the underlying surface finish (as seen in Figure 12.7). Ductile fatigue fracture is the typical mode of solder joint fracture in personal and enterprise office equipment, in such

Chip Scale Package (CSP)   Quad Flat-pack No-Lead (QFN)

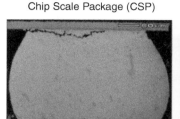 

Figure 12.5 Chip Scale Package (CSP) is an example of an area array joint and quad flat no-leads (QFN) is an example of a peripheral array joint. Both joints have been subjected to thermal cycling and cracks are visible in the solder.

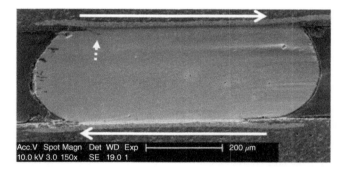

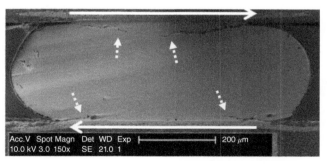

Figure 12.6 Ductile crack growth after a (a) 15% load drop (top) and (b) 40% load drop (bottom) from isothermal fatigue experiments at 25 °C and 4×10^{-5} s^{-1}. The large arrows indicate that the loading is in shear and the small dashed arrows show where cracking has occurred [8].

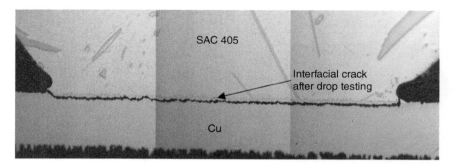

Figure 12.7 Brittle interfacial crack from high strain rate drop test on SAC405 solder on a Cu substrate with an ENIG surface finish [7].

applications as desktop computers, medical diagnostic equipment, and servers. This type of failure results from thermomechanical fatigue. Differing coefficients of thermal expansion between the silicon die and PCB induce damage, leading to fracture in the solder joints as the electronics are thermally cycled during use (see Figure 12.8a). In use, temperature extremes can be low as −40 °C and as high as 125 °C, with even higher temperatures for underhood automotive electronics.

Ductile fatigue fracture occurs by progressive damage during mechanical or thermomechanical cycling which is characterized by crack formation and

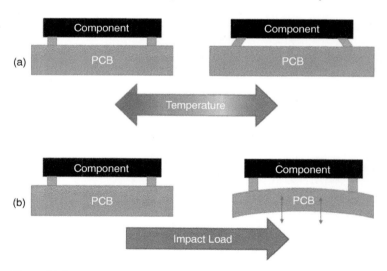

Figure 12.8 Schematic representation of the strain induced on the solder joints due to (a) the PCB having a higher coefficient of thermal expansion than the component, which means that the PCB will expand more for a given change in temperature and (b) the PCB bending up and down during impact loading.

propagation, grain coarsening in the vicinity of the crack front, and load drops with progressive thermal cycling. At the other end of spectrum, portable electronics, such as laptops/tablets and cell phones, are susceptible to impact loading when dropped. Sudden impact causes the PCBs to flex, with the stiffness of the components determining the flexural behavior [9]. As seen in Figure 12.8b, this flexure leads to a tensile load on the solder joints. It is important to note the fracture behavior under these two conditions is different due to the inherent rate-dependent nature of the mechanical response of solder. These two limiting fracture mechanisms of metals are often referred as ductile fracture and cleavage fracture [10]. Ductile fracture is associated with lower loading rates, and is deformation-controlled. Cleavage (or brittle) fracture occurs at higher loading rates and is stress controlled. Cleavage crack growth occurs when the stress reaches a critical level before plastic flow can accommodate the increasing stress.

Being tin-based, solder alloys operate at high homologous temperatures (a high fraction of their melting temperature), with room temperature at a homologous temperature of approximately 0.6 for pure Sn (melting temperature of 232 °C) and for SAC305 (eutectic temperature of 217 °C). This means that yielding occurs at low applied stresses and creep and other processes requiring diffusion are rapid. As a result, solder alloys exhibit both strain-rate-independent behavior expected of solids as well as strain-rate-dependent (plastic) flow behavior that is typical of fluids [11]. A schematic of a typical BGA joint along with the simplified mechanical system in tension is shown in Figure 12.9a. The mechanical behavior of the bulk alloy may be modeled as a combination of a spring and dashpot (Figure 12.9b) that

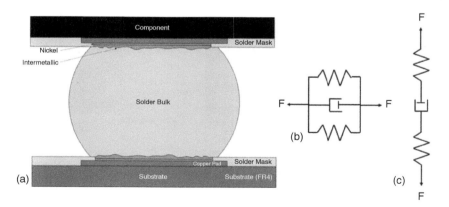

Figure 12.9 (a) A typical BGA solder joint connecting a component to a PCB. Along with two simplified mechanical models for loading in (b) shear and (c) tension under a given load, F. The intermetallic is represented by springs and the solder bulk is represented by a dashpot.

captures its viscous and velocity (strain rate) dependent response. The IMC, on the other hand, being mostly elastic in its response, may be modeled as a spring.

The initial response of a dashpot-spring combination is nearly rigid, but with time, the dashpot will extend in the load direction. The longer-term response will be largely determined by the extension of the dashpot. This is a reasonable first-order mechanical model of the strain-rate-dependent behavior of bulk solder: at low loading rates the solder alloy has enough time to deform plastically under the applied load. At higher loading rates, however, the solder alloy behavior is stiffer since it is unable to extend instantaneously under the applied load, and instead transfers the load onto the IMC layer. Thus, depending on the applied loading rate, a typical solder joint will undergo ductile or brittle failure.

During ductile fatigue fracture, the crack initiates at a location of high stress, and a stable crack slowly progresses across the solder joint [8, 12–14]. The fracture is driven by irreversible dissipation of energy resulting in accumulated damage in the joint that is highest in the region of the crack tip and eventually causes the crack to advance. It is expected that the stable ductile fracture growth will absorb more energy due to dissipation of inelastic energy compared to brittle fracture, which is often unstable. Brittle fracture occurs when an unstable crack zips across the entire joint. This occurs at high strain rates where the intermetallic layer experiences a high stress causing fracture either along the interface between the intermetallic layer and bulk solder or entirely through the intermetallic interface. Brittle fracture has a statistical nature related to the sizes of defects inherent in the material, making it very difficult to predict. In experiments, as the loading rate increases, the fracture will go from ductile fracture to mixed to brittle as can be seen in Figure 12.10.

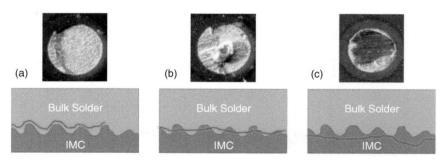

Figure 12.10 Schematic of crack path and images of actual ball shear tests at different strain rates. (a) Low strain rate loading where the crack stays primarily in the ductile phase material; (b) Intermediate strain rate loading where the crack goes through both the ductile bulk solder and brittle intermetallic; (c) High strain rate loading where the crack goes entirely through the brittle intermetallic layer [10].

This strain rate at which the ductile to brittle transition occurs shifts to slower and slower strain rates as the intermetallic layer thickness grows. At long annealing times, failure can be brittle at all strain rates. The implication of this is that joints become more fragile as the IMC grows with high temperature annealing and thermal cycling, with them being able to support very little load before fracture at some point. Thus, thinner IMC layers, higher fracture toughnesses of the IMC, and stronger interfaces are all desirable in reducing brittle fracture. In many scientific studies, the crack propagation behavior is followed by systematically changing the IMC thickness and using scanning electron microscopy (SEM) cross-sections and fracture surfaces to determine the crack path as a function of strain rate. Fluorescent staining of crack surfaces of non-failed joints, known as "die and pry," and X-ray/CT (computed tomography) scans can also provide information on crack growth rate and morphology. Specific component/solder joint/PCB systems are typically characterized by drop tests, the minimum data from which are the number of drops to failure per component to generate a Weibull plot. Additional information on the microstructure and the crack path can also be obtained from the characterization methods listed above. This effect is described below (Section 12.3.1.2) for SnAgCu/Cu solder/surface finish combinations.

Microbumps and small-scale solder joints in 2.5D and 3D packages contain a significant fraction of IMC after reflow. In addition to the amount of IMC expected in single pass reflow, many of these joints will experience as many as six reflows in component manufacturing and circuit board assembly, causing further IMC growth. Figure 12.11 shows the relative scale of a microbump compared to a standard BGA cross-section. Joints with a small standoff height, such as under the

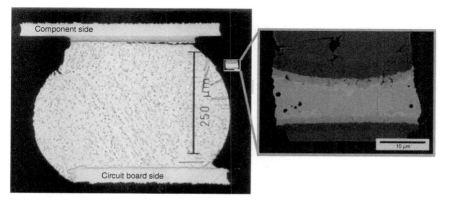

Figure 12.11 The size of a microbump compared to a BGA. The left images are at the same size scale. The higher magnification image of the micro-joint shows a greater amount of IMC relative to the joint size.

bottom termination of gull wings leads, castellations, and passive components, can also have regions of the joint where the IMC makes up a large portion of the joint. The reliability of these joints is then dependent on the total solder volume, the geometry of the component leads and pad, and material properties of the inter-metallics that form. Increasing research is being focused on understanding the mechanical and thermomechanical behavior of intermetallic-rich microbumps, but the contribution of each factor has still not been quantified.

These are the basics of the mechanical and thermomechanical behavior of sol-der joints. However, this is only the first step. Solder joint microstructures are not static: during aging, IMCs in the bulk of the solder joint coarsen and the inter-facial IMCs thicken and often flatten during aging. These changes during aging can have a profound effect on the reliability of solder joints. As the IMCs in the bulk coarsen, the solder softens, i.e. yield strength and the UTS (Ultimate Tensile Strength) decrease. In addition, the transition from ductile fatigue fracture and brittle fracture occurs at lower stresses and lower stain rates as the interfacial IMC thickens and the interface between the IMC and the solder becomes flatter.

These are the basic phenomena. Additional IMC-related phenomena also affect reliability. For example, a weak P-rich layer can form during Ni-Sn IMC formation in the presence of P-containing Ni layers in ENIG surface finishes. This can lead to brittle fracture in the P-rich layer between the IMC and the remaining electroless Ni-P layer. In Section 12.3, the roles of IMC formation and microstructure evolu-tion on reliability are described for specific systems using the principles presented above.

As already indicated, solder joints will typically fail under thermal and mechani-cal loads, either through ductile creep-fatigue fracture or brittle fracture. The crack path is strain-rate-dependent, and the application of the circuit board can deter-mine the types of strain rates the solder joints may encounter during the product lifetime. The microstructure at the solder-surface finish interface has a significant effect on the reliability, even more so with microbumps and other small-scale sol-der joints.

12.3 Common Lead-Free Solder Alloy Systems

Intermetallics form in lead-free solder joints during reflow, aging, and thermal cycling. The types of IMC, their morphologies and compositions, and their effects on reliability depend on the systems which are reviewed here. Table 12.3 details the solder alloy/substrate-surface finish combinations examined in this chapter. The phases formed within the solder and the IMCs forming at the interfaces between the solders and the substrates are listed in the as-solidified and aged states. The morphologies of the IMCs along the interface are described, which illustrate the

Table 12.3 Alloy/substrate systems covered in this chapter.

Solder joint systems	SnCu/Cu	SnAgCu/Cu and SnAg/Cu	SnAg/Ni(P)[a]	SnAgCu/Ni(P)[a]	Cu/SnAgCu/Ni(P)[a]	High-lead/Cu
Intermetallic phases	Solder: Cu_6Sn_5	Solder: Ag_3Sn, Cu_6Sn_5	Solder: Ag_3Sn	Solder: Ag_3Sn, Cu_6Sn_5	Solder: Ag_3Sn, Cu_6Sn_5	Solder: N/A
	Interface: Cu_3Sn, Cu_6Sn_5	Interface: Cu_3Sn, Cu_6Sn_5	Interface: Ni_3Sn_4, Ni_3P[a]	Interface: Ni_3Sn_4, Ni_3P[a]	Interface: Cu_3Sn, Cu_6Sn_5, Ni_3Sn_4, Ni_3P[a]	Interface: Cu_3Sn
	Solder: Cu_6Sn_5	Solder: Ag_3Sn, Cu_6Sn_5	Solder: Ag_3Sn	Solder: Ag_3Sn	Solder: Ag_3Sn	Solder: N/A
	Interface: Cu_3Sn, Cu_6Sn_5	Interface: Cu_3Sn, Cu_6Sn_5	Interface: Ni_3Sn_4, Ni_3P[a]	Interface: $(Cu,Ni)_6Sn_5$, $(Ni,Cu)_3Sn_4$, Ni_3P[a]	Interface: $(Cu,Ni)_6Sn_5$ $(Ni,Cu)_3Sn_4$, Ni_3P[a]	Interface: Cu_3Sn
Liquidus temperature	227°C	217–221°C	221°C	217°C	217°C	260–310°C
Use temperature	<150°C	<150°C	<150°C	<150°C	<150°C	<150°C
Reliability issues	Fast IMC growth	Fast IMC growth compared to other systems, spalling	Black pad, formation of weak high-P interlayer	Black pad, formation of weak high-P interlayer	Spalling, black pad, formation of weak high-P interlayer	Spalling
Interfacial microstructures and morphologies	Scalloped	Scalloped	Needle, reentrant angle	Needle, reentrant angle	Scalloped/needle	Planar
	Scalloped	Scalloped	Reentrant angle, equiaxed and faceted	Scalloped	Scalloped/scalloped	Planar

For each category in the left-hand column, there are two conditions listed: the darker gray rows are for as-reflowed microstructure, the lighter gray rows are aged microstructures. Note that if the electrolytic Ni(P) surface finish is replaced by electrolytic Ni, all the P-containing phase are no longer present.

a) The Ni_3P phase will be discussed later in the chapter.

variation of solder IMC morphology as a function of composition. The liquidus temperature determines the reflow temperature, and the use/application temperature typically tracks the liquidus temperature. Additional phenomena that affect reliability, such as spalling, formation of a brittle P-rich layer between Ni and Ni_3Sn_4, and black pad, can also occur in some of these systems.

Several of the most common intermetallic phases, such as Cu_6Sn_5, Cu_3Sn, Ni_3Sn_4, Ag_3Sn, $AuSn_4$, and $AuSn_2$, are present in these systems. These IMCs, their Young's moduli, coefficient of thermal expansion (CTE), and several properties relevant to solder joint reliability are listed in Table 12.4, in comparison to the Sn, Cu, Ni, Au, and Ag that they are made of. Given the similarity in Young's modulus and CTE between a surface finish and its corresponding IMC, internal shears are not expected in the solder joint during thermal changes.

Table 12.4 Properties of intermetallics relevant to solder joint reliability.

Phase	Young's modulus (GPa)	Coefficient of thermal expansion ($\times 10^{-6}$ K^{-1}) (RT-60 °C)	Crystal structure/space group	
β-Sn	44[a]	23, with a range of 15–30	Body-centered tetragonal	A5
Si	133–188 [15]	2.6	Diamond cubic	Fd_3m
FR-4	21–24	12–14	—	—
Cu	117	16.5	Face-centered cubic	Fm_3m
Cu_3Sn	108 ± 4.4 [16] 134 [17]	18.2 [18] 19.0 ± 0.3 [16]	Orthorhombic	Cmcm
Cu_6Sn_5 (η)	85.56 ± 1.65 [16] 112.3 ± 5.0 [17]	18.3 [18] 16.3 ± 0.3 [16]	Monoclinic	C12/c1
Ni	200	13.4	Face-centered cubic	Fm_3m
Ni_3Sn_4	133.3 ± 5.6 [16]	14.6 [18]	Monoclinic	C12/m1
Au	78	14.2	Face-centered cubic	Fm_3m
$AuSn_4$	71 [19] 39 ± 4 [20]	15.5[b][18]	Orthorhombic	Aba2
$AuSn_2$	103 ± 9 [20]	13.6[b][18]	Orthorhombic	Pbca
Ag	83	18.3	Face-centered cubic	Fm_3m
Ag_3Sn	78.9 [17]	21 [21]	Hexagonal	P6$_3$/mmc Pmmn

a) The β-Sn Young's modulus has a large range, which will be discussed.
b) The CTE values for $AuSn_2$ and $AuSn_4$ are estimated to be the same as $PdSn_2$ and $PdSn_4$.

Additional mechanical, thermal, and physical properties of Cu_6Sn_5, Cu_3Sn, and Ni_3Sn_4 IMCs are available from NIST (National Institute of Standards and Technology); these include Vicker's Hardness and fracture toughness as a function of temperature, Poisson's ratio, thermal diffusivity, heat capacity, resistivity, density, and thermal conductivity [16]. The differences in thermal expansion coefficients between the IMCs are not large, generally within the range exhibited by Sn (15–30) and a factor of 5–7 greater than Si. The Young's moduli of the IMCs are in the range of 80–134 GPa, in the range of Ag and Cu. Note that although the modulus of tin is given as 44 GPa, tin-based solders creep rapidly at room temperature, and therefore never behave elastically. The IMCs tend to be brittle due to their ordered arrangement of atoms in complex crystal structures with typically few active slip systems.

Since lead-free solder joint systems have more than two components, it is important to take into account the interaction between Ni, Cu, Ag, and Au, as these interactions can lead to dramatically different microstructures and properties. Binary intermetallics can also deviate from stoichiometry, such as Ni_3Sn_4 as shown in Figure 12.24. Table 12.5 shows both the deviations from stoichiometry and solubilities for the most common intermetallics in solder joint systems.

Figure 12.12 shows a portion of a Sn-Cu-Ni ternary phase diagram at 240 °C, indicating the solubility of Cu in Ni_3Sn_4 and Ni in Cu_3Sn and Cu_6Sn_5 during reflow. The blue lines in Figure 12.12b show the compositional bounds of each intermetallic phase, in contrast with the top of the triangle where liquid tin has low solubility of Ni and Cu and is represented as a point rather than a region. This system will be explored in more depth in Section 12.3.2, and other relevant solubilities will be discussed later as necessary.

Several common intermetallic phases, such as Cu_6Sn_5, Cu_3Sn, Ni_3Sn_4, Ag_3Sn, and $AuSn_4$ are present in each solder alloy/substrate system, and their properties are important to solder joint reliability. Many of the elements found in solder joint systems are soluble in these common intermetallics, and can form ternary intermetallic phases.

Table 12.5 Intermetallic compounds and their deviations from stoichiometry for both binary and ternary systems.

	at% Ag	at.% Cu	at.% Au	at.% Ni
Ag_3Sn	75–76 (240 °C) [23]	Negligible [24]	0–11 (250 °C) [25]	Negligible [26]
Cu_3Sn	Negligible [24]	76–77 (240 °C) [27]	12 (360 °C) [28]	55 (240 °C) [29]
Cu_6Sn_5	Negligible [24]	55–57 (240 °C) [27]	15 (170 °C) [28]	22 (240 °C) [29]
$AuSn_4$	Negligible [25]	Negligible [28]	20 [30]	11 (150 °C) [31]
Ni_3Sn_4	Negligible [26]	10 (240 °C) [29]	1 (150 °C) [31]	53–57 (240 °C) [32]

Shaded boxes show the binary deviation from stoichiometry, and white boxes show the ternary solubilities of other elements in the IMCs.

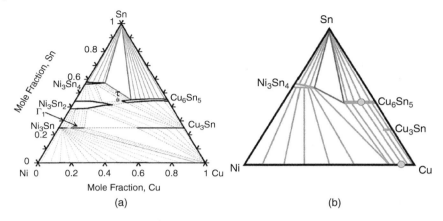

Figure 12.12 (a) Calculated metastable equilibria in the Sn-Cu-Ni systems at 240 °C. The possible τ-phase is indicated and dotted lines in the Cu-Ni side of the diagram beyond the formation of $(Cu,Ni)_6Sn_5$ and $(Ni,Cu)_3Sn_4$ reflect the uncertainties in the phase equilibria Source: Plot reproduced from [33]. (b) Schematic reaction diagram represents the phases observed in diffusion couples composed of Cu-Ni high temperature phases and Sn around 240 °C. The gold circles indicate the final phases remaining after complete consumption of the Sn for a Cu-10 wt% Ni high temperature alloy [34].

12.3.1 Solder Joints Formed Between Sn-Cu, Sn-Ag, and Sn-Ag-Cu Solder Alloys and Copper Surface Finishes

12.3.1.1 Sn-Cu Solder on Copper

The Cu-Sn binary phase diagram (Figure 12.13) shows the three important IMCs in the system: ε (Cu_3Sn), and η (high temperature Cu_6Sn_5 –hexagonal) and η' (low temperature Cu_6Sn_5 – monoclinic). The phases observed to form at the Cu-solder interface during reflow are, in physical order: $Cu{:}\varepsilon(Cu_3Sn){:}\eta(Cu_6Sn_5){:}\beta$-Sn, with the relative thicknesses of the two intermetallic layers depending on reflow profile and subsequent aging conditions. For a eutectic Sn-Cu solder at 20 °C above the eutectic temperature of 227 °C, a typical reflow temperature, Cu dissolves into the liquid up to approximately 1 wt% Cu. Upon cooling to room temperature, the alloy solidifies into tin with negligible solubility of Cu in tin and Cu_6Sn_5. The η to η' phase transformation in the Cu_6Sn_5 occurs at approximately 187.5 °C, with a volume contraction of 2.15% during cooling from the reflow temperature just to the phase transformation alone [36]. Additions of Ni as low as 0.05 wt% Ni have been found to stabilize the η phase, and thus eliminate cracking associated with the η to η' transformation [37]. Some of the IMC physical properties important to the performance of the resulting solder joints are shown in Table 12.4.

Note that there are also several other η phases that have been observed, including a new monoclinic crystal structure of Cu_6Sn_5, designated as η^{4+1}, which formed in

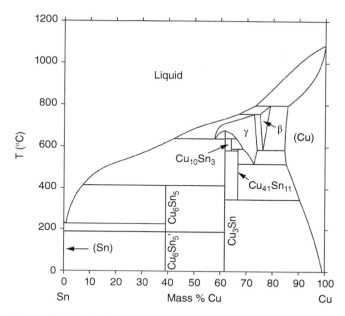

Figure 12.13 Sn-Cu binary phase diagram [35].

a directly alloyed stoichiometric sample [38]. It is not clear at this point whether phases other than the η and η' phases are relevant to solder joints.

Typical scalloped microstructures of the two Cu-Sn intermetallics forming at the solder-Cu interface are seen in Figure 12.14. Microvoids sometimes form near the Cu:Cu_3Sn interface when there are significant impurities incorporated into the Cu during electrodeposition. Based on the movement of inert markers, these appear to be Kirkendall voids, but they only form when bath additives and their decomposition products are present in the electrolyte and act as nucleation sites for the microvoids [39]. The Cu-Sn IMCs grow by multiple pathways: by bulk/lattice diffusion through the grains, by grain boundary diffusion, and by coarsening when the IMC particles in the bulk dissolve and re-precipitate onto the interface IMC. It has been observed that the in-plane grain size of the Cu_6Sn_5 increases with increased exposure to reflow temperatures and during aging. This means that the growth rates of the IMC may slow down as the total grain boundary area decreases. The morphologies and the directional growth of Cu_6Sn_5 may change significantly as a function of time at reflow temperature as shown in Figure 12.15 for times between one second and one hour at 250 °C. These experiments by Zhang et al. [40] on Cu single crystals revealed that the initial nucleation even on a single crystal produced the scalloped structure typical of growth on polycrystalline Cu substrates. For longer annealing, even at times as

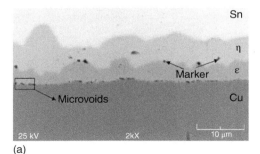

(a)

(b)

Figure 12.14 Microvoids formed in (a) between Cu_3Sn and Cu. (b) has fewer microvoids, due to a higher purity of Cu substrate [39].

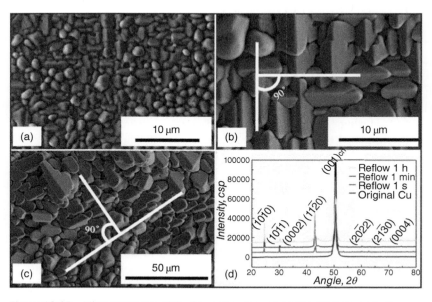

Figure 12.15 Microstructures of Cu_6Sn_5 grains formed on the (001) Cu pad after soldering at 250 °C for (a) one second, (b) one minute and (c) one hour, and (d) the corresponding XRD patterns. Additionally, the elongations of Cu6Sn5 grains are marked by the white lines in (b) and (c) [40].

short as one minute, growth became highly anisotropic with two orientations observed, and even greater growth anisotropy developing for longer annealing times. Such anisotropic structures may occur when the grain size of the Cu surface finish is large, though this situation occurs infrequently.

As already indicated, the Cu-Sn binary system contains three important IMCs, Cu_3Sn and both high temperature (η) and low temperature (η') Cu_6Sn_5. The transformation between η and η' involves a volume change during cooling. Microvoids may form at the interface between Cu and Cu-Sn IMCs due to impurities in the Cu substrate, which can reduce reliability. The morphologies of Cu_6Sn_5 are susceptible to changes in reflow time and temperature.

12.3.1.2 Sn-Ag and Sn-Ag-Cu Solder Alloys on Copper

The lead-free Sn-Ag-Cu (SAC) solder family spans from extremely low Ag compositions (0.1 wt%) to as high as 4 wt% Ag, and from Cu concentrations of 0.5 to 1.0 wt% Cu. Note that compositions are given in wt%, using the standard terminology "SAC305" for an alloy with 3.0 wt% Ag, 0.5 wt% Cu, and remainder tin and "SAC105" for an alloy with 1.0 wt% Ag, 0.5 wt% Cu, and remainder tin. According to IPC J-STD-006, solder compositions can deviate from their nominal compositions as follows: "Except where otherwise indicated, the component elements in each alloy shall deviate from their nominal mass percentage by not >0.10% of the alloy mass when their nominal percentage is ≤1.0%; by not >0.20% of the alloy mass when their nominal percentage is >1.0% to ≤5.0% or by not >0.50% when their nominal percentage is >5.0%" [41]. The nominal alloy composition of SAC305 in commercial alloys is Ag: 2.8–3.2 wt% Ag and Cu: 0.4–0.6 wt% Cu, with the remainder tin, as allowed by the standard.

Here we are using the equilibrium binary and ternary phase diagrams as starting points to understand what phases, including IMCs, will be in the system both for the solder alloys themselves and for the alloys in contact with copper at the reflow temperature and during cooling.

The binary Sn-Ag phase diagram is shown in Figure 12.16, with the IMC phases between Ag and Sn being ζ-Ag and Ag_3Sn. At reflow temperatures, the Ag_3Sn phase is the only phase that is observed between Ag and tin. The typical solder compositions in this system are between 1 wt% Ag and 4 wt% Ag, with the solidified solder containing only Sn and Ag_3Sn. By examining the binary phase diagram, one can see that even with a low initial concentration of Ag in the alloy, heating the alloy in contact with Ag leads to dissolution of Ag up to the solubility of Ag in Sn, which for a reflow temperature of 240 °C is approximately 4 wt%. When Sn-Ag solders come into contact with copper surface finishes, Cu dissolves into the liquid; this transforms the solder into a Sn-Ag-Cu ternary alloy, the ubiquitous lead-free ternary solder system.

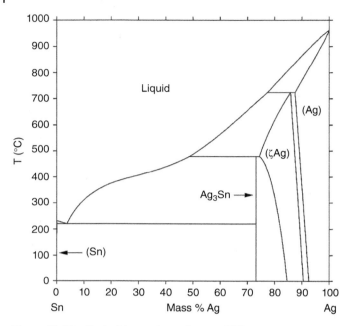

Figure 12.16 Sn-Ag binary phase diagram [42].

The tin-rich corner of the liquidus projection of the SnAgCu ternary phase dia-gram is needed to understand what IMCs are present (Figure 12.17). The bottom left corner is 100% Sn, and the axes show increasing concentrations of Cu and Ag, with different scales on the two axes to reflect the different eutectic compositions of Sn-Ag [5]. The liquidus projection is a contour map of the lowest tem-perature where a liquid still exists as a function of composition and shows the first phase that forms at equilibrium below that temperature. The binary and ternary eutectic temperatures are: Sn-Cu (227 °C), Sn-Ag (221 °C), and Sn-Ag-Cu (217 °C). Individual solder alloy compositions are points on this graph, with Sn3.5Ag (star), SAC105 (circle), and SAC305 (square) as indicated in Figure 12.17. For Sn3.5Ag (the eutectic composition) not in contact with Cu surface finish, β-Sn and Ag_3Sn are the two phases that form from the melt. For both SAC105 and SAC305 not in contact with a Cu surface finish, the first phase predicted to form during cool-ing is the solid β-Sn primary phase field in the phase diagram, with Ag_3Sn and Cu_6Sn_5 forming at the eutectic. If these solders are in contact with Cu, the sol-der alloys can dissolve additional copper up to their solubility limits at that tem-perature. For a reflow temperature of 240 °C, the dashed arrows in the liquidus

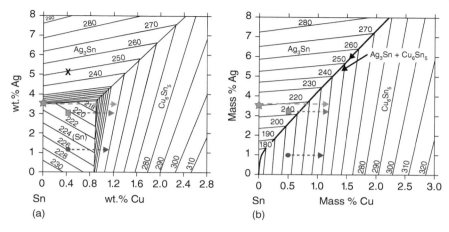

Figure 12.17 Two projections of the liquidus surface of the Sn-Ag-Cu ternary system with three initial compositions marked Sn-3.5Ag (star), SAC305 (square), and SAC 105 (circle). (a) The equilibrium liquidus projection with the arrows indicating the starting compositions and the ending compositions after dissolution of Cu into the alloys up to the solubility limits of the alloys at 240 °C. Dissolution of Cu puts all three alloys in the Cu_6Sn_5 primary phase field. (b) The metastable liquidus shows the phase fields when β-Sn is slow to nucleate. Source: ternary liquidus projections modified from [5, 6].

projections in Figure 12.17a start from the compositions of the original alloys to the copper-saturated alloy compositions at over 1 wt% Cu. What this means is that the amount of Cu_6Sn_5 in the solder joint will be greater than in the original alloy, and the amount of Cu_6Sn_5 in the joint will increase as the reflow temperature increases. For Sn-3.5Ag in contact with Cu, the final composition can have slightly more copper dissolved than SAC305 given the shape of the liquidus surface, and will be indistinguishable from SAC305 in contact with copper.

The IMCs in the Sn-Ag-Cu system form relatively easily during solidification, however, it is well known that, during solidification of tin-rich alloys, β-Sn is difficult to nucleate and will not form until significantly lower temperatures. (This phenomenon is known as "undercooling".) This delay in β-Sn nucleation can be represented by a metastable phase diagram. Figure 12.17b shows the metastable extension of formation of Ag_3Sn and Cu_6Sn_5 far below their equilibrium phases. That is, if the β-Sn phase does not form, the IMCs will continue to form well below the eutectic temperature. For the SAC305 alloy alone, i.e. not in contact with copper, this means that the first phase to form is Ag_3Sn followed by Cu_6Sn_5. If SAC305 is in contact with a copper surface finish, the first phase to form is Cu_6Sn_5 followed by Ag_3Sn. At sufficient undercooling, β-Sn will form as dendrites, often with 1 to

3 twin-related orientations per solder joint. There is essentially zero solubility of Ag and Cu in β-Sn, of Cu in Ag_3Sn, and of Ag in Cu_6Sn_5 (Figure 12.18).

As mentioned in Section 12.2.1, the creep behavior of the solder is affected by composition and the presence of intermetallic particles. Within a single composition, however, the creep behavior can change with temperature as well. The Mukherjee-Bird-Dorn equation (Eq. (12.1)) describes the relationship between the minimum creep rate, $\dot{\gamma}$, and the applied stress, τ. D_v is the diffusivity, k_B is the Boltzmann constant, A is a dimensionless constant, T is absolute temperature, G is the shear modulus, and n is the stress exponent. Figure 12.19 shows creep data for a Sn-3.5Ag solder under a range of temperatures. The applied stress necessary for creep decreases with increasing temperature, and a clear change in stress exponent designates low stress and high stress regimes.

Equation (12.1): Mukherjee-Bird-Dorn equation, describing the minimum creep rate $\dot{\gamma}$ related to applied stress τ [45].

$$\dot{\gamma} = A \left(\frac{D_v Gb}{k_B T} \right) \left(\frac{\tau}{G} \right)^n \qquad (12.1)$$

Several size scales of Ag_3Sn and Cu_6Sn_5 intermetallics form in the bulk solder during solidification, including at size scales smaller than seen in Figure 12.3. In Figure 12.20, approximately 200 nm diameter particles of Ag_3Sn are observed to act as effective pinning sites for dislocations in β-Sn in the Sn-3.5Ag binary alloy not in contact with copper surface finish. The presence of small precipitates inhibits plastic deformation and causes an overall strengthening of the solder. The strengthening effect is proportional to the volume fraction and inversely proportional to the size of the intermetallics dispersed in the solder matrix. That is, for the same volume fraction of particles, small particles act as more effective pinning sites than larger particles [44].

The strengthening effect of the IMC particles changes with time after solidification and during elevated temperature annealing. During aging of SnAgCu solder joints, whether at room temperature or elevated temperature, coarsening of the IMC particles and the tin dendrites occurs. The volume fraction of IMCs remains constant. However, coarsening means that their average particle size and spacing increase with annealing time. Chavali quantified the microstructural coarsening during aging of Sn3Ag0.5Cu solder joints as a function of aging temperature and time, as seen in Figure 12.21 [46]. Coarsening of the IMC particles leads to a decrease in the effectiveness of the IMCs as dislocation pinning sites, a decrease in the yield stress, and an increase in the creep strain rate at a given load, i.e. it leads to a softening of the solder in the joint [47].

The roughness of the Cu_6Sn_5/solder interface and IMC thickness have been observed to have a pronounced effect on the transitions between modes of failure: ductile fatigue failure, brittle failure and mixed mode as a function of

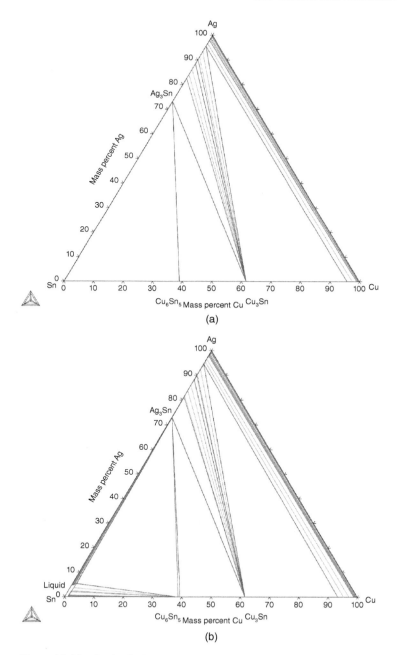

Figure 12.18 Sn-Ag-Cu isothermal phase diagram at (a) 150 °C and (b) 240 °C. There is no solubility of Cu in Ag_3Sn or Ag in Cu_3Sn and Cu_6Sn_5 at either temperature [43].

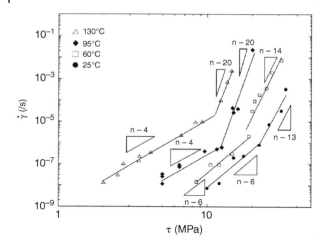

Figure 12.19 Minimum creep rate versus applied stress for Sn-3.5Ag at 25, 60, 95, and 130 °C. Note the change in stress exponent n in the low stress regime with increasing temperature [44].

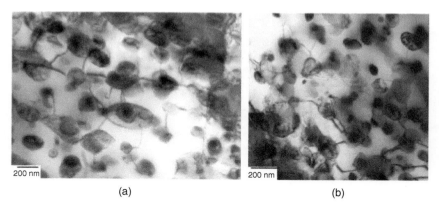

(a)   (b)

Figure 12.20 Microstructure of Sn-3.5Ag after creep deformation (22 MPa at 60 °C, with a strain rate in the order of 10^{-4} s^{-1}, ⟨110⟩ zone axis). (a) Dislocations pinned by Ag$_3$Sn particles in the eutectic mixture after creep deformation. (b) Pinning of dislocation substructure by Ag$_3$Sn particles [44].

aging time and strain rate. That is, at a given strain rate, the joint will transition from ductile, to mixed ductile/brittle to brittle failure as the intermetallic becomes thicker. An and Qin [48] examined the interfacial IMC microstructure evolution in Sn3Ag0.5Cu/Cu surface finish joints annealed at 125 °C and tested at strain rates from 2×10^{-4} s^{-1} to 2 s^{-1}. This transition was characterized in SEM cross-section as well as by the changes in IMC surface morphologies and

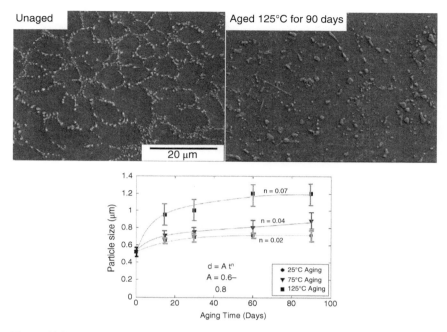

Figure 12.21 Microstructure of Sn3Ag0.5Cu solder alloy showing dark β-Sn dendrites in a networked arrangement of Ag_3Sn or Cu_6Sn_5 precipitates, as reflowed and after aging at 125 °C for 90 days. The graph shows Ag_3Sn particle size as a function of aging time at three different temperatures. The second micrograph corresponds to the upper right data point [46].

grain sizes revealed by etching away the β-Sn, leaving a view of the IMC surface morphology (Figure 12.22). The in-plane grain size of the IMC increases as the layers thicken, meaning that the grain structures of the IMC layers are coarsening by grain boundary migration which also may have an effect on load transfer in the joint.

As the IMC grows during solid-state annealing, the IMC-solder interfaces become flatter and a transition is observed between fully ductile fatigue failure in the solder joint and brittle fracture as a function of strain rate. This can be seen by comparing the characteristics of fracture surfaces in Figure 12.23 with the categorization of the behavior into the three failure modes. At the lowest strain rate, the fracture surfaces reveal cup-and-cone fracture patterns characteristic of the final stages of necking and cavity link-up in ductile failure for all aging times. At the intermediate strain rate, there is ductile fatigue failure for the as-fabricated joints, which transitions to a combination of mixed ductile and brittle failure with increase aging time. The fraction of the fracture area displaying ductile

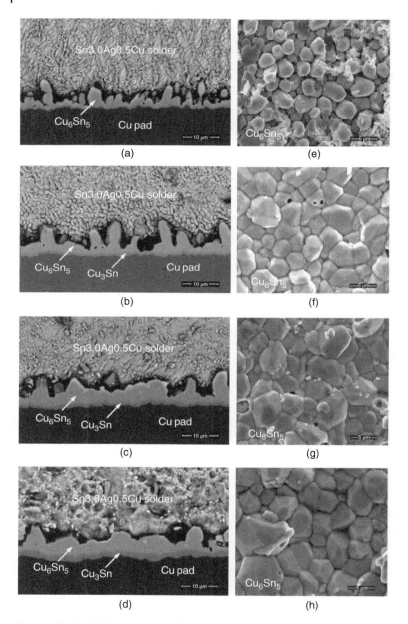

Figure 12.22 SEM images of Sn3.0Ag0.5Cu/Cu interface aged at 150 °C for (a) 0 hour, (b) 72 hours, (c) 288 hours, (d) 500 hours and the top views of the IMC layer aged at the corresponding times [48].

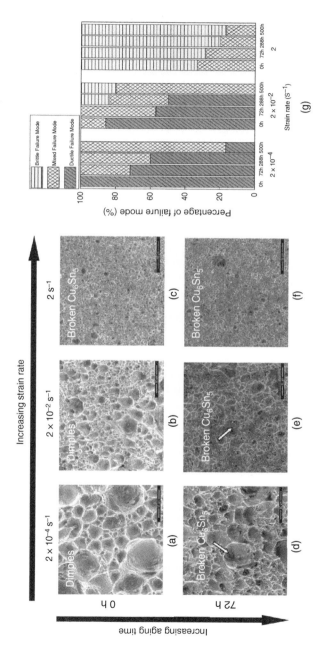

Figure 12.23 Fracture surfaces of unaged and aged samples as a function of aging time at 150 °C and strain rate. Distribution of the three failure modes occurred in the solder joints for different aging times and tested at different strain rates [48].

cup-and-cone deformation patterns decreases as the aging time increases, with fractured Cu_6Sn_5 grains taking up an increasing fraction of the fracture surface. Based on their microstructure-based fracture criteria, no regions of ductile fatigue fracture were observed at the longest aging times and the intermediate strain rate. At the highest strain rate, solder joint behavior is characterized by mixed mode fracture and brittle fracture, with the fraction of mixed mode decreasing with increasing aging time. These interactions are the reasons why limiting IMC growth is critical to maintaining solder joint reliability.

Sn-Ag solder alloys in contact with copper substrates form a Sn-Ag-Cu solder alloy due to the solubility of Cu in Sn. These Sn-Ag-Cu systems range in composition and microstructure, with two intermetallics formed, Cu_6Sn_5 and Ag_3Sn. The morphologies of these IMCs affects the creep behavior of the solder.

Both Sn-Ag and Sn-Ag-Cu solder alloys in contact with Cu substrates will form Cu_6Sn_5 and Ag_3Sn intermetallics in the bulk with Cu_3Sn and Cu_6Sn_5 at the interface. Ag_3Sn particles with average diameter of about 200 nm strengthen the solder by pinning dislocations, although the intermetallics coarsen with time and annealing. At the interface, Cu_6Sn_5 forms a more uniform layer with annealing. This changes the failure mode from ductile to brittle.

12.3.2 Solder Joints Formed Between Sn-Cu, Sn-Ag, and Sn-Ag-Cu Alloys and Nickel Surface Finishes

12.3.2.1 Ni-Sn

As with Section 12.3.1, we start with the binary phase diagrams Sn-Ni and move into the ternary Sn-Ag + nickel surface finish and Sn-Cu + nickel surface finish systems, ending with the Sn-Ag-Cu + nickel surface finish quaternary system. The Ni-Sn binary phase diagram is shown in Figure 12.24. In soldering, the temperatures of interest are below 250 °C, so only three intermetallics in the phase diagram are relevant: Ni_3Sn, Ni_3Sn_2, and Ni_3Sn_4. Although other IMCs have been reported after long duration annealing of Sn-Ni, Ni_3Sn_4 is generally the dominant Ni-Sn IMC observed in solder joints [49]. This is due to the slow growth kinetics of Ni_3Sn and Ni_3Sn_2. The eutectic composition has been reported to be Sn-0.16 wt% Ni, with the solubility of nickel in liquid tin being extremely low as a function of temperature. From the phase diagram, the binary eutectic is expected to be Sn and Ni_3Sn_4. However, as reviewed in a 2016 paper by Belyakov and Gourlay on phase formation in the Ni-Sn system during soldering and after aging and thermal cycling, $NiSn_4$ was observed both in the binary eutectic and after aging [50, 51]. The presence of $NiSn_4$ in a solder joint is of concern for two reasons: (i) $NiSn_4$ grows very quickly, having itself a serious effect, therefore, on the reliability of the joints, and (ii) the formation of $NiSn_4$ also rapidly consumes Sn at the Sn:Ni ratio of 4:1, which also reduces the amount of ductile phase in the joint. Although there was no Au in

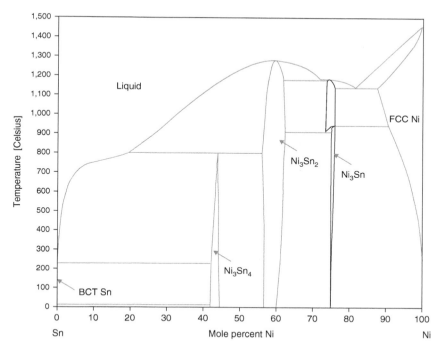

Figure 12.24 Ni-Sn binary phase diagram [22].

the joints studied by Belyakov and Gourlay, it should be noted that having Au in the solder increases the volume fraction of $NiSn_4$ in the system due to the high solubility of Au in $NiSn_4$. This is discussed further in Section 12.3.3.

The interfacial intermetallic Ni_3Sn_4 has several different morphologies at the interface, including needle-type, reentrant angle-type, scallop-type, and equiaxed-type. Changes in the interfacial IMC morphology can affect the mechanical properties of the solder joint, particularly the change from a discontinuous to a continuous layer. Needle-type Ni_3Sn_4 grows from the interface as long, thin needles in a discontinuous layer. Reentrant angle-type Ni_3Sn_4 may be twinned grains which coarsen into equiaxed-type grains. Figure 12.25a shows both needle-type and reentrant angle-type morphologies. The scallop-type morphology is similar to Cu_6Sn_5 morphology (Figure 12.26). Equiaxed- and scallop-type morphologies form a continuous layer of intermetallic.

The Ni-Sn binary system contains one important IMC, Ni_3Sn_4, which has several interfacial morphologies due to the low solubility of Ni in liquid solder. Some of these morphologies do not form a continuous layer of intermetallic at the interface. The growth of Ni-Sn IMCs from electroless Ni substrates can cause a brittle P-rich layer, as discussed in Section 12.3.2.4.

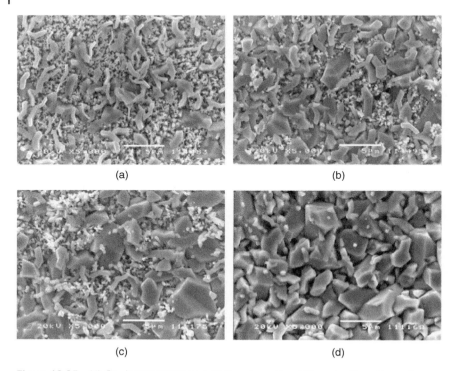

Figure 12.25 Ni_3Sn_4 intermetallic in a Ni-P system after different reflow times at 251 °C: (a) 30 seconds; (b) 90 seconds; (c) 180 seconds; and (d) 600 seconds. With increased reflow time, the morphologies of the Ni_3Sn_4 change from needle- and reentrant angle-type to scallop- and equiaxed-type [52].

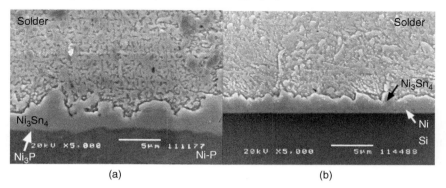

Figure 12.26 Sn-3.5Ag solder reflowed for 600 seconds at 251 °C with (a) electroless Ni-P and (b) sputtered Ni. The Ni_3Sn_4 morphology consists of reentrant angle and chunk-type in (a) and scallop-type in (b). Both interfaces of the Ni_3Sn_4 are flatter with sputtered Ni than with Ni-P [52].

12.3.2.2 Sn-Ag Solder Alloys on Nickel

As mentioned above, the solubility of nickel in liquid tin is extremely low, and thus Ni-Sn intermetallics are only found at the interface. In the Sn-Ag/Ni system, the interfacial intermetallic composition does not change during aging, since the solubility of Ag in Ni_3Sn_4 is negligible. The interfacial intermetallics do grow and change morphology. During a five-second reflow at 251 °C, needle-type Ni_3Sn_4 grains were found to be the primary morphology at the interface, with significant spacing between the intermetallics [52]. With increased reflow time, the Ni_3Sn_4 morphology consisted of more reentrant angle and chunk-type intermetallics, as shown in Figure 12.26 [52, 53]. With a pure Ni substrate, the Ni_3Sn_4 grains are scallop-type [52].

During aging, the growth of the Ni_3Sn_4 intermetallics follows an approximate $t^{1/3}$ dependence. Growth occurs through several methods, including grain boundary diffusion and grain coarsening [52]. When there is spacing between the needles, direct Ni-Sn diffusion occurs as well until the intermetallic coarsens. The addition of Ag to a Sn/Ni system has also been shown to reduce voiding and slow the growth of Ni_3Sn_4 intermetallic layers [54]. The role of P in Ni_3Sn_4 formation as seen in Figure 12.26 will be discussed in Section 12.3.2.4. Similar to the Sn-Ag-Cu system (Figure 12.18), there are only binary Ag-Sn and Ni-Sn intermetallics present (Figure 12.27).

In the Sn-Ag-Ni system, the intermetallics present are binary Ag_3Sn and Ni_3Sn_4 due to the low solubility of Ni and Ag, respectively. As with the Ni-Sn binary system, increased reflow time results in a continuous layer of intermetallic at the interface.

12.3.2.2.1 Sn-Ag-Cu and Sn-Cu Solder Alloys on Nickel and Ni-Cu-Sn Intermetallics

The Sn-Ag-Cu/Ni system has several interfacial intermetallics present throughout the aging process. As the nickel solubility in liquid tin is very low, Ni-containing IMCs are only found at the interface, as with the Sn-Ag/Ni system. During reflow, the copper from the solder comes into contact with the nickel surface finish and forms the ternary IMC $(Ni,Cu)_3Sn_4$ (Figures 12.12 and 12.28). Cu_6Sn_5 particles migrate from the bulk to the interface during annealing, and slowly change both the composition and morphology of the interfacial intermetallic from a needle-like Ni_3Sn_4-based to scalloped Cu_6Sn_5-type intermetallic, as shown in Figure 12.29.

The critical amount of copper necessary to change the $(Ni,Cu)_3Sn_4$ to $(Cu,Ni)_6Sn_5$ is shown in Figure 12.29b, and is less than 1 wt% Cu. Snugovsky et al. [56] posited a new ternary phase, $Cu_{33}Ni_{23}Sn_{44}$, which would form in between Ni_3Sn_4 or Cu_6Sn_5 (Figure 12.30). This phase is in equilibrium with both binary intermetallics.

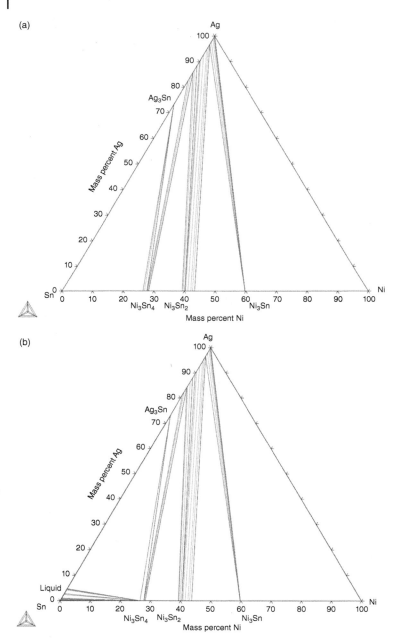

Figure 12.27 Sn-Ag-Ni isothermal phase diagram at (a) 150 °C, and (b) 240 °C. There is no solubility of Ni in Ag$_3$Sn or Ag in Ni$_3$Sn, Ni$_3$Sn$_2$ and Ni$_3$Sn$_4$ at either temperature [43].

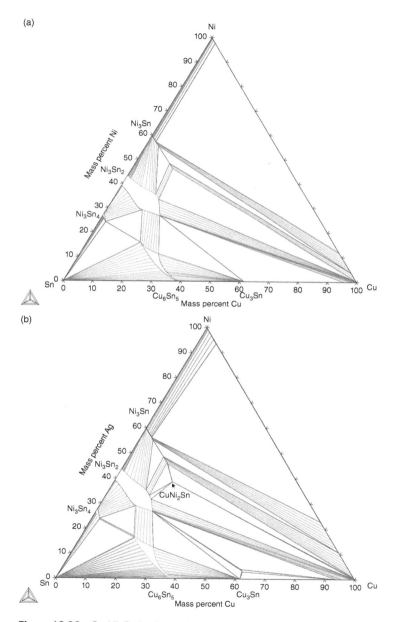

Figure 12.28 Sn-Ni-Cu isothermal phase diagrams at (a) 150 °C and (b) 240 °C. There is solubility of Cu in Ni₃Sn, Ni₃Sn₂ and Ni₃Sn₄ and of Ni in Cu₆Sn₅ at both temperatures [43].

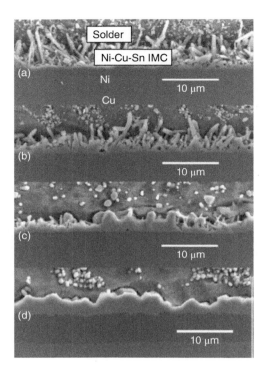

Figure 12.29 Etched micrographs of SnAgCu/Ni interface during isothermal aging at 125 °C. (a) 0 hour, (b) 119 hours, (c) 262 hours, (d) 480 hours. The interfacial intermetallics change morphology from needle-like to scalloped [55].

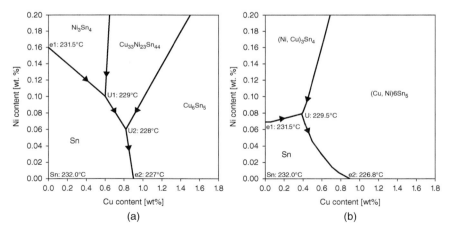

Figure 12.30 Liquidus projections for tin-rich corner of the Sn-Cu-Ni phase diagram. Source: (a) based on data by [56] and (b) based on data by [33].

Smaller amounts of nickel in the solder joint have been shown to stabilize the high temperature, hexagonal $(Cu,Ni)_6Sn_5(\eta)$, even at lower temperatures where the monoclinic η' phase is known to be stable in the binary, such as in Figure 12.13 [37]. Nogita and Nishimura determined that 9 at% Ni will stabilize the high temperature $(Cu,Ni)_6Sn_5$ phase at lower temperatures [37].

It is expected that the Sn-Cu/Ni solder system contains the same interfacial intermetallics as a SnAgCu/Ni system due to the insolubility of Ag in other intermetallics, and due to the higher weight percent of copper present in the solder, the reaction seen in Figure 12.29 may progress at a faster rate. Vuorinen et al. found that the addition of copper to a Sn/Ni diffusion couple at 240 °C increased the intermetallic growth rate, with a peak at a Cu:Ni ratio of 9 : 1 [33].

The stable intermetallics in the Sn-Cu-Ni system change over time, due to the solubility of copper in Ni_3Sn_4 and nickel in Cu_6Sn_5. The change from $(Ni,Cu)_3Sn_4$ to $(Cu,Ni)_6Sn_5$ also changes the morphology of the interfacial IMCs. Small amounts of nickel in the solder have been shown to stabilize the high temperature $(Cu,Ni)_6Sn_5$. The spalling of the intermetallics caused by the change in equilibrium phases will be discussed in the following section.

12.3.2.3 Spalling

Spalling is the separation of intermetallic layer from the substrate, and can weaken the joint significantly. During reflow and annealing, an intermetallic layer forms at the interface between the surface finish and solder. Yang et al. posit that for spalling to take place, the local equilibrium must be sensitive to one or more elements present in the intermetallic, and one of those elements must be limited [57]. Figure 12.31b shows spalling at the interface between a SnAgCu solder and nickel substrate. According to the phase diagram in Figure 12.28, with 0.27 wt%

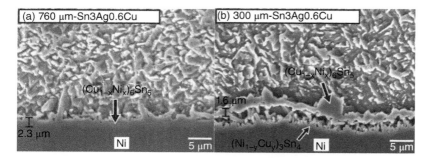

Figure 12.31 Micrographs showing the interface of solder joints of composition Sn-3Ag-0.6Cu on a 375 μm Ni pad. (a) Cu concentration 0.57 wt%, (b) Cu concentration 0.27 wt%. Spalling only occurs in (b) due to the limited concentration of Cu in the system [57].

Cu and less than 0.08 wt% Ni in the solder, $(Cu,Ni)_6Sn_5$ forms. As the intermetallic grows, Cu is depleted from the Sn, and Ni continues diffusing into the intermetallic. The local equilibrium changes such that $(Ni,Cu)_3Sn_4$ is in equilibrium with the Ni instead of $(Cu,Ni)_6Sn_5$. The new $(Ni,Cu)_3Sn_4$ is not in equilibrium with $(Cu,Ni)_6Sn_5$, and Sn forms between them, causing spalling of the $(Cu,Ni)_6Sn_5$ layer.

Yang et al. soldered Sn3.0Ag0.6Cu solder balls with varying diameters to a 375 μm-diameter nickel pad to view the effect of residual copper concentration on interfacial intermetallics [57]. Figure 12.31 shows the interface for a 760 and 300 μm solder ball. The composition of residual copper in the solder decreased due to the larger volume fraction of intermetallic formed, and the copper was limited enough to change the local equilibrium and promote spalling.

In the short term, the presence of massive spalling increased the ductility of the solder joint due to the reduced intermetallic thickness at the interface [58]. With additional aging, new intermetallic grows and leads to a brittle joint.

The addition of palladium seems to reduce spalling effects and improve mechanical properties of the joints due to suppression of intermetallic growth [59]. Therefore, ENEPIG (Electroless Nickel Electroless Palladium Immersion Gold) may be a better choice of surface finish for lead-free solder joints to reduce spalling defects.

Spalling occurs in solder systems in which Cu or Ni are limited, and the composition of the intermetallic changes in contact with solder over time due to the reduced availability of the limited element in the solder. In lead-free solder joints, this can be mitigated with an ENEPIG surface finish, or by having very high Cu concentrations in the solder.

12.3.2.4 Effects of Phosphorus Concentration in ENIG on Solder Joint Reliability

Pure nickel, Ni-P (Electroless Nickel), and ENIG (Electroless Nickel Immersion Gold) surface finishes change the evolution of the interfacial microstructure despite the similar phases in the system. ENIG substrates essentially behave as Ni-P substrates, as the gold dissolves into the tin to form a tin solid solution. Few if any Au-Sn intermetallics are visible when the only source of gold comes from an ENIG finish.

Electroless Ni also contains 8–10% phosphorus [60]. There are no current standards for phosphorus content in electroless nickel. During Ni-Sn intermetallic formation, the nickel diffuses out of the Ni-P layer, leaving a phosphorus-rich nickel layer that has been discussed in several studies. Figure 12.32 shows layers at the interface of a Sn-3.5Ag/Ni-P solder joint held at 251 °C for one hour. The lowest layer is the amorphous Ni-P coating from the electroless nickel plating. The next layer is the phosphorus-rich layer, which was identified as Ni_3P, in contact with Ni_3Sn_4 [52, 53]. Wojewoda et al. found $Ni_{12}P_5$ instead of Ni_3P [61].

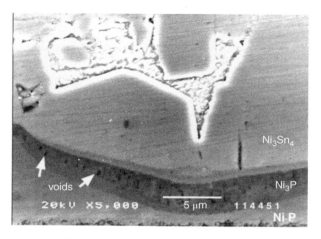

Figure 12.32 Voids forming in Ni_3P layer after reflow at 251 °C for one hour [52].

Due to the thinness of the phosphorus-rich layer, it is difficult to determine an exact composition, and it is also possible that Ni_3P in contact with amorphous Ni-P changes composition slightly. Wojewoda et al. also found a thin Ni-Sn-P layer that was determined to be Ni_2SnP with TEM [61]. Ni_3Sn_4 and $(Cu,Ni)_6Sn_5$ were also seen at the interface. Hung et al. posited that the growth of Ni_3Sn_4 assisted the crystallization of Ni_3P, the composition of the phosphorus-rich layer [53].

Figure 12.26 shows the interface between Ni_3Sn_4 intermetallic and solder with Ni-P and pure Ni. Both interfaces are flatter with a pure nickel substrate than with Ni-P. He et al. posited that crystallization of Ni_3P increased the mobility of nickel through phosphorus-rich layer into Ni_3Sn_4 due to a release of internal energy from crystallization and could facilitate a rougher interface [52].

The voids present in the Ni_3P layer in Figure 12.32 are due to the diffusion of nickel into the solder. During growth of Ni_3Sn_4, nickel diffuses out of the Ni-P layer into the solder, but tin is unable to diffuse into the phosphorus-rich or Ni_3P layer. Therefore, there is an unequal flux at the Ni_3P interface and voids form to equilibrate the system.

There are no current standards for phosphorus content in electroless nickel surface finishes. During the formation of Ni-Sn IMCs at the interface, a phosphorus-rich layer forms between the electroless nickel and the intermetallic, which has not been fully characterized.

12.3.3 Au-Sn

Any discussion of intermetallics and their effects on reliability would be incomplete without addressing the Au-Sn system. Gold-tin intermetallics are ubiquitous in solder joints due to the high solderability of gold. These intermetallics, particularly $AuSn_4$, can cause significant negative impacts on the mechanical reliability of solder joints, as noted by the IPC standard [62] that limits gold concentrations to ≤5 wt% in both lead-free and SnPb solder joints. In this section, an overview of the Au-Sn system and its IMCs is given, followed by a summary of research on Au-Sn solid-state diffusion and intermetallic growth; finally, the problem of gold embrittlement is discussed, not only for the Au-Sn system but for Au-Ni-Sn system as well.

The Au-Sn binary phase diagram is shown in Figure 12.33. Six IMCs may be present in the system: β ($Au_{10}Sn$), ζ and ζ' (Au_5Sn), AuSn, $AuSn_2$, and $AuSn_4$. Two areas of the phase diagram are particularly relevant for soldering applications: the higher-gold area surrounding the eutectic point at 20 wt% Sn, and the lower-gold area including compositions from about 85 to 100 wt% Sn. The higher-gold intermetallics ζ' and AuSn appear in applications utilizing Au-20Sn with gold metallization [64], while the lower-gold intermetallics are more commonly observed in tin-based solders. Although Au-20Sn solders are used in specific applications such as microwave components and hermetic packaging, when their high temperature performance justifies the higher cost [64], this discussion will focus on the more widespread case of tin-based solder joints containing relatively low concentrations of gold.

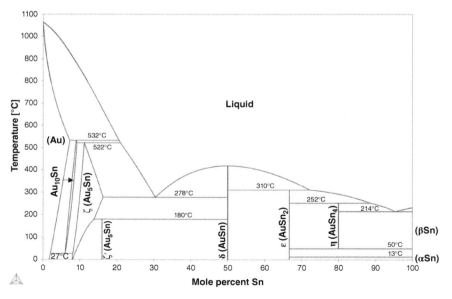

Figure 12.33 Au-Sn binary phase diagram. Source: Adapted from [63].

In a solder joint, these IMCs may precipitate during the solidification portion of the reflow process, and grow via lattice diffusion, grain boundary diffusion and coarsening. Diffusion in real or simulated solder joints is highly complex and usually involves other components in addition to gold and tin; however, the simplest case of diffusion and IMC growth in the Au-Sn system would be a binary diffusion couple of pure gold and pure tin.

Significant research has been conducted on the growth of the IMCs via diffusion in the binary Au-Sn system with thermal aging. Yamada et al. fabricated Sn/Au/Sn sandwich diffusion couples, annealed them in a silicone oil bath at 433 K (160 °C) over a range of times from 1 to 127 hours, and measured the resulting thickness of each intermetallic layer using a differential interference contrast (DIC) optical microscope; composition across the diffusion couples was measured using electron probe microanalysis (EPMA) [65]. The resulting concentration profile is shown in Figure 12.34.

For the range of annealing times investigated, the ratio between intermetallic layer thicknesses was found to be consistent at approximately 4:1:1 for $AuSn_4$, $AuSn_2$, and AuSn, respectively; the remaining intermetallics were not observed to grow at the annealing temperature chosen [65]. Yamada found that the effective interdiffusion between gold and tin was strongly dependent on grain size, indicating that grain boundary diffusion was a significant factor [65].

Baheti et al. compared Au-Sn bulk diffusion couples aged at 125–200 °C with electroplated diffusion couples aged at 25–200 °C. At 200 °C, all intermetallics except $Au_{10}Sn$ were seen to develop; at 150 °C, Au_5Sn was no longer present. The electroplated diffusion couples only developed $AuSn_2$ and $AuSn_4$ at 75 °C.

Figure 12.34 Concentration profile of gold in the Sn/Au/Sn binary diffusion couple annealed at 160 °C for 127 hours $(4.57 \times 10^5 \text{ s})$ [65].

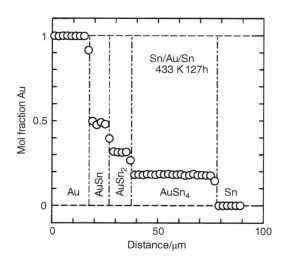

Composition was measured using EPMA, and the resulting composition profile is shown in Figure 12.35 along with the bulk diffusion couple micrograph [66].

Using electroplated diffusion couples, room temperature aging of $AuSn_4$ was observed after aging for 30 and 912 days (2.5 years) as shown in Figure 12.36. Substantial growth of the intermetallic phases can be seen even with aging at room temperature. The elongated, needle-like morphology of the $AuSn_4$ phase is not observed to become smoother over time [66].

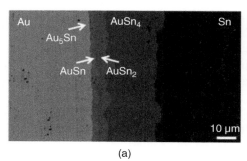

(a)

Figure 12.35 Au/Sn bulk diffusion couple annealed at 200 °C: (a) for 4 hours, BSE image and (b) for 400 hours, composition profile of the interdiffusion zone [66].

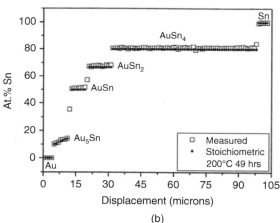

(b)

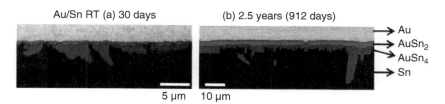

Au/Sn RT (a) 30 days   (b) 2.5 years (912 days)

→ Au
→ $AuSn_2$
→ $AuSn_4$
→ Sn

5 μm   10 μm

Figure 12.36 BSE images of the Au/Sn electroplated diffusion couple after storage at room temperature for (a) 30 days and (b) 912 days, i.e. 2.5 years [66].

The growth of Au-Sn intermetallics in low-temperature, tin-based solders used to attach gold-plated components causes the common reliability problem known as gold embrittlement. Gold, present in component and connector platings, dissolves readily in liquid tin solder during reflow. From the binary phase diagram in Figure 12.33, more than 10 wt% Au can dissolve into pure tin at 250 °C, and the presence of other elements may further increase the gold solubility. As the solder joint solidifies, Au-Sn IMCs form. The precipitates are dispersed throughout the bulk of the solder joint upon solidification, and coarsen with thermal aging. Significant IMC formation is correlated with decreased joint strength [67, 68].

The amount of gold necessary to cause embrittlement in different solder types is unclear. IPC J-STD-001 specifies that at least 95% of gold be removed from most solderable surfaces to prevent embrittlement [62]. The corresponding guide IPC-HDBK-001 sets 3–4 wt% gold in a solder joint as an upper limit to prevent embrittlement [69]; however, several studies and failure analyses have cast doubt on the requirements in the standard. Hare noted embrittlement failure at a calculated 1.65 wt% Au [67]; analyses by Vianco found that the double-tinning process used to remove excess gold was not always sufficient to prevent failure by gold embrittlement [70]. Studying embrittlement in SnAgCu solders, Glazer et al. found a gradual decrease in strength with additional gold, but did not find a singular composition where the loss of strength became precipitous [71]. Mei et al. observed brittle fracture in aged PBGAs due to gold embrittlement at a concentration of only 0.1 wt% Au in the joint [72].

Gold-containing solder joints with an ENIG surface finish or an electrolytic Au coating on Ni also form $(Au,Ni)Sn_4$ at the solder-Ni interface rather than or in addition to Ni_3Sn_4, even when the Au layer in the ENIG is thin[73]. During reflow of Sn or Sn-Ag solders in contact with ENIG or electrolytic Au, the Au dissolves in the solder and then forms $AuSn_4$ in the bulk solder during solidification. During aging, the $AuSn_4$ in the bulk dissolves and reprecipitates at the interface as $(Ni,Au)Sn_4$ with the substrate being the source of the Ni through solid state diffusion at the $AuSn_4$-Ni interface. The activation energy for growth of $(Au,Ni)Sn_4$ is significantly lower than Ni_3Sn_4 and therefore forms rapidly with the presence of Ni, Au, and Sn. This does not change with the presence of Pb or Ag because both are insoluble in the Au-Ni-Sn intermetallics. However, with the presence of Cu, a quaternary intermetallic based on Cu_6Sn_5 may form instead [51].

In summary, understanding and controlling the growth of Au-Sn and Au-Ni-Sn intermetallics are critical for ensuring mechanical reliability of solder joints. These intermetallics occur frequently in tin-based solders in contact with gold. Upon thermal aging, the intermetallics grow; this can lead to brittle fracture of the solder joints. Although current industry standards assert that an upper limit of 3 wt%

gold prevents gold embrittlement, multiple studies and failure analyses have cast doubt on the validity of this limit in particular small amounts of gold with ENIG. Although further data are needed to validate or revise the current standard, the possibility of gold embrittlement should always be carefully considered during the design process.

12.4 High Lead – Exemption

Solders with more than 85% lead are exempt from the European Union RoHS legislation, and are typically used in combination with copper substrates. The High-Lead/Cu system contains one intermetallic: Cu_3Sn. The Cu_3Sn forms mainly at the interface, but some copper does dissolve in the solder during reflow, allowing the intermetallic to form in the bulk as well.

Figure 12.37 shows the growth of Cu_3Sn at the interface between the copper substrate and two high-lead solders. Spalling of the intermetallic is seen in 95Pb5Sn solder but not in 90Pb10Sn solder. This is most likely due to the limited amount of tin in 95Pb5Sn. Spalling is discussed further in Section 12.3.2.3.

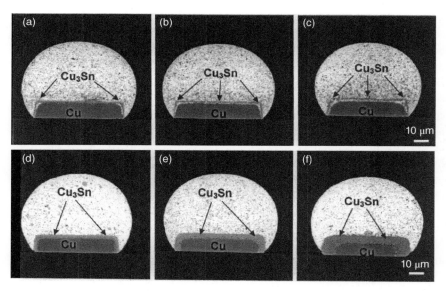

Figure 12.37 Cross-section backscattered SEM images of 95Pb5Sn solder bumps after annealing at (a) 170 °C for 500 hours, (b) 170 °C for 1000 hours, (c) 170 °C for 1500 hours, and 90Pb10Sn after annealing at (d) 170 °C for 500 hours, (e) 170 °C for 1000 hours, (f) 170 °C for 1500 hours. 95Pb5Sn shows spalling of Cu_3Sn, while 90Pb10Sn shows a continuous layer of intermetallic. Source: Adapted from [74].

12.5 Conclusions

The chapter reviewed the roles that intermetallics play in determining solder joint reliability, and how those roles change as a result of aging or damage induced by thermal cycling. It looked at the performance of common lead-free solder alloys in combination with metallizations and surface finishes to help understand what to expect in the specific systems discussed as well as the problems that may arise when combining new solder alloys and surface finishes/metallizations and the methodologies that can be used to separate out the different possible root causes.

Various effects were discussed in relation to intermetallic growth and brittle fracture, gold embrittlement, spalling, and phosphorus effects in ENIG on solder joint reliability. Although there has been good progress in these areas to understand the effects and interactions occurring, more research and development work is needed to develop electronic products with improved reliability.

References

1 Pang, J.H., Xu, L., Shi, X. et al. (2004). Intermetallic growth studies on Sn-Ag-Cu lead-free solder joints. *Journal of Electronic Materials* 33 (10): 1219.

2 Spinelli, J.E. and Garcia, A. (2013). Microstructural development and mechanical properties of hypereutectic Sn-Cu solder alloys. *Materials Science and Engineering A* 568: 195–201.

3 Li, D., Liu, C., and Conway, P.P. (2005). Characteristics of intermetallics and micromechanical properties during thermal ageing of Sn-Ag-Cu flip-chip solder interconnects. *Materials Science and Engineering A* 391: 95–103.

4 Zeng, G., Xue, S., Zhang, L. et al. (2010). A review on the interfacial intermetallic compounds between Sn-Ag-Cu based solders and substrates. *Journal of Materials Science: Materials in Electronics* 21: 421–440.

5 Moon, K.-W., Boettinger, W., Kattner, U. et al. (2000). Experimental and thermodynamic assessment of Sn-Ag-Cu solder alloys. *Journal of Electronic Materials* 29 (10): 1122–1136.

6 Swenson, D. (2007). The effects of suppressed beta tin nucleation on the microstructural evolution of lead-free solder joints. *Journal of Electronic Materials* 18 (1–3): 39–54.

7 Suh, D., Kim, D.W., Liu, P. et al. (2007). Effects of Ag content on fracture resistance of Sn-Ag-Cu lead-free solders under high-strain rate conditions. *Materials Science and Engineering A* 460–461: 595–603.

8 Tucker, J., Chan, D., Subbarayan, G., and Handwerker, C. (2014). Maximum entropy fracture model and its use for predicting cyclic hysteresis in

Sn3.8Ag0.7Cu and Sn3.0Ag0.5Cu solder alloys. *Microelectronics Reliability* 54: 2513–2522.

9 Wong, E., Seah, S., van Driel, W. et al. (2009). Advances in the drop-impact reliability of solder joints for mobile applications. *Microelectronics Reliability* 49: 139–149.

10 Kaulfersch, E., Rzepka, S., Ganeshan, V. et al. (2007). Dynamic mechanical behavior of SnAgCu BGA solder joints determined by fast shear tests and FEM simulations. In: *International Conference on Thermal, Mechanical and Multi-Physics Simulation Experiments in Microelectronics and Micro-Systems,* 1–4.

11 Chan, D.K. (2012) A maximum entropy fracture model for low and high strain-rate fracture in SnAgCu alloys. Dissertation Abstracts International 74-03B(E).

12 Towashiraporn, P., Subbarayan, G., McIlvanie, B. et al. (2002). Predictive reliability models through validated correlation between power cycling and thermal cycling accelerated life tests. *Soldering and Surface Mount Technology* 14 (3): 51–60.

13 Bhate, D., Mysore, K., and Subbarayan, G. (2012). An information theoretic argument on the form of damage accumulation in solids. *Mechanics of Advanced Materials and Structures* 19 (1–3): 184–195.

14 Chan, D., Subbarayan, G., and Nguyen, L. (2011). Maximum-entropy principle for modeling damage and fracture in solder joints. *Journal of Electronic Materials* 41 (2): 398–411.

15 Hopcraft, M., Nix, W., and Kenny, T. (2010). What is the Young's modulus of silicon? *Journal of Microelectromechanical Systems* 19 (2): 229–238.

16 Fields, R., Low III, S. and Lucey, G., Jr (1991) Physical and mechanical properties of intermetallic compounds commonly found in solder joints. Metal Science of Joining, TMS Symposium, Cincinnati.

17 Deng, X., Koopman, M., Chawla, N., and Chawla, K. (2004). Young's modulus of (Cu,Ag)-Sn intermetallics measured by nanoindentation. *Materials Science and Engineering A* 364: 240–243.

18 Jiang, N., Clum, J., Chromik, R., and Cotts, E. (1997). Thermal expansion of several Sn-based intermetallic compounds. *Scripta Materialia* 37 (12): 1851–1854.

19 Ghosh, G. (2004). Elastic properties, hardness, and indentation fracture toughness of intermetallics relevant to electronic packaging. *Journal of Materials Research* 19 (5): 1439–1454.

20 Chromik, R., Wang, D.-N., Shugar, A. et al. (2005). Mechanical properties of intermetallic compounds in the Au-Sn system. *Journal of Materials Research* 20 (8): 2161–2172.

21 Xian, J., Zeng, G., Belyakov, S. et al. (2017). Anisotropic thermal expansion of Ni_3Sn_4, Ag_3Sn, Cu_3Sn, Cu_6Sn_5 and beta-Sn. *Intermetallics* 91: 50–64.

22 Okamoto, H. (2006). Ni-Sn (nickel-tin). *Journal of Phase Equilibria and Diffusion* 27 (3): 315.

23 Gierlotka, W., Huang, Y., and Chen, S. (2008). Phase equilibria of Sn-Sb-Ag ternary system. (II): calculation. *Metallurgical and Materials Transactions A: Physical Metallurgy and Materials Science* 39: 3199–3209.

24 Lee, T., Choi, W., Tu, K. et al. (2002). Morphology, kinetics, and thermodynamics of solid-state aging of eutectic SnPb and Pb-free solders (Sn-3.5Ag, Sn-3.8Ag-0.7Cu and Sn-0.7Cu) on Cu. *Journal of Materials Research* 17: 291–301.

25 Wang, J., Liu, H., Liu, L., and Jin, Z. (2007). Thermodynamic description of the Sn-Ag-Au ternary system. *CALPHAD: Computer Coupling of Phase Diagrams and Thermochemistry* 31: 545–552.

26 Hsu, H. and Chen, S. (2004). Phase equilibria of the Sn-Ag-Ni ternary system and interfacial reactions at the Sn-Ag/Ni joints. *Acta Materialia* 52: 2541–2547.

27 Li, M., Du, Z., Guo, C., and Li, C. (2009). Thermodynamic optimization of the Cu-Sn and Cu-Nb-Sn systems. *Journal of Alloys and Compounds* 477: 104–117.

28 Bochvar, N. and Liberov, Y. (1995). Gold-Copper-Tin. In: *Ternary Alloys*, vol. 12 (eds. G. Effenberg, F. Aldinger and A. Prince), 401–409.

29 Li, C. and Duh, J. (2005). Phase equilibria in the Sn-rich corner of the Sn-Cu-Ni ternary alloy system at 240°C. *Journal of Materials Research* 20: 3118–3124.

30 Grolier, V. and Schmid Fetzer, R. (2007). Thermodynamic evaluation of the Au-Sn system. *International Journal of Materials Research* 98: 797–806.

31 Song, H., Ahn, J., Minor, A., and Morris, J.J. (2001). Au-Ni-Sn intermetallic phase relationships in eutectic Pb-Sn solder formed on Ni/Au metallization. *Journal of Electronic Materials* 30: 409–414.

32 Schmetterer, C., Flandorfer, H., Richter, K. et al. (2007). A new investigation of the system Ni-Sn. *Intermetallics* 15: 869–884.

33 Vuorinen, V., Yu, H., Laurila, T., and Kivilahti, J. (2008). Formation of intermetallic compounds between liquid Sn and various CuNix metallizations. *Journal of Electronic Materials* 37 (6): 792–805.

34 Reeve, K., Holaday, J., Choquette, S. et al. (2016). Advances in Pb-free solder microstructure control and interconnect design. *Journal of Phase Equilibria and Diffusion* 37 (4): 369–386.

35 National Institute of Standards and Technology (1996) Cu-Sn System: calculated phase diagram.

36 Ghosh, G. and Asta, M. (2005). Phase stability, phase transformations, and elastic properties of Cu6Sn5: Ab initio calculations and experimental results. *Journal of Materials Research* 20 (11): 3102–3117.

37 Nogita, K. and Nishimura, T. (2008). Nickel-stabilized hexagonal $(Cu,Ni)_6Sn_5$ in Sn-Cu-Ni lead-free solder alloys. *Scripta Materialia* 59: 191–194.

38 Wu, Y., Barry, J., Yamamoto, T. et al. (2012). A new phase in stoichiometric Cu_6Sn_5. *Acta Materialia* 60: 6581–6591.

39 Kumar, S., Handwerker, C., and Dayananda, M. (2011). Intrinsic and Inter-diffusion in Cu-Sn System. *Journal of Phase Equilibria and Diffusion* 32 (4): 309–319.

40 Zhang, Z., Li, M., Liu, Z., and Yang, S. (2016). Growth characteristics and formation mechanisms of Cu6Sn5 phase at the liquid-Sn0.7Cu/(111)Cu and liquid-Sn0.7Cu/(001)Cu joint interfaces. *Acta Materialia* 104: 1–8.

41 IPC J-STD-006 (2013) *Requirements for electronic grade solder alloys and fluxed and non-fluxed solid solders for electronic soldering applications*. IPC International.

42 National Institute of Standards and Technology (1996) Ag-Sn System: calculated phase diagram.

43 Thermo-Calc Software, Thermo-Calc Database TCSLD3.

44 Kerr, M. and Chawla, N. (2004). Creep deformation behavior of Sn-3.5Ag solder/Cu couple at small length scales. *Acta Materialia* 52: 4527–4535.

45 Mukherjee, A., Bird, J., and Dorn, J. (1968). Creep of metals at high temperatures. *ASM Transactions Quarterly* 61 (4): 697–698.

46 Chavali, S.C. (2012) Diffusion driven microstructural evolution and its effect on mechanical behavior of SnAgCu solder alloys, Dissertation Abstracts International 74-07B(E).

47 Chavali, S., Singh, Y., Kumar, P. et al. (2011) Aging aware constitutive models for SnAgCu solder alloys. Electronic Components and Technology Conference 2011.

48 An, T. and Qin, F. (2014). Effects of the intermetallic compound microstructure on the tensile behavior of Sn3.0Ag0.5Cu/Cu solder joint under various strain rates. *Microelectronics Reliability* 54: 932–938.

49 Baheti, V., Kashyap, S., Kumar, P., Chattopadhyay, K., and Paul, A. (2017). Solid-state diffusion-controlled growth of the intermediate phases from room temperature to an elevated temperature in the Cu-Sn and the Ni-Sn systems. Journal of Alloys and Compounds, vol. 727, pp. 832–840.

50 Belyakov, S.A., and Gourlay, C. (2016). The Influence of Cu on Metastable NiSn4 in Sn-3.5Ag-xCu/ENIG Joints. Journal of Electronic Materials, vol. 45, no. 1, pp. 12–20.

51 Li, M., Lee, K., Olsen, D. et al. (2002). Microstructure, joint strength and failure mechanisms of SnPb and Pb-free solders in BGA packages. *IEEE Transactions on Electronics Packaging Manufacturing* 25 (3): 185–192.

52 He, M., Lau, W., Qi, G., and Chen, Z. (2004). Intermetallic compound formation between Sn-3.5Ag solder and Ni-based metallization during liquid state reaction. *Thin Solid Films* 462–463: 376–383.

53 Hung, K., Chan, Y., and Tang, C. (2000). Metallurgical reaction and mechanical strength of electroless Ni-P solder joints for advanced packaging applications. *Journal of Materials Science: Materials in Electronics* 11: 587–593.

54 Chuang, H., Yu, J., Kuo, M. et al. (2012). Elimination of voids in reactions between Ni and Sn: a novel effect of silver. *Scripta Materialia* 66: 171–174.

55 Xu, L., Pang, J. and Che, F. (2005) Intermetallic growth and failure study for Sn-Ag-Cu/ENIG PBGA solder joints subject to thermal cycling. Electronic Components and Technology Conference 2005.

56 Snugovsky, L., Snugovsky, P., Perovic, D., and Rutter, J. (2006). Phase equilibria in Sn rich corner of Cu-Ni-Sn system. *Materials Science and Technology* 22 (8): 899–902.

57 Yang, S., Ho, C., Chang, C., and Kao, C. (2007). Massive spalling of intermetallic compounds in solder-substrate reactions due to limited supply of the active element. *Journal of Applied Physics* 101 084911-1.

58 Chen, H., Tsai, Y.-L., Chang, Y.-T., and Wu, A. (2016). Effect of massive spalling on mechanical strength of solder joints in Pb-free solder reflowed on Co-based surface finishes. *Journal of Alloys and Compounds* 671: 100–108.

59 Yoon, J.-W., Noh, B.-I., and Jung, S.-B. (2011). Comparative study of ENIG and ENEPIG as surface finishes for a Sn-Ag-Cu solder joint. *Journal of Electronic Materials* 40 (9): 1950–1953.

60 Sade, W., Proenca, R., de Oliveira Moura, T., and Branco, J. (2011). Electroless Ni-P coatings: preparation and evaluation of fracture toughness and scratch hardness. *ISRN Materials Science* 2011: 1–6.

61 Wojewoda-Budka, J., Huber, J., Litynska-Dobrzynska, L., Sobczak, N., and Zieba, P. (2013). Microstructure and chemistry of the SAC/ENIG interconnections. *Materials Chemistry and Physics* 139: 276–280.

62 IPC J-STD-001G (2005) *Requirements for soldered electrical and electronic assemblies.* IPC International.

63 Liu, H., Liu, C., Ishida, K., and Jin, Z. (2003). Thermodynamic modeling of the Au-In-Sn system. *Journal of Electronic Materials* 32 (11): 1290–1296.

64 McNulty, C. (2008). Processing and reliability issues for eutectic AuSn solder joints. In: *41st International Symposium on Microelectronics*, 909–916.

65 Yamada, T., Miura, K., Kajihara, M., Kurokawa, N., and Sakamoto, K. (2004). Formation of intermetallic compound layers in Sn/Au/Sn diffusion couple during annealing at 433 K. *Journal of Materials Science* 39: 2327–2334.

66 Baheti, V., Kashyap, S., Kumar, P., Chattopadhyay, K., and Paul, A. (2018). Solid-state diffusion-controlled growth of the phases in the Au-Sn system. *Philosophical Magazine* 98 (1): 20–36.

67 Hare, E. (2010). *Gold Embrittlement of Solder Joints*. Snohomish, WA: SEM Lab, Inc.

68 Hillman, C., Blattau, N., Arnold, J., Johnston, T., Gulbrandsen, S., Silk, J., and Chiu, A. (2013) Gold embrittlement in lead-free solder. Surface Mount Technical Association International Conference 2013.

69 IPC (2016). *IPC-HDBK-001F: Handbook and Guide to Supplement J-STD-001*. IPC International.

70 Vianco, P. (1993). Embrittlement of surface mount solder joints by hot solder-dipped, gold-plated leads. In: *Surface Mount Technical Association International Conference*, 1–19.

71 Glazer, J., Kramer, P., and Morris, J. (1992). Effect of gold on the reliability of fine pitch surface mount solder joints. *Circuit World* 18 (4): 41–46.

72 Mei, Z., Kaufmann, M., Eslambolchi, A., and Johnson, P. (1998). Brittle interfacial fracture of PBGA packages soldered on electroless nickel/immersion gold. In: *Proceedings of the 48th Electronic Components Technology Conference*, 952–961.

73 Liu, Y., Chen, Y., Gu, S., Kim, D., and Tu, K. (2016). Fracture reliability concern of $(Au,Ni)Sn_4$ phase in 3D integrated circuit microbumps using Ni/Au surface finishing. *Scripta Materialia* 119: 9–12.

74 Ramanathan, L., Jang, J., Lin, J., and Frear, D. (2005). Solid-state annealing behavior of two high-Pb solders, 95Pb5Sn and 90Pb10Sn, on Cu under bump metallurgy. *Journal of Electronic Materials* 34 (10): L43–L46.

13

Conformal Coatings

Jason Keeping

Celestica Inc., Toronto, Canada

13.1   Introduction

This chapter will discuss the industry updates in the use of conformal coatings and their use in electronics manufacturing and their effect on reliability.

For most traditional manufacturing processes, conformal coating and environmental protection are not a planned process, minus applications that were designed for harsh environmental conditions from the start. With the transition within the industry for higher reliability and protection from environmental conditions, conformal coating and other ruggedization processes have been developed and applied over the years.

In that, once a printed circuit assembly (PCA) has been fully assembled, tested and inspected, and if ruggedization is required, this is the point in the process where protections from the environment are applied. This protection from the environment, needs to be completed for the end-use environment where the product will ultimately be used. Typical environmental aspects include moisture, salt, dirt, fungus, sulfur, mechanical shock, and vibration, along with other factors that will cause failure on an unprotected assembly.

The application of a ruggedization process and materials are the PCA's last line of defense against the elements that provides a thin layer of coating that conforms to the shape of all the part leads, solder joints, conductive surfaces, and other complex features on the completed PCA. This conformal coating can be applied in many forms – liquid, atomized spray, and/or vapor deposition to the fully assembled PCA – and is cured in place to form a protective layer of insulation for the assembly within the given cure conditions of the material that are selected in terms of air, heat, moisture, ultraviolet, sublimation, and combinations.

Lead-free Soldering Process Development and Reliability, First Edition. Edited by Jasbir Bath.
© 2020 John Wiley & Sons, Inc. Published 2020 by John Wiley & Sons, Inc.

This chapter will discuss various aspects of conformal coatings including:

- Environmental Health and Safety (EHS) requirements
- An overview of the five basic conformal coating types, and new emerging materials
- Preparatory steps necessary to ensure a successful coating process
- Various methods of applying conformal coating
- Aspects for cure, inspection, and demasking
- Repair and rework processes
- Design guidance on when and where coating is required, and which physical characteristics and properties are important to consider

13.2 Environmental, Health, and Safety (EHS) Requirements

The application of a conformal coating to a printed circuit assembly (PCA) can involve brushing, dispensing, spraying, or otherwise applying a solution in which the conformal coating resin can be dissolved in a solvent onto the assembly. Another method of coating application is vacuum deposition. The solvent then evaporates/volatilizes leaving the resin coating on the PCA. This conformal coating process involves the following EHS issues:

1. Employee exposure to the solvent, resin, or other precursor materials. Risks include corrosive materials, solvent inhalation, skin irritation.
2. Fire safety considerations when the solvent is flammable or combustible.
3. Air emissions from the volatilized solvent (e.g. VOCs, Hazardous Air Pollutants [HAPs], etc.); precursor materials can also be pyrophoric.
4. Waste management from used manual tooling (brushes, wipes, etc.), personal protective equipment (PPE), other applicators, dispenser cleanup/maintenance mixtures, unused coating solution, waste coating material, and used masking materials (tape, dots, covers, etc.).
5. Material content declaration (MCD) requirements.
6. Greenhouse gas (GHG) reporting.
7. Equipment design and construction.

Each of these EHS issues must be addressed for compliant conformal coating operations, both internally and externally, with all applicable regulatory agencies.

13.3 Overview of Types of Conformal Coatings

Conformal coatings are polymeric or other materials used to protect electronic assemblies from a wide variety of life cycle contaminants. Conformal coatings provide a high degree of insulation protection and can be resistant to many types of solvents and harsh environments encountered in the product life cycle. The coating materials also act to immobilize various types of particulates on the surface of the assembly and function as protective barriers to the various devices on the board.

They are resistant to moisture and humidity, which may reduce the potential of leakage currents, "cross talk," electrochemical migration (ECM), Conductive anodic filament (CAF), dendrite growth, and arcing. These issues are becoming more critical with the reduction in component size, pitch, circuitry spacing, laminate thickness and voltage plus the rise in speed (frequency) of signals as well as the transition of electronics closer to harsh environments and/or new designs from indoor to outdoor environments.

13.3.1 Types of Conformal Coatings

Traditionally, conformal coating materials were available in five basic chemistries, each with multiple application methods and several drying/curing options. As such, no one material will fit every application, but the wide variety will allow the designer to select a material that meets most technical, budgetary, and manufacturing needs.

The traditional five basic chemistries are:

1. *Acrylic.* Acrylics are easy to apply and remove, due to their reaction with solvents.
2. *Urethane.* This coating has good moisture and chemical resistance and good electrical insulation.
3. *Epoxy.* This coating is tough, durable, and very chemically resistant.
4. *Silicone.* Being soft and having good adhesion, silicone is useful over a wide temperature range.
5. *Para-xylylene.* This very thin, pin-hole-free coating is vapor deposited instead of being applied as a liquid.

There are two new emerging chemistries which are:

1. *Synthetic Rubber (SC).* Similar to silicone, with a wide temperature range, but they are organic based.
2. *Ultra-thin (UT).* Similar to para-xylylene, but are applied at an even thinner thickness.

More information on these and other less common types of conformal coating may be found in IPC-HDBK-830 [1].

13.3.1.1 Acrylic Resins (Type AR)

Acrylic coatings (AR) are usually available as one-part compounds at a given solids percentage within a solvent transportation medium. They provide excellent moisture resistance and electrical and flexibility properties.

Acrylic coatings are relatively easy to apply and remove, due to their reaction with their base solvents. Typical application is 0.001–0.005 in. (25–125 μm) thick. Acrylic coatings are often used in consumer and aerospace electronics.

13.3.1.1.1 Chemistry

Acrylic coatings are usually supplied as dissolved pre-polymerized acrylic chains. The acrylic chemistry does not cure by polymerization and cross-linking as the other coating materials do, but instead hardens gradually as the solvent evaporates. Acrylics are also available in heat-curable and UV-cured formulations.

13.3.1.1.2 Properties

Acrylics are easy to apply, and are the easiest coating to remove, since relatively mild solvents soften and dissolve the acrylic coating while leaving the epoxy encapsulated parts and printed wiring board (PWB) unharmed. They can be cured quickly.

13.3.1.1.3 Advantages and Disadvantages

The greatest strengths of acrylic coating are the ease of rework and the fast room temperature cure. Acrylics provide good moisture resistance and fluoresce easily, aiding inspecting under UV lamps. Since they are so easy to rework, acrylics are also susceptible to inadvertent chemical attack from solvent splash during hand cleaning of solder joints elsewhere in the assembly. The high emission of solvent inherent in the acrylic process makes them less environmentally friendly than other materials. As the solvent evaporates, the coating shrinks and exerts stress on the components, so acrylics may not be suitable for all low-temperature applications.

Acrylics dry rapidly, reaching tack-free condition in a short period of time, are fungus resistant, and provide long pot life. Furthermore, acrylics give off little or no heat during cure eliminating damage to heat-sensitive components, have no further shrinkage once the carrier solvent has evaporated off, and have good humidity resistance. Acrylic coatings exhibit low glass transition temperatures (Tgs). Above the Tg, the large expansion can result in damaging effects.

As for rework, acrylics are easy to apply, and the dried film can be removed using the same solvent method from their application. Similarity for spot removal of the coating to repair a solder joint or replace a component this can also be easily accomplished by localized solvent application.

For other removal methods please refer to Section 10 of the IPC-HDBK-830 Conformal Coating Handbook [1].

13.3.1.2 Urethane Resins (Type UR)

Polyurethane coatings (UR) are available as either single or two-component formulations. Both provide good humidity and chemical resistance, plus higher sustained dielectric properties.

Urethanes are relatively easy to apply with dip, spray, or brush. Cure times (other than UV-curable) range from a few hours for two-part heat-curable, to several days for a single-part room temperature cure. Typical application is 0.001–0.005 in. (25–125 μm) thick. Urethane coatings are often used for space applications due to their low outgassing characteristics.

13.3.1.2.1 Chemistry

Urethanes are based on a diisocyanate and polyol backbone. They are available in solvent evaporative cure, heat cure, and UV cure formulations.

13.3.1.2.2 Properties

Since urethanes are polymerized and cross-linked in place, they have excellent resistance to chemical, moisture, and solvents. They are available in hardnesses ranging from tough, abrasion-resistant varieties to low-modulus versions suitable for extreme temperature ranges.

13.3.1.2.3 Advantages and Disadvantages

Urethanes have good adhesion to most materials, including epoxy part bodies, metals, and ceramics. As such, the coating process is fairly robust. Since urethanes are chemically resistant, they are also difficult to remove except by thermal or mechanical means. Urethanes can be soldered through, although this often results in a brownish discoloration that must then be removed.

Early polyurethane compounds exhibited instability or reversing of the cured film to a liquid under high humidity and temperature conditions. Some newer formulations, however, eliminate this phenomenon.

Single component polyurethanes, while easy to apply, sometimes require days at room temperature to reach their full properties. It is advised to refer to their technical data sheets for more information.

Two-component formulations, on the other hand, reach optimum cure properties at elevated temperatures within hours, with pot lives up to three hours.

As for rework, their chemical resistance is an advantage, however, can be a major drawback for rework and can make this process become difficult and costly.

To repair or replace a component, it is advised to refer to Section 10 of the IPC-HDBK-830 Conformal Coating Handbook for processes and methods [1].

13.3.1.3 Epoxy Resins (Type ER)

Epoxy coatings (ER) are usually available as two-part compounds. They provide reasonable humidity resistance and good abrasive and chemical resistance.

Epoxy coatings are chemically stable and very resistant to chemical attack. Typical application thicknesses are 0.001–0.005 in. (25–125 μm) thick. Epoxy coatings are useful in extreme environments where chemical vapors or high temperatures are present, as seen in various industrial applications.

13.3.1.3.1 Chemistry
Epoxy coatings are based on epoxy resin systems and come in four types: solvent evaporation, heat curing, UV curing, and catalyzed.

13.3.1.3.2 Properties
Epoxies have a low coefficient of thermal expansion (CTE) that matches well with the PWB epoxy resin, since they share very similar chemistry. They have a higher Tg than most of the other coating materials. They are also very tough and abrasion-resistant, so that rework is very difficult; epoxy coating can form the basis of an anti-tampering coating.

13.3.1.3.3 Advantages and Disadvantages
Epoxies are useful at moderately high temperatures, up to about 150 °C (302 °F). Because of their strength, they also provide mechanical support for components. The disadvantages of epoxy coatings are their usually pungent odor and the possibility of skin irritation. They are difficult to rework. Some formulations of epoxies are chemically delicate and do not cure properly in the presence of inhibiting compounds. Cure shrinkage is also of concern for fragile components; a softer buffer coating should be applied locally before the epoxy, particularly for assemblies that need to withstand wide swings in temperature.

When most epoxies are applied, a "buffer" material should be used around fragile components to prevent their damage from film shrinkage during polymerization.

Curing of epoxy systems takes place in up to three hours at an elevated temperature or up to seven days at room temperature.

Two-part epoxies begin to chemically react as soon as they are mixed. Consequently, there is a finite pot life and processing window. The processing properties of the epoxy mixture will change throughout the processing window.

Single-part epoxy resin coatings with temperature-activated hardeners are also available. These coatings require cure temperatures higher than 66 °C (150 °F). Single-part UV-curable coatings are available, which minimize the need for exposure to elevated temperatures.

Single-part vinyl-modified epoxy compounds are available for special applications that require higher thermal performance properties. These compounds are also less brittle.

As for rework, epoxies are virtually impossible to remove chemically for rework since any stripper that will remove the coating may vigorously attack epoxy-potted components as well as printed circuit board laminate materials. Other effective ways to repair a board or replace components are to burn through the epoxy coating with a knife or a soldering iron or use of abrasive media.

To repair or replace a component, it is advised to refer to Section 10 of the IPC-HDBK-830 Conformal Coating Handbook for processes and methods [1].

13.3.1.4 Silicone Resins (Type SR)

Silicone coatings (SR) are usually available as 100% solid coatings that are very flexible, and stay that way over a wide range of temperatures.

Silicone coatings have good adhesion to a variety of surfaces, but contaminate the surface and, once applied, prevent other materials from adhering. Typical application is 0.002–0.008 in. (50–200 μm) thick. Silicone coatings are useful in high temperature and moisture-resistant environments, as seen in various automotive electronic applications.

13.3.1.4.1 Chemistry

The silicone polymer chain is based on an alternating silicon-oxygen backbone. Silicone coatings are available in three types, room temperature vulcanizing (RTV), UV cure, and catalyzing (addition) cure.

13.3.1.4.2 Properties

Silicones have relatively stable properties from −55 to +200 °C (−67 to 392 °F). Their CTE is higher than that of urethanes, but this is mostly offset by their lower modulus, so that the level of stress exerted on the parts is still relatively low. They have high resistance to moisture and humidity, as well as polar solvents.

13.3.1.4.3 Advantages and Disadvantages

Silicones are usable over a wide temperature range and are relatively easy to remove via mechanical or thermal means. If mishandled, silicones may contaminate the work area, causing adhesion problems on other Circuit Card Assemblies (CCAs). Careful process sequencing and process separation is necessary to prevent this.

Silicone coatings provide high humidity resistance along with good thermal endurance, making them desirable for assemblies with heat dissipating components such as power electronics. For high-impedance circuitry, silicones offer a low dissipation factor and offer good resistance to polar solvents.

Cross-contamination factors stemming from the use of silicones and the effects on other production processes no longer pose major concerns with the new solvent-free, low volatility chemistries that have been developed along with the usage of proper housekeeping practices.

Secondary cure for the UV-curable versions is accomplished with an effective ambient moisture mechanism. It should also be considered that high temperature protection may generally demand that the silicone coating be cured at or near to the maximum temperature it is designed to withstand. Silicone coatings can be applied at large thicknesses (depending on viscosity).

To repair or replace a component, it is advised to refer to Section 10 of the IPC-HDBK-830 Conformal Coating Handbook for processes and methods [1].

13.3.1.5 Para-xylylene (Type XY)

Para-xylylene coatings (XY) are unique, since they are applied by vapor deposition rather than as a liquid. They provide excellent humidity, electrical, and chemical resistance.

Para-xylylene coatings are chemically stable and very resistance to chemical attack. Typical application thicknesses are 0.0005–0.002 in. (12.5–50 μm) thick. Para-xylylene coatings are often used in biomedical devices due to the inert character and uniform application of the coating.

13.3.1.5.1 *Chemistry*

Para-xylylene coating is supplied as a dimer, a polymer chain only two units long. During the vacuum deposition process, the dimer powder is gradually vaporized and is condensed onto the CCA, polymerizing with the material already deposited. Typical application is 0.0005–0.002 in. (12.5–50 μm) thick.

13.3.1.5.2 *Properties*

This coating is chemically inert and moisture-resistant. It is very thin, uniform layers are possible, with no pin-holes or voids. This coating also has very high dielectric strength. Due to the nature of the deposition process, no volatiles are generated.

13.3.1.5.3 *Advantages and Disadvantages*

This coating is the highest performing coating in many regards. It has the highest dielectric strength, and has the lightest weight, lowest outgassing, is mostly chemically inert, and most moisture-resistant. It provides the most uniform coating over, under, and inside parts where liquid coatings cannot be applied.

For this coating type, the coating process must be performed in a batch mode, using specialized coating equipment. This process will also require unique masking processes as the application is done under a vacuum. It may be impossible to mask adjustable components such as potentiometers adequately.

Both material characteristics and application technique yield XY conformal coatings that are uniquely different from the liquid applied coating types. The obtained coating film yields consistent thickness with true conformance to the assembly contour and is pin-hole and bubble-free. The XY film is also characterized by properties such as good dielectric, low thermal expansion, good abrasion resistance and outstanding chemical resistance, among others. This makes XY coatings a good choice for protecting circuits against effects from harsh environments, notably high humidity with condensation, intermittent immersion, salt fog, atmospheric pollutants and exposure to aggressive solvents.

This type of coating is frequently used in Food and Drug Administration (FDA) approved devices for medical and biomedical applications.

Additional staking or adhesive bonding of parts is required, since the thin coating provides little mechanical support.

A fluorinated version is also available that can maintain its properties at temperatures in excess of 400 °C (752 °F), has increased UV stability, and has a lower dielectric constant.

As for rework, this is very difficult; since the thin coating cannot be peeled, a micro-abrasion process is usually required to remove the coating. However, the true question is how to apply material to patch the area or should the assembly be re-coated.

13.3.1.6   Synthetic Rubber (Type SC)

Synthetic Rubber (SC) coatings are usually available as one-part compounds at a given solids percentage within a solvent transportation medium as an organic option to silicone coatings, providing a very flexible coating which stays that way over a wide range of temperatures.

Synthetic Rubber (SC) coatings are considered excellent moisture and environmental protection coatings with a conformal coating typical application being 0.001–0.005 in. (25–125 μm) thick. There are only a few coatings available as SCs, and are a new technology coating providing new material options on an ongoing basis.

The materials used in the application of these coatings, as well as the application methods, are similar to other solvent-based coating types.

Synthetic Rubber coatings can also be applied with traditional methods such as dip or spray coating.

13.3.1.6.1   Advantages and Disadvantages

As this classification of conformal coating is fairly new, there is limited information available to provide as deep a list as other conformal coating types, however, as per evaluations that have been completed, here is some information.

Synthetic rubber (SC) has properties similar to natural rubber, but with greater resistance to abrasion, wear, and water. Additional advantages over natural rubber include: low-cost alternative material to natural rubber, and thus it has a variety of market applications that this conformal coating material can be designed into for usage.

Other advantages as compared to natural rubbers are a superior low temperature flexibility, very good heat resistance and heat-aging qualities.

The only true disadvantage that is known at this time, is that not enough information has been reviewed that can either provide further advantages and/or concerns for this type of conformal coating as compared to other coating types.

13.3.1.7 Ultra-Thin (Type UT)

Ultra-Thin (UT) coatings cover the largest range of coatings available as UTs, with different chemistries, coating methods, and applications, depending on the type of coating required.

Ultra-Thin coatings are commonly applied similar to XY type materials, providing excellent coverage and moisture protection and are classified as any conformal coating below 0.0005 in. (12.5 μm) thick.

The materials used in the application of these coatings, as well as the application methods, can vary greatly from other coating types. While coatings less than 0.0005 in. (12.5 μm) can be deposited with the typical, solvent-based methods mentioned very thin coatings can be deposited using vacuum based methods, e.g. plasma, Chemical Vapor Deposition (CVD), Atomic Layer Deposition (ALD), etc. Vacuum deposition methods are highly beneficial when a conformal coating is required, as the obtained coatings have true conformance to the substrate structure, a relatively consistent thickness across 3D structures, and tend to not show pin-holes or defects. There are a wide range of materials for these coatings, such as monomers, oligomers, gases, etc., leading to significant variance in the type of coating formed.

Ultra-Thin coatings can also be applied with traditional methods such as dip or spray coating.

Due to the broad nature of the UT coatings and the application methods, many types of coating chemistries can be applied as UTs, e.g. amorphous hydrocarbon, fluoropolymer, silicon hydrides/carbides, etc. These chemistries can also be tailored for different coating properties, and as such, a defined list of chemical and physical properties can be very large. However, coatings can be selected based on the desired properties, such as strength, toughness, water repellency, thermal protection, moisture protection, chemical resistance, electrical resistance, and dielectric constant, with a range of applications just as broad.

For vapor deposited UTs, the lack of solvents and a coating curing step can make these coatings very useful, and combined with the wide range of applications and coating chemistries, they can be very beneficial when a conformal coating is required. Another benefit of UT coatings is that some do not require masking.

13.3.1.7.1 Advantages and Disadvantages

As this classification of conformal coating is fairly new, there is limited information available. However, due to the similarity of powder coatings that have been used for decades, there are various similarities that can be compared, here is some information.

Ultra-Thin (UT) are a very large group since the main driver of distinction for this group is the thickness that these materials are applied, so they can cover several chemical types. Due to this wide range of usage, dependent on the specific chemistry used, these may slightly differ, but the following should be common among the group, which cover the following aspects.

Environmental As most UT coatings emit zero or near zero volatile organic compounds, produce less hazardous waste than conventional liquid coatings, generally have fewer appearance differences between horizontally coated surfaces and vertically coated surfaces than liquid coated items, and are solvent-free, no additional time is required for the flash-off of solvents.

Application As UT coatings can produce much thinner coatings than conventional liquid coatings without running or sagging this also ensures complete coverage, even on complex shapes.

Technical Some UT coatings due to their application and chemistry aspects will provide better corrosion resistance, mechanical and chemical performance and electrical insulation capabilities as compared to other coating types.

Productivity and Costs For some UT coatings their overspray can be recycled and thus it is possible to achieve nearly 100% usage of the material, which is cost-efficient on small batches or single items. It is thus ideal for job coaters and is ready to be used (no material preparation of stirring, mixing, or thinning) by being directly fed from the incoming containers.

The only true disadvantage that is known at this time, would be due to the thickness that this coating type has, it could be sensitive to abrasion, otherwise there is not enough information that has been reviewed that can either provide further advantages and/or concerns for this type of conformal coating as compared to other coating types.

13.4   Preparatory Steps Necessary to Ensure a Successful Coating Process

Most coating problems can be traced to improper preparation. The four preparation steps are:

- Assembly: cleaning to remove flux, oils, and contaminates
- Assembly: masking to ensure coating is applied only where desired
- Preparing the surface or priming to promote adhesion of the coating
- Baking out to prevent moisture from being trapped under the coating

13.4.1   Assembly Cleaning

Thoroughly cleaning the CCA before conformal coating is the best way to ensure proper adhesion of the coating. There are various cleaning chemistries which can be used to remove different kinds of fluxes.

Besides removing flux, the cleaning process removes any contamination such as mold release agents on plastic part bodies, fingerprints, etc. With the design and release of No-Clean, "low-solid based" soldering chemistries, there has been a reduction in cleaning processes, but with recent years, the cleaning of no-clean is gaining interest again.

No-clean fluxes reduce the need for waste water and waste solvent treatment. For boards that require conformal coating, the benefits of a no-clean system without assembly cleaning must be balanced with the risks of poor coating yields. A minor change in flux or coating chemistry could render the coating incompatible with the no-clean flux being used, causing delamination, cure inhibition, or other defects. In general, coating results are not as good with no-clean flux as with standard flux plus cleaning, yet in some cases may still be adequate to do the job.

13.4.2   Assembly Masking

To control the areas where coating is applied, and areas to be free of coating, some kind of masking is usually needed. Masking tapes or adhesive circles may be used to keep flat areas and holes free of coating. Since the tape needs to be removed completely after the coating operation, the tape adhesive must be capable of withstanding the cure temperature without peeling or becoming permanently bonded to the surface. To mask irregular surfaces, latex masking material is used. This material is applied with automatic or manual dispensing equipment and is cured in an oven before the coating operation is performed. It then peels off easily after coating. For high-quantity production, boots or covers can be molded to fit over the areas that do not require coating. These are reusable, easy to install, and produce well-defined repeatable results. Rather than being masked, areas may be sealed with a permanent bead of a thixotropic (non-runny) adhesive. This eliminates the demasking step, and is useful for complicated geometries such as the back of connectors and the perimeter of ball grid arrays (BGAs). The only aspect is that this is a permanent mask and should be reviewed with the design and usage of the assembly in the field to ensure there are no concerns.

13.4.3 Priming and Other Surface Treatments

The adhesion of some coatings – notably epoxies, silicones, and para-xylylenes – will be improved if the CCA surface is primed before coating. Primers are applied as a very thin coat using dip or spray (generally the thinner the better) and oven-dried to drive any remaining solvent from crevices, vias, and such. If the primer fails to wet all areas of the board, the coating will not wet either but will easily delaminate instead. Such surfaces are not compatible with the primer, and an adjustment in primer chemistry is required. For difficult surfaces, such as large fluoropolymer parts or substrates, a more aggressive measure such as plasma-etch or mechanical abrasion may be needed to promote adhesion. Mechanical abrasion (grit blasting) may be used to roughen troublesome surfaces to promote coating adhesion. It may be easier to do this step as the parts are fabricated (in the case of mechanical parts) or as part of the make-ready process before assembly. Micro-abrasion can be used with proper care on the finished assembly as long as electrostatic discharge is controlled by the design and use of the abrasion equipment, and is monitored daily to ensure that it is working properly. Micro-abrasion is also a viable method for removing coating.

Another aspect that can impact the ability of a coating to adhere to a substrate is the surface energy of the substrate. In general, a rough substrate, having high surface energy, will exhibit better wetting and coating adhesion than a smooth substrate, having low surface energy, all other factors being equal (which of course, seldom happens). A good example is comparing the adhesion of a conformal coating to the surface of a ceramic DIP component versus a plastic bodied DIP of the same configuration. Smooth plastic surfaces, often compounded by the presence of fluorinated (e.g. Polytetrafluoroethylene [PTFE]) or silicone mold release agents from the injection molding process, may show a higher incidence of coating delamination in thermal cycling due to surface energy conditions.

13.4.3.1 Measuring Surface Energy

There are a number of general methods for measuring or estimating the energy of a surface prior to coating application.

1. *Dyne pens.* These are liquid-filled pens, similar to highlighting markers (Figure 13.1). Each pen body contains a solution with a calibrated

Figure 13.1 Standard dyne pen samples.

surface tension. The most commonly available commercial sets start with 30 dynes cm^{-1} and end with 44 dynes cm^{-1}. In use, the marker tip is drawn across the substrate surface. If the solution does not immediately "bead up" or dewet, then the next higher pen is tried in an adjacent area. If a solution of 34 does not dewet and a solution of 36 does dewet, then the surface energy is judged to be 35 dynes cm^{-1}. This method is relatively inexpensive, very quick to use, but is variable and requires operator judgment.

These commercially available pens are very economical and it is often easier to purchase a set than to make your own solutions. However, if solutions are desired for measuring outside the range of 30–44 dynes cm^{-1}, individual solutions can be formulated using the information found in ASTM-D-2578, titled "Wetting Tension of Polyethylene and Polypropylene Films" [2]. This test method focuses on measuring the surface energy of polyethylene and polypropylene films, but the test method includes the procedures on mixing the necessary solutions.

The methods in ASTM-D-2578 use combinations of reagent grade formamide (HCONH$_2$) and reagent grade ethyl cellosolve (CH$_3$CH$_2$OCH$_2$CH$_2$OH) to create solutions ranging from 30 to 56 dynes cm^{-1}. It is advised that both ethyl cellosolve and formamide are toxic and pose exposure risks.

As a general rule of thumb when examining the surface energy of a substrate with dyne pens, the following general guidelines can be used (Table 13.1).

Table 13.1 General guidelines when using dyne pens.

Surface energy, dynes cm^{-1}	Expected result
Above 40	Adhesion expected to be good
35–40	Adhesion generally good, with some intermittent delamination under severe conditions
30–35	Adhesion generally poorer, with increasing incidence of delamination under severe conditions
Below 30	Coating adhesion poor in any climatic testing

Many factors affect coating adhesion, but all other factors being equal, this scale can be used to determine an overall degree of risk for coating delamination during environmental stress screening tests.

To show how surface energy can have an impact in production and reliability, the following case studies are provided for the conformal coating process.

13.4.3.1.1 Case Study 1

Company A used a solder mask coated coupon as part of its process control methodology for the conformal coating process. Each day, operators would coat the coupon using methods similar to those being used on production hardware. Following the standard cure methods, the coupons would be allowed to sit for one hour. The coupons would then have a $1'' \times 1''$ (2.5 cm × 2.5 cm) X scribed into the coating using a sharp razor blade. A standard tape was applied to the surface of the scribed coating and then pulled off. The coating was to have no loss of adhesion in the scribed area. While the company had numerous qualified solder masks, surface energy was never part of the qualification protocol, and the coupons were standardized on the most common solder mask used. Failures in the tape test were seldom encountered, but always passed in a subsequent retest.

Then, during the same week, multiple failures began to occur in the process control coupons at multiple manufacturing sites. While processes between sites were similar, it was highly unusual for such failures to occur. The quality assurance policies of company A required that hardware be visually examined after product burn-in testing. No evidence of any coating delamination was found for the hundreds of complex assemblies examined. The loss in productivity and delays in shipment was high.

After extensive root cause analysis, it was found that the solder mask on the control coupons had been changed. The drawings allowed for any solder mask that was on the approved list, and only tribal knowledge was used in ordering the coupons. The mask was always assumed to be the same. The previous surface energy of the solder mask was in excess of 44 dynes cm^{-1} when using the dyne pens. The surface energy of the new solder mask was 30 dynes cm^{-1}. The inventory control system for the coupons was done in large lots to keep the per part prices low. The production problems began when the supply of the old coupons was exhausted and the supply of new coupons began.

Fearing that this meant a large-scale quality problem, a sampling of unpopulated circuit boards was made for all vendors and all solder masks, again using the dyne pens. All bare boards were found to have surface energies in excess of 40 dynes cm^{-1}, which was deemed to be acceptable. Circuit boards go through a much wider range of chemical processes during board fabrication, such as plating and etching, and so have a higher surface energy as the mask surface is etched at the microscopic scale. In contrast, the coupon processing methods were very simple and did not have such exposure, creating the wide disparity in surface energy between coupon and bare board, even for the same solder mask.

13.4.3.1.2 Case Study 2

As noted, one of the elements which can impact the ability of a coating to adhere to a substrate is the surface energy of the substrate. In general, a rough coating, having high surface energy, will exhibit better coating adhesion than a smooth coating, having low surface energy, all other factors being equal.

To validate this surface energy adhesion, there are a couple of methods which can be used, such as the IPC TM-650(2.4.1.6) adhesion test [3] and the water-break test [4]. The adhesion test is a destructive test method that is completed by etching a cured conformal coated surface into a grid structure and placing a section of tape with a calibrated adhesion strength and measuring the amount of material loss. Once this test is passed and there is no adhesion loss noticed, the assumption is that there is full adhesion with no potential for conformal coating peeling. However, this is not fully true. If the IPC adhesion test is passed, and the surface energy below the conformal coating material is below 40 dynes cm^{-1} there can still be a field failure for conformal coating adhesion as noted earlier within the table of surface energies (Table 13.1). This potential field failure aspect is only presented as a cured conformal coated material on an assembly when environmentally tested within a cycle consisting of high and low temperatures. In that, as the assembled product is cycled the conformal coating material pulls away from the product board surface area during the heating process and starts to form areas of peeling/cracking within the coating area during cooling (Figures 13.2 and 13.3).

One relatively inexpensive qualitative method for examining the surface energy of a substrate is a test commonly called a "water-break" test. This test is

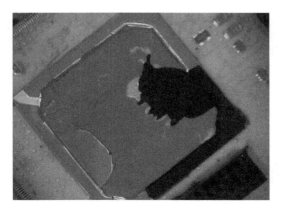

Figure 13.2 Cracked coated areas.

Figure 13.3 Peeling coated areas.

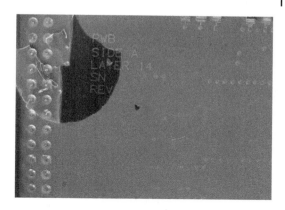

commonly used in the printed wiring board fabrication industry to examine the surface energy of a substrate prior to the application of solder mask. The test is also commonly used prior to plating and anodizing operations. The formal test method can be found in ASTM-F-22, "Standard Test Method for Hydrophobic Surface Films by the Water-Break Test" [4].

This test method covers the detection of the presence of hydrophobic (non-wetting) films on surfaces and the presence of hydrophobic organic materials in processing ambients. When properly conducted, the test will enable detection of molecular layers of hydrophobic organic contaminants. On very rough or porous surfaces, the sensitivity of the test may be significantly decreased.

The test methodology indicates the following:

- Take a cleaned and dried part and set it in a vertical position.
- Use a spray bottle containing distilled water.
- Spray the part two to three times from at least 6″ (15 cm) away.
- If the part is clean and free of oily residue, the water spray should sheet off.
- If some oily residue remains, the water will tend to adhere in droplets or patchy, non-uniform slicks.

Alternatively, several drops of distilled water are applied to the cleaned surfaces. If the surface is inadequately cleaned, the spherical form of the drop is largely retained, and the surface must be cleaned once more. If the water runs on the treated surface, then wetting has been satisfactory and the part is ready for subsequent operations.

It is cautioned that this test will not work with certain wetting agents, which will cause a continuous liquid film to spread over the metal surface even though the surface still has grease or oil on it.

This difficulty can be overcome in certain cases by dipping the parts in a dilute sulfuric or hydrochloric acid solution (the acid inhibits or destroys the "spreading

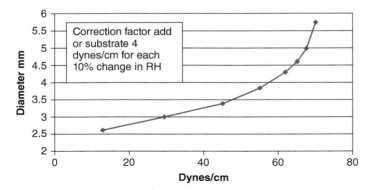

Figure 13.4 Critical surface tension for 5 μl of deionized water drop at 50% RH.

power" of the wetting agent so a more correct test result is obtained) and then rinsing again. It should be kept in mind that even with a good water-break test the surface still may contain soils/contaminants.

A practical method for direct surface energy measurement and A to B comparison revolves around the simple concept of placing a known volume of a liquid on a surface and then simply measuring the resulting drop diameter. With a micro-liter pipette, it is fairly easy to place many small drops on a substrate and then measure the resulting drop diameter by any number of methods such as optically using a microscope, or mechanically with a caliper. If one desires to go through a detailed geometric proof, the drop diameter can be directly related to the contact angle, which in turn can be related to the surface energy. Figure 13.4 shows this relationship for a 5 μl DI water droplet at 50% RH. It is fairly quick and low cost to place a number of 5 μl drops on a surface, measure the diameter and the relative humidity, convert drop diameter to surface energy and calculate a fairly accurate estimation of the substrate surface energy. The application of this test method is not limited to just DI water. Correlations for other fluids and/or curing materials can be obtained.

13.4.3.1.3 Case Study 3
The following example of applied surface energy shows how these test methods [3, 4] may be applied to a silicone conformal coating as well. This example shows the interaction between materials (flux, solder mask) and process (the wave soldering process). Just like the previous DI water example (Figure 13.4) this example applies a known and controlled volume of a conformal coating and then measures the resulting drop diameter after the conformal coating cures.

A 5-μl drop of a conformal coating was applied to various substrates which were previously exposed to different fluxes (same basic flux formulation with

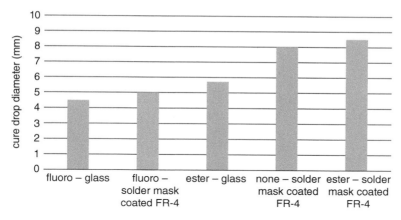

Figure 13.5 The effect of flux surfactant type on conformal coating wetting.

different surfactants). This example shows that a no-clean flux formulation with a fluoro-surfactant had a major impact on the conformal coating wetting (Figure 13.5). The fluoro-surfactant significantly reduced wetting when compared to another potential flux ester based surfactant. Again, the amount of wetting directly corresponds to the relative drop diameters (i.e. a large drop diameter indicates improved wetting). In this example, to solve the coating wetting issue, the production wave soldering flux formulation was changed.

13.4.3.2 Water Drop Contact Angle

One quantitative test method for the determination of surface energy is to measure the contact angle of a droplet of distilled or deionized water on the substrate surface (Figure 13.6). A substrate with a high surface energy will have a large contact angle (water will bead up). A substrate with a low surface energy will have a low contact angle (water wets out). This method is more quantitative and repeatable than the dyne pen method, but requires a greater capital investment.

Figure 13.6 Diagram of the pertinent measurements for contact angle.

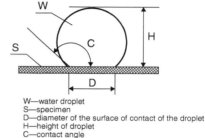

W—water droplet
S—specimen
D—diameter of the surface of contact of the droplet
H—height of droplet
C—contact angle

ASTM-D-7334, titled "Standard Test Method for Measurement of the Surface Tension of Solid Coatings, Substrates, and Pigments using Contact Angle Measurements," [5] outlines one method used to determine contact angle and a calculation of resulting surface energy.

ASTM-D-5725, titled "Standard Test Method for Surface Wettability and Absorbency of Sheeted Materials Using an Automated Contact Angle Tester," [6] shows an automated (though capital intensive) method of determining contact angle.

A manual method of measurement makes use of a device called a goniometer. An example of such an instrument is shown in Figure 13.7. A substrate is placed on the center pedestal. A calibrated syringe is used to dispense a droplet of distilled/deionized water on the surface of the substrate. A light source behind the water droplet illuminates the water droplet. A front viewer is set level with the surface of the substrate. Reticules in the eyepiece are adjusted to measure the contact angle of the drop. A hydrophobic surface will have a high bulbous shape. A hydrophilic surface will have a flat and wide droplet shape.

13.4.4 Bake-Out

Since the coating is intended to be waterproof, it is imperative that all moisture be baked out before the coating is applied, especially for hydroscopic substrates such as polyimide. If moisture is left in place, it can cause corrosion of traces and/or parts and will promote dendritic growth between conductors, as well as enabling

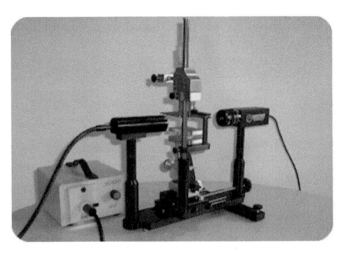

Figure 13.7 Goniometer. Source: Courtesy of Ramé-Hart Instrument Company.

the growth of CAF along the glass fibers in the weave of the PWB. A bake of 93 °C (200 °F) for four hours will be sufficient to drive out moisture from the CCA. With certain component designs, a lower temperature and different qualification time may be required.

13.5 Various Methods of Applying Conformal Coating

Conformal coatings can be applied by a variety of methods, including

- Manual coating with a brush
- Dipping the board into the coating
- Hand spraying with an aerosol can or handheld spray gun
- Automatically spraying the board with coating as it passes by on a conveyor
- Selectively applying coating to certain areas of the board with a robotic nozzle
- Vapor depositing para-xylylene onto the board with specialized equipment

13.5.1 Manual Coating

Manual coating allows coating to be applied wherever desired, with no setup time and low equipment cost. It is still the preferred method for touching up thin spots in a bulk applied coating and for coating replaced components. It can be used to reach areas where spray or automatic nozzles cannot reach, and works well for smaller, one-of-a-kind items. Hand painting is the most variable of all methods.

13.5.2 Dip

Dip coating works best for coatings that have a long pot life at room temperature, such as solvent evaporation cured coatings and one-part materials. Since the unused coating remains in the dip tank, a long (or infinite) pot life keeps the waste down to a minimum. The viscosity of the material in the bath should be checked daily, using a flow cup viscometer or other method, and solvent added to keep the viscosity of the material within the acceptable range. For a given dip speed, a lower viscosity will produce a thinner coating. The dip machine (see Figure 13.8) will have bars or hangars on which to hang the board to be coated. Insertion speed should be slow enough to prevent the formation of bubbles as the coating flows around and under surface mounted parts, etc. Withdrawal speed should range between 1 and 6 in. min^{-1} (2.5–15.2 cm min^{-1}) and will dictate the finished coating thickness; a slower withdrawal will produce a thinner coating. Dip coating cannot be used multiple times to build up thicker layers of acrylic coating, since solvent

Figure 13.8 Dip coating equipment.
Source: Courtesy of SCS, a KISCO Company.

in the bath will loosen or dissolve previously applied layers of acrylic. Any areas that are not to be coated must be completely masked before dipping. Assemblies can be partially dipped to leave the top edge of the board uncoated if the board is designed so that all components that need coating are below the level of the bath.

13.5.3 Hand Spray

Hand spraying is used for all types of liquid coatings. The boards to be coated are sprayed several times; the angle is changed 90° each time so that complete coverage is achieved. Tall components may shadow or block the spray from lower profile components behind them, so care must be taken to ensure all areas of the board are fully coated. A witness strip is often sprayed at the same time for use in measuring the coating thickness. As with the dip method, viscosity control is a key factor to reducing bubbles, cobwebs, and improper coating thickness. Two-part materials are mixed immediately prior to filling the spray gun container, or are mixed in line if a continuously fed mixing gun is used. Solvents may also be used to modify the viscosity of the coating being applied. Single-part materials (particularly acrylics) are also available in aerosol cans, eliminating mixing altogether.

13.5.4 Automatic Spray

For high-volume coating application, a semi-automatic spray machine is used to eliminate the dependency on the operator and allow in-line coating. The boards enter the machine on a paper-covered conveyor and are sprayed by a rotating or reciprocating spray head. The motion of the spray head(s) is designed to cover a given width of the belt completely from all angles. Shadowing is of greater concern with semi-automatic spray equipment, since special treatment cannot be given to any one area of the board. Full masking is required to keep the coating out of areas that should not be coated. Once the board is coated on one side, it is partially cured (or fully cured), turned over, and coated on the other side.

13.5.5 Selective Coating

The selective coating process utilizes a robot similar to the automated pick and place machine to dispense coating only where it is needed. Four- or five-axis machines can be used to coat the sides of tall parts and around corners. The greatest benefit to selective coating is that it drastically reduces the amount of labor needed for masking and mask removal; areas that need to be free of coating simply have no coating applied there. Coating utilization is highest of all the methods, since almost no coating is lost in continuous production. Spray patterns may be atomized, spraying droplets onto the board and providing a thinner coating (see Figure 13.9), or continuous, applying a stream or sheet of liquid to the board (Figure 13.10), producing a well-defined boundary between the coated and uncoated areas and eliminating the need for masking. In the same way as the semi-automatic sprayer, the board is tack-cured and then turned over and coated on the other side. An automated board flipper may be used to accomplish this to automate the entire coating process.

13.5.6 Vapor Deposition

Vapor deposition is only used with para-xylylene (XY) coatings, and requires specialized coating equipment. In the evaporator, a measured amount of dimer is placed in a crucible. The coating cycle is then started, and the chamber is pumped down to an almost complete vacuum (see Figure 13.11). During the deposition process, the dimer powder is vaporized and condenses onto the batch of CCAs,

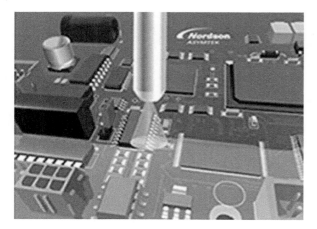

Figure 13.9 Atomized spray applicator. Source: Courtesy of Nordson-Asymtek.

Figure 13.10 Film applicator nozzle. Source: Courtesy of Nordson-Asymtek.

Figure 13.11 Vapor deposition equipment. Source: Courtesy of Comelec.

forming a uniform coating covering all corners and edges evenly. The vapor deposition process drives the vapor into any crevices and openings. Vapor deposition has the most stringent masking requirements, since leaks in the masking will still admit vapor that will contaminate the surface behind, and air pockets inside masking boots or tape may dislodge the boot in the vacuum and create leaks as the air escapes.

13.6 Aspects for Cure, Inspection, and Demasking

The finishing steps require most of the labor involved in the conformal coating process. Three finishing operations are:

- Curing the coating
- Inspecting under UV light to verify complete coverage and touch-up
- Removing mask materials – demasking

13.6.1 Cure

There are five different formulation types that liquid chemistries are available to be cured within, with each formulation having different cure mechanisms, as shown in Table 13.2.

13.6.1.1 Solvent Evaporation

Systems that are cured by solvent evaporation can be air dried, or heated slightly to drive the solvent out faster. Care must be taken not to overheat the freshly coated assemblies; the coating will skin over and cause bubbles. Long-term exposure to the unevaporated solvents trapped in the coating may also be deleterious to the materials in the parts and PWB. Drying of solvent cure assemblies must be done under a ventilation hood or other properly ventilated area.

Table 13.2 Cure mechanisms (liquid coatings).

Cure format	Time	Comment
Solvent evaporation	Hours	Requires ventilation
Room temperature vulcanization (RTV)	Hours	Requires humidity/moisture
Heat cure	Minutes	Requires oven. Requires ventilation
UV cure	Seconds	Requires ultraviolet source. Requires ventilation
Catalyzed	Minutes	Requires mixing process

13.6.1.2 Room Temperature Vulcanization (RTV)

Some silicone-based coatings use the RTV cure process, which consumes moisture from the surrounding environment as part of the cure process. Some means of humidity control is required; if cure is attempted in a dry oven, the material will not cure properly, if at all.

13.6.1.3 Heat Cure

Oven-curable systems use an extended bake, ranging from 15 minutes to several hours or longer, for the material to cure fully. Thermally activated systems require the increased temperature to start the curing process, whereas thermally accelerated systems will cure eventually at room temperature but are baked to speed up the process and improve material properties. Generally, the hotter the oven, the quicker the cure, and the harder the finished material. However, care must be considered to ensure that there are no thermally sensitive components that are added prior to the coating process when higher temperatures are used to expedite the cure.

13.6.1.4 UV Cure

For high-volume production, UV-curable materials are used. Immediately after the coating machine is a booth with high-intensity ultraviolet lamps to illuminate the assembly. After several seconds, the assembly has cured enough to be able to be handled at the next stage of assembly, such as coating the opposite side. UV-curable materials do not cure fully under components, where the coating is not illuminated. Therefore, most UV-curable materials also contain a secondary curing mechanism, such as heat or moisture cure, so that at the end of the line the CCA can be fully cured in a single baking or moisture step or naturally come to full cure in ambient conditions inside the finished product. Since different coatings are sensitive to different wavelengths of UV light, it is important to coordinate the purchase of the light source with the supplier of the coating.

13.6.1.5 Catalyzed

The catalyzed process involves mixing two parts together just before application, to initiate polymerization and cross-linking. This process occurs slowly at room temperature but may be accelerated by elevating the temperature in an oven. Catalyzed coatings may be affected by contaminates present in the PWB or parts being coated, resulting in uncured areas or peeling coatings. Since the pot life of catalyzed materials tends to be short, it is imperative that the mixed material be cleaned from sprayers, tubing, and so on, before it thickens and hardens.

Figure 13.12 Conformal coating AOI bubble capture. Source: Courtesy of Nordson-YESTech.

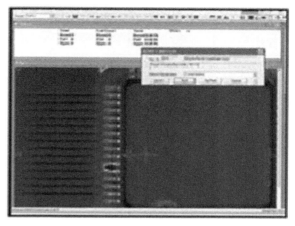

13.6.2 UV Inspection

Most conformal coatings contain a fluorescent dye that will glow under UV light and show where the coating has (or has not) been applied. Areas without coating can be touched up at this time. The area most susceptible to insufficient coating are corners of parts, where the coating can flow away and leave little behind; the protruding ends of through-hole part leads; and shadowed areas under or behind larger components. If large flat areas are uncoated and show beads of coating, there is a wetting problem due to improper surface preparation or a material incompatibility. The coating should also be inspected for improper thickness, bubbles, trapped material or debris, and incomplete curing (see Figure 13.12).

Recently, with the migration of conformal coating into new market segments, new suppliers have developed camera-based automated optical inspection (AOI) systems that can measure the presence of the coating, detect defects such as insufficient coverage and voids and calculate the coating thickness of the flat, unencumbered cured surface. Completely transparent coatings yield a low optical signature, the signal the AOI system detects under ambient lighting. The resolution of the AOI is frequently not high enough to see the clear film.

Fluorescent coatings provide an enhanced optical signature when illuminated, making AOI inspection easier. AOI systems utilize filters to eliminate background light not emanating from the fluorescent coating. This can be a fast solution for in-line verification of coverage.

13.6.3 Demasking

Mask materials should be removed carefully; to prevent peeling, it may be desirable to score or cut the coating at the edge of the mask so that the coating tears along the correct line. It is necessary to ensure that the mask is removed entirely and that no residue has been left behind on the board. Latex mask that has been overheated will be sticky and difficult to remove completely, as will overheated masking tapes.

13.7 Repair and Rework Processes

Since the different varieties of available coatings vary widely in their properties, removal methods must be evaluated and developed on practice boards first before being used on deliverable CCAs. The coating can then be removed as easily as possible, with the least risk to the CCA. Several methods to remove conformal coating are:

- *Chemical.* Using a solvent to loosen or dissolve the coating
- *Thermal.* Using a hot soldering iron or hot-air nozzle to solder through the coating while simultaneously removing the electrical part
- *Mechanically.* Cutting, abrading, or picking away the coating
- *Abrasion.* Etching the coating away using rarefied gas plasma (typically for para-xylylene [XY] coating only)

13.7.1 Chemical

The easiest conformal coatings to remove are acrylics, since they do not react when they are applied. For local coating removal, the area to be repaired can be cotton swabbed several times with the proper solvent. Care must be taken to keep the solvent only in areas that are to have the coating removed. The entire board may be stripped of coating with a solvent soak. This is most useful on acrylic coatings, but is also possible for some urethanes and silicones, which swell and loosen when soaked. It should be noted that this process may damage the components or solder joints on the board, which will be stressed as the expanding coating pushes up the components. The solvent may attack the components and PWB itself. Epoxy coatings cannot be removed with solvent or chemical strippers, since many parts and the PWB are also made of epoxy.

13.7.2 Thermal

Thin coatings may be loosened from solder joints with a soldering iron or a hot-air reflow tool, while the faulty part is being desoldered. The area is then cleaned of charred and/or damaged coating, and the repair completed. Although this method is simple, it usually produces toxic gases, and must be performed with adequate ventilation. The decomposed coating is also difficult to clean off from the board and soldering iron/reflow tool.

13.7.3 Mechanical

Mechanical means such as scraping and picking are necessary to remove all coatings, even if the bulk has been removed via one of the preceding methods. Softer coatings such as polyurethane coatings (UR) or silicone coatings (SR) can be scraped away with a wood implement, so as not to cause damage. A soft rotary tool may be used for larger areas. The board may be heated slightly to make the job easier. For harder coatings, a hot knife is used to cut through the material.

13.7.4 Abrasion (Micro-Abrasion)

The principle of micro-abrasive blasting is based on projecting a small particle onto the coating surface with enough force to mechanically abrade the surface and eventually remove the coating. Balance must be achieved between the abrasion of the coating surface and abrasion of the printed wiring board surface beneath the coating.

This coating removal method uses a micro-abrasive blasting system and a very fine soft abrasive powder. The powder is propelled through a small nozzle toward the area where the coating needs to be removed. The nozzle is positioned around the base of component leads to selectively remove the coating. This technique requires practice in order to not damage the substrate beneath the coating along with contamination by the abrasion media to the surrounding board areas.

Micro-abrasive blasting will generate substantial static charges. Proper ESD measures need to be used in order to ensure that static charges are handled appropriately.

13.7.5 Plasma Etch

An extended plasma etch will remove para-xylylene (XY) coatings from large areas, since the plasma erodes the surface evenly. Since the plasma will chemically change the surface (leaving behind oxides, reduction products, and ash), compatibility of the plasma-stripped surface with the new coating must be verified.

13.8 Design Guidance on When and Where Conformal Coating is Required, and Which Physical Characteristics and Properties are Important to Consider

When designing a conformal coating system into the product, the designer needs to consider a number of questions:

- Is the coating required for this product, or should it be omitted for cost savings or other reasons?
- What material properties are most important for this application, and which materials will provide the best balance?
- What areas should not be conformally coated?

13.8.1 Is Conformal Coating Required?

One of the basic questions to consider is whether or not to coat the particular assembly. Generally, this will be a trade-off between reliability and cost.

13.8.1.1 Why Use It?

Customers may flow down requirements for conformal coating for industries where it is commonly used such as military or aerospace electronics. The following are some of the reasons to consider the use of conformal coating:

- The end products are used in humid or condensing atmospheres, such as electronics boxes for military and ship board applications. It should be noted that although conformal coatings are moisture-resistant, they are not waterproof, and should not be used for items that will see extended or continuously wet conditions. For those applications, a watertight enclosure should be used with the boards conformally coated inside. Conformal coating will protect against condensation, such as when the product is brought from a cold environment into a warm humid one.
- Clean conformal coating does not support mold and fungus growth.
- The end products can become contaminated by dirt and debris. Conformal coating is especially useful in space applications, since loose debris will float readily from area to area inside the electronics box.
- Conformal coating greatly improves the board's capability to withstand high voltage, especially at altitude or low pressure, since an electrical arc now has to travel through two layers of insulation rather than simply jumping from one trace to its neighbor.
- Conformal coating protects against tin whiskers. Pure tin plating has the tendency to extrude long (>1 mm) whiskers of pure tin, which can connect to neighboring part leads and cause short circuits. Conformal coating provides protection against short circuits caused by tin whiskers. Even if a tin whisker can force its way through a layer of coating on the way out, it cannot force its way back through the coating on the neighboring lead without bending and buckling
- Conformal coating provides mechanical support for components during shock and vibration. Components weighing more than 1/4 oz (7 g) per lead should be supported by other means as well, such as staking with a bead of adhesive along the side.

- Conformal coating may be made opaque, and thus provide a level of security. A thick opaque epoxy coating will prevent the observer from determining what is under the coating.

13.8.1.2 Why Not Use Conformal Coating?

Most of the reasons for not coating on an assembly stem from the recurring cost of cleaning, masking, demasking, touch-up, etc. The target cost of the product may not support the extra cost incurred by the coating, and if reduced reliability of the uncoated assembly is still acceptable, the rational decision is not to coat the assembly. Since the coating is a dielectric material, it changes the impedance of high-speed traces on the surface of the PWB. Boards with micro-strip are usually not coated for this reason. Radio frequency (RF) PWBs for severe environments can still be coated if the traces are buried micro-strip or strip-line, which are unaffected by the additional coating on the board surface.

13.8.2 Desirable Material Properties

When selecting a conformal coating, the entire suite of material properties must be considered in order to choose the best coating for the application. Unfortunately, there is no "one size fits all" solution, but many different coatings, each suited to different applications, can be compared in various tables and formats, one example which is shown in Table 13.3 (with a ranking of 1 for good and a ranking of 5 for poor).

The following are some of the properties to consider:

- *Ease of application and producibility.* The intended production environment may determine the cure process and application process that is best for the job, and will narrow the field of alternatives.

Table 13.3 Standard conformal coating material comparison.

Coating classifications Coating properties	Acrylic (AR)	Polyurethane (UR)	Epoxy (ER)	Silicone (SR)	Para-xylylene (XY)
Abrasion resistance	4	3	1	5	2
Humidity resistance	1	1	4	1	4
Temperature resistance	3	4	4	1	2
Mechanical strength	4	1	1	4	1
Thermal conductivity	2	2	2	2	3
Reworkability	1	3	3	2	5

Ranking: 1–5.

- *Compatibility*. The coating must not be inhibited from curing by any material on the completed assembly. Often the best way to address this is to ask the manufacturer and try out the coating in question on a sample of the materials that are going to be used.
- *Moisture resistance*. Sometimes called moisture and insulation resistance (M&IR), is a measure of how well the insulation characteristics are maintained when exposed to elevated conditions of temperature and humidity. A material with a low M&IR value would be a poor choice to protect a circuit in a high humidity end-use environment.
- *Dielectric constant*. For RF and high-speed circuits, a material with a low dielectric constant will change circuit performance less than a material with a higher one.
- *Dielectric strength*. For high-voltage applications, materials with a high dielectric strength should be used. For a valid comparison, tests must be performed on samples of equivalent thickness, since thin samples will break down at a higher Volts/mil than thick samples of the same material.
- *Insulation resistance*. Also known as resistivity, insulation resistance is of concern in high-voltage or high-impedance circuits, where the resistance in the circuit is very high (>10 MΩ) and is significant compared to the resistance of the coating and PWB laminate. For these applications, a coating should be chosen with high (>10^{15} Ω-cm) resistivity and excellent moisture resistance.
- *Hardness*. For products that need to withstand cold temperatures <−10 °C [<14 °F], a soft coating is recommended so that at the cold temperatures the coating will not exert excessive force on fragile components (see Figure 13.13). For products that need good abrasion resistance and protection from external handling damage, a harder coating must be used.
- *Glass transition temperature (T_g)*. For assemblies that need to withstand cold environments, a coating should be selected that either has a T_g below the coldest temperature required or a T_g of 25 °C (77 °F) and above. Coatings that have a low T_g in the range of operation will suddenly get hard at cold temperatures and impose high stress on the surrounding components.
- *Elongation*. Materials with high elongation are resistant to cracking and abrasion.
- *Reworkability*. Acrylic coatings (AR) are the easiest to rework, whereas para-xylylene (XY) and epoxy coatings (ER) are the hardest.
- *Abrasion resistance*. This is the ability of the coating to withstand damage from handling and later assembly operations, as well as in the field.
- *Flammability*. For assemblies that need to withstand prolonged exposure to sources of heat, a coating should be selected that has the correct flammability certification standards published by Underwriters Laboratories (ULs) which may be a requirement. Coatings that do not have this certification, if used within this condition could fail and cause a danger to the public.

Figure 13.13 Coating cracking (due to hardness).

13.8.3 Areas to Mask

During the design, the designer needs to consider all the areas and parts on the board that need to be uncoated and identify them on the assembly drawing, factory instructions, or other paperwork that the coating engineer will use. Although it is obvious to the design engineer that an expensive infrared (IR) sensor element should not be coated, it may not be so obvious to the engineer holding a drawing that indicates, "Coat the entire board," and nothing else. Being specific on the drawing clarifies what should be coated and how.

The following areas should be considered:

- Coating on connector pins can cause open circuits, and may interfere with the full mating of the connector. It is necessary to mask the back of the open back connectors as well as the front, to keep the coating from flowing along the back of the pin into the active area. Also, pin alignment depends on a certain amount of looseness on the pins, and filling the connector shell with coating will lose this feature.
- Coating should be kept from the edges of plug-in CCAs where the card guides travel. The coating will interfere with proper operation of the card guide and may prevent proper grounding of the card.
- Mounting holes for screws should be masked to keep the coating from filling or reducing the size of the hole and to ensure that the PWB is properly grounded.
- Test points should be masked so that they can be reliably accessed during testing.
- The adjustment screws of adjustable components and if the adjustable components have openings in them, the openings should be masked also.

- If the product contains sensors for air temperature, pressure, humidity, etc., the coating should not block the air from entering the sensor.
- Optical devices, light-emitting diodes (LEDs), infrared sensors, and light sensors may have their operation affected by the coating. This is especially true for UV and IR devices since the coating may be opaque to UV and IR wavelengths of light. The coating may also darken, cloud, or turn brown as it ages, eventually affecting the operation of the circuit. Connectors for fiber-optic cable should also remain clear of coating for the same reason.
- Components mounted in sockets should have the sockets masked and the components inserted after coating.
- Areas under BGA packages should not be underfilled with conformal coating but with an underfill adhesive particularly suited for that purpose. If no underfill is used, the edges of the BGA should be masked, or dammed with a permanent adhesive barrier to prevent the coating from entering under the BGA. Due to the shrinkage of conformal coating, during thermal cycling the BGA balls will be crushed and the reliability of the BGA will be adversely affected.

13.9   Long-Term Reliability and Testing

In general, the subject of long-term reliability and testing is a detailed area that the reader is advised to review the following references for further information [7–33].

There are various materials that are used, within various market segments, that each have their own advantages as well as obstacles to overcome. There is a large scope of end-use environments that are currently being investigated which will continue in the near future.

13.10   Conclusions

The chapter has covered a wide range of topics needed to understand a conformal coating process such as:

- EHS requirements, not just for the materials, but also the various application requirements.
- A detailed review of the five basic conformal coating types, and new emerging materials.
- The various preparation steps necessary to ensure a successful coating process, covering application, cure, inspection, demasking, repair, and rework processes.
- Design guidance on when and where coating is required, and which physical characteristics and properties are important to consider.

Overall, the topic of conformal coating is not as simple as applying a paint polish, but is a very interconnected and complicated subject that if approached for the first time, industry expert discussion and collaboration is an excellent starting point.

13.11 Future Work

The conformal coating sector is a dynamic and it is continuing to change to meet the various new aspects that the electronics industry is looking into now and in the future.

With this in mind, there was an industry round robin study that was just completed, that was one of the first of its kind that was a true blind test study. The study has also generated and confirmed what the industry produced that many individuals were either not aware of or had overlooked.

From the completion of this work, there are various aspects and cross-functional teams that are now looking at how to use this data to review the various industry aspects that need to be covered for the future; such as tin whisker mitigation, the transition to lead-free as well as how to mitigate from the various harsh environmental conditions that electronics need to work in [34–37].

References

1 IPC-HDBK-830A (2013). *Conformal Coating Handbook*. IPC International.

2 ASTM D2578-04a (2004). *Standard Test Method for Wetting Tension of Polyethylene and Polypropylene Films*. West Conshohocken, PA: ASTM International.

3 IPC-TM-650 2.4.1.6 (1995). *Adhesion, Polymer Coating*. IPC International.

4 ASTM-F-22 (2004). *Standard Test Method for Hydrophobic Surface Films by the Water-Break Test*. West Conshohocken, PA: ASTM International.

5 ASTM-D-7334 (2004). *Standard Test Method for Measurement of the Surface Tension of Solid Coatings, Substrates and Pigments Using Contact Angle Measurements*. West Conshohocken, PA: ASTM International.

6 ASTM-D- 5725 (2004). *Standard Test Method for Surface Wettability and Absorbency of Sheeted Materials Using an Automated Contact Angle Tester*. West Conshohocken, PA: ASTM International.

7 Wickham, M., Lewis, A., and Clayton, K. (2017). *1000 Days of Testing Tin Whiskering PCB Assemblies to Determine the Suitability of Conformal Coatings to Mitigate Against Shorting*, vol. 30-4. National Physical Laboratory.

8 Cho, J., Meschter, S.J., Maganty, S. et al. (2014). *Polyurethane Conformal Coatings Filled with Hard Nanoparticles for Tin Whisker Mitigation*, vol. 27-3. Binghamton University (SUNY), BAE Systems, Henkel.

9 McKeown, S.A., Meschter, S.J., Snugovsky, P., and Kennedy, J. (2015). *SERDP Tin Whisker Testing and Modeling: Simplified Whisker Risk Model Development*, vol. 28-1. BAE Systems and Celestica Inc.

10 Meschter, S.J., Snugovsky, P., Kennedy, J. et al. (2014) Strategic environmental research and development program (SERDP) tin whisker testing and modeling: thermal cycling testing. *ICSR (Soldering and Reliability) Conference Proceedings*.

11 Wickham, M., Lewis, A., Clayton, K. (2017) 1000 days of testing tin whiskering PCB assemblies to determine the suitability of conformal coatings to mitigate against shorting. *SMTAI Conference*.

12 Hou, W. and Salvaraski, B. (2013) 3M electronic grade conformal coating for moisture and corrosion protection of PCB and electronic components. *ICSR (Soldering and Reliability)*.

13 Kumar, R. (2014) A high temperature vapor phase conformal coating for electronics applications. *High-Performance Cleaning and Coating*.

14 Hannafin, J. (2002) A novel approach to thermal management and EMI shielding via a metallic conformal coating on a plastic housing. *Telecom Hardware Solutions*.

15 Kinner, P., Urquhart, J., and Hunt, C. (2016) A new selective conformal coating process for the increased ruggedization of printed circuit assemblies with particular emphasis on condensing environments. *SMTA International*.

16 Gonsolin, M. (2017) Air plasma at atmospheric-pressure for improving conformal coating reliability. *High-Performance Cleaning and Coating*.

17 Sit, O. and Fonseca, H.L. (2017) Automated conformal coating inspection and thickness measurement. *SMTA International*.

18 Hindin, B. (2014) Behavior of conformal coatings in sulfur vapor-bearing environments. *High-Performance Cleaning and Coating*.

19 Guene, E. and Puechagut, C. (2015) Case study: tools to assure compatibility of conformal coatings and no-clean lead-free solder pastes. *SMTA China*.

20 Cho, J., Meschter, S.J., Maganty, S. et al. (2013) Characterization of hybrid conformal coatings used for mitigating tin whisker growth. *ICSR (Soldering and Reliability)*.

21 Hillman, D. (2017) Characterization of the capability of conformal coating to inhibit tin whisker growth. *ICSR (Soldering and Reliability)*.

22 Xia, C. and Priore, S. (2013) Conformal coating process for networking and communication devices in harsh environments. *High-Performance Cleaning and Coating*.

23 Jason Keeping, P. (2011) Critical considerations for conformal coating. *ICSR (Soldering and Reliability)*.

24 Jason Keeping, P. (2008) Evaluating the manufacturability and operational costs for new conformal coating processes. *Pan Pacific Symposium*.

25 Woodrow, T.A., and Ledbury, E.A. (2006) Evaluation of conformal coatings as a tin whisker mitigation strategy, Part II. *SMTA International.*

26 Jason Keeping, P.E. (2007) Process development and optimization using a newly designed conformal coating test vehicle. *SMTA International.*

27 Cho, J., Meschter, S.J., Maganty, S. et al. (2014) Polyurethane conformal coatings filled with hard nanoparticles for tin whisker mitigation. *ICSR (Soldering and Reliability).*

28 Meschter, S., Cho, J., Maganty, S. et al. (2016) Strategic environmental research and development program (SERDP) layered nanoparticle enhanced conformal coating for whisker mitigation. *ICSR (Soldering and Reliability).*

29 McKeown, S., Meschter, S., Snugovsky, P. et al. (2015) Strategic environmental research and development program (SERDP) nanoparticle enhanced conformal coating project: coating modeling for tin whisker mitigation. *SMTA International.*

30 Meschter, S., Cho, J., Maganty, S. et al. (2014) Strategic environmental research and development program (SERDP) nanoparticle enhanced conformal coating for whisker mitigation. *SMTA International.*

31 Wettermann, B. (2018). *Webinar: Consider the options – How should I mask for Conformal Coating?* BEST, Inc.

32 Jason Keeping, P. (2011). *Webinar: Critical Considerations for Conformal Coating Reliability.* Celestica Inc.

33 Pauls, D. (2014). *Webinar: Evaluating the Performance of Conformal Coatings.* Rockwell Collins.

34 National Physical Laboratory (NPL) tin whisker investigation.

35 Woody, L. and Fox, B. (2014) Tin whisker risk management by conformal coating. IPC APEX EXPO Conference Proceedings.

36 Pb-free Electronics Risk Management Council (PERM, IPC Committee 8-81) (2018) Mitigation of pure tin risk by tin-lead SMT reflow – results of an industry round-robin – final report., WP-022.

37 SERDP (2017) Novel whisker mitigating composite conformal coat assessment. SERDP WP2213 Final Report. A collaboration between BAE Systems, Celestica, Bayer Material Science, Henkel, and Binghamton University evaluated enhancing polyurethane coating properties to obtain improved tin whisker mitigation.

Index

Lead-free Soldering Process Development and Reliability, First Edition. Edited by Jasbir Bath.
© 2020 John Wiley & Sons, Inc. Published 2020 by John Wiley & Sons, Inc.